AF593485

CLINICAL ENZYMOLOGY SYMPOSIA

CLINICAL ENZYMOLOGY SYMPOSIA

Proceedings of the 7th International Symposium on Clinical Enzymology held in Venezia, Fondazione G. Cini, April 5-6-7th, 1976

Edited by
A. BURLINA and L. GALZIGNA

1

PICCIN MEDICAL BOOKS – PADUA

The Symposia on Clinical Enzymology are organized by Professor Angelo Burlina, M.D., under the auspices of the "Ministero della Sanità", "Ministero della Pubblica Istruzione" and "Consiglio Nazionale delle Ricerche"

Since 1969 the "International Symposia on Clinical Enzymology" have been a Forum for a wide interdisciplinary survey of the frontiers of the biomedical research in the field.

One of the main results of the initiative, which has facilitated contacts and the exchange of problems among scientists of different countries, was the creation in 1974 of an International Society for Clinical Enzymology.

The Society was officially founded in Venice on the occasion of the 7th International Symposium on April 7th, 1976.

The Founding Members elected a Board comprised of the following: J.H. Wilkinson (President), A. Burlina (Vice-President), M. Roth (Secretary), E. Kaiser (Treasurer), D.M. Goldberg, W. Gerhardt and G. Szasz.

The International Symposia on Clinical Enzymology can now be considered a traditional event which is made possible by the assistance of a Scientific Advisory Board composed of: P. Boivin (Paris), D.M. Goldberg (Toronto), W. Gerhardt (Helsingborg), E. Kaiser (Wien) and G. Szasz (Giessen), and a Standing Organizing Committee composed of: G. Astaldi, P. Boccato, L. Galzigna, U. Lippi, P.G. Paleani-Vettori, F. Porta, E. Salvidio, G. Torlontano and L. Tentori, Chairman A. Burlina.

The help and technical assistance of a number of Italian firms and particularly that of Boehringer-Biochemia, Ames-Miles Italiana and General Diagnostics made this achievement possible.

The present volume of Proceedings is published by M. Piccin, Padova and constitutes the first publication of the International branch called Piccin Medical Books. We wish a good success to this publication and to all future publications of Piccin Medical Books.

A.B.

Symposia on Clinical Enzymology:

1st Symposium: Conegliano, November 22-23d, 1969 Proceedings published by "Quaderni Sclavo di Diagnostica", Siena

2d Symposium: Conegliano, October 24-25th, 1970 Proceedings published by "Quaderni Sclavo di Diagnostica", Siena

3d Symposium: Conegliano, November 26-28th, 1971 Proceedings published by "Quaderni Sclavo di Diagnostica", Siena

4th Symposium: Conegliano, October 6-7th, 1972 Proceedings published by "Quaderni Sclavo di Diagnostica", Siena

5th Symposium: Venezia, October 18-20th, 1973 Proceedings published by "Edizioni LAB", Milano

6th Symposium: Venezia, October 24-25th, 1974 Proceedings published by "Kurtis Editore", Milano

7th Symposium: Venezia, April 5-7th, 1976 Proceedings published by "Piccin Medical Books" Padova

CONTENTS

Part 2: Clinical and Methodological Aspects

G.F. GALLANTI PRIZE FOR CLINICAL ENZYMOLOGY

On the occasion of the 7th International Symposium on Clinical Enzymology, to honour the memory of Dr. G.F. Gallanti, supporter and sponsor of the Symposia from the very first meetings, a "G.F. Gallanti International Prize" has been instituted with the following Statute:

Statute

1. Boehringer-Biochemia (Diagnostic Division) has established a Prize to honour the memory of Dr. G.F. Gallanti.
2. The Prize consisting of a memorial plaque and a sum of $ 2000 shall be awarded to a scientist or a group of scientists who have contributed in a significant way to the advancement of Clinical Enzymology.
3. Boehringer-Biochemia has decided that the Prize will be given on the occasion of the International Symposia on Clinical Enzymology organized in Italy every two years.
   The Prize will be presented by the Honorary President or by the President of the Organizing Committee of the Symposium. The recipient is expected to present his most important results in a short lecture.
4. The Prize can be awarded to any scientist or group of scientists who have contributed in a significant way to the progress of Clinical Enzymology. The relevant paper(s) should be published or accepted for publication in the interval between two Symposia. In particular cases a group of papers published over a longer period of time might be taken into consideration. The paper(s) considered for the Prize should not have been considered for other Awards. The details on the awarding of the Prize shall be contained in the Proceedings of the Symposium.
5. The winner will be chosen by an *Award Committee* composed of:
   a. The President of the Standing Organizing Committee of the International Symposium on Clinical Enzymology.

b. Two members of the Scientific Advisory Board of the Symposium.
c. One member of the Standing Organizing Committee.
d. One representative of Boehringer-Biochemia.
Members referred to as b) and c) shall be designated by the Standing Organizing Committee in a Meeting preceeding the Symposium.

6. The Award Committee will consider papers and candidates presented by members of Scientific Societies and Scientific Committees known at international level or by members of the Committee itself.
7. The Secretariat of the Symposium will coordinate the Meetings of the Award Committee and Boehringer Biochemia will financially support such Meetings.
8. The recipient of the Prize shall be designated by ballot by the Award Committee. The winner will be notified in advance so that his name can be communicated in the Final Program of the Symposium and he will have sufficient time to prepare his lecture.

During the 7th Symposium an "Ad hoc Committee" unanimously decided to attribute the 1st G.F. Gallanti Prize to Professor J.H. Wilkinson (London), for his original contributions and pioneering work in the field of Clinical Enzymology.

The Prize was presented to Professor Wilkinson on April 6th, 1976 and Professor Wilkinson generously donated half of the sum he received to the newly-formed International Society for Clinical Enzymology.

The "Ad hoc Committee" that decided on the awarding of the prize during the 7th Symposium was formed by the following: A. Burlina, L. Galzigna, W. Gerhardt, D.M. Goldberg, J. Homolka, E. Kaiser and M. Roth.

OPENING REMARKS

A. Burlina

Ladies and Gentlemen,

It is a great pleasure to welcome you here in Venice and see the take off of this 7th edition of the International Symposium on Clinical Enzymology. I hope that you will enjoy your stay in this town and find the Symposium profitable.

Only an act of faith made it possible to continue the organisation of this Symposium in these difficult times and, besides faith, perhaps a true belief in Medicine and Research at the service of the suffering people.

This year the Scientific Program consists of Invited Lectures and a large number of Communications given in the form of Posters. The topics include pathophysiology of enzymes, their actual or potential diagnostic use, and methodology. In this introduction I can just list the most important contributions.

These begin with the fundamental researches of *Wilkinson* on the catabolism of plasma enzymes. The use of labeled LAD_5 and LAD_1 isoenzymes made it possible to study their intravascular inactivation, their intestinal absorption, their hydrolysis to aminoacids. *Moss* presents an essential contribution on identification of the origin of circulating enzymes showing differences between in vivo and in vitro behaviors, and differences in the distribution among the various types of cells which constitute an organ. A study on thyroid hormones in the mitochondrial synthesis of RNA and fatty acid chain elongation might explain some enzymatic alterations observed in hypothyroid states (*Quagliariello*). The pathophysiological role of collagenase is discussed on the basis of its activity on λ and k chains of immunoglobulins (*Coletti-Previero*); the characteristics of human hypoxantine-guanine phosphoribosyltransferase (*Salerno*) are also illustrated and the relationships between lipase and lipoproteinlipase are discussed (*Galzigna*). A critical review on the principles of cholinesterase determination is presented by *Augustinsson* and the progress in the determination of substrates with kinetic methods and fast analyzers in shown by *Bergmeyer* and his associates. *Castelli* analyzes the technical factors which influence the determination of serum transaminases, *Copeland* presents a turbidimetric reference standard for the determination of amylase and *Werner* illustrates the methodologies and the clinical usefulness of urinary enzymes determination after gel-filtration of

Laboratorio di Biochimica Clinica e di Ematologia – Centro Ospedaliero di Borgo Trento 37100 Verona – Italy

urines. *Persijn* presents a new substrate for GGT which gives higher normal values than those obtained with the Szasz method. *Freise* shows the present difficulties in considering the various GGT fractions as true isoenzymes and *Henderson* demonstrates the presence of MB isoenzyme of CPK in normal serum while *Chemnitz* illustrates the diagnostic use of CPK isoenzymes in heart disease and extracardial conditions. *Gerhardt* discusses the differential diagnostic use of enzymes and isoenzymes in myocardial infarction and *Statland* shows the increase of enzyme activities such as CPK, AST and LAD in serum after muscular exercise.

The potential diagnostic use of new computer-aided techniques is proposed by *Gunzer* and the promising results on glutathione reductase allow *Goldberg* to propose its determination in the screening of patients with tumors. *Tettamanti* illustrates the assay and clinical value of lysosomal enzymes in serum and *Bardslay* presents new kinetic methods which can be used for the determination of diamino-oxidase in pregnancy. *Gregolin* analyzes the correlation between LCT, lipoprotein catabolism and atherosclerosis while *Kaiser* considers the same enzyme in hepatic diseases; *King* compares normal and atypical cholinesterases from the standpoint of sensitivity to anti-cholinesterase agents, *Roth* discusses the relationship between angiotensin converting enzyme and α_1-antitrypsin in pulmonary embolism and *Griffiths* illustrates the use of glyceraldehyde-3-phosphate dehydrogenase in the diagnosis of acute myocardial infarction.

The present program appears to follow on logically from that of 1974, as will appear from the Proceedings, and this makes an explanation necessary. In the Proceedings of the 1974 Symposium we proposed a system of standardized abbreviations for the names of the enzymes of diagnostic use. This time we did not use the same system because we had many negative reactions and criticisms. In the present book, we have therefore avoided any modification of the abbreviations proposed by the authors. Nevertheless, we feel that the system proposed in 1974, based on a three capital letters notation, should be given serious consideration for its suitability in computer-aided data-elaboration and usefulness in avoiding confusion with other existing abbreviations.

Before concluding these remarks, with my best wishes for a fruitful 1976 edition of our "Enzyme Symposium", I wish to remind you that the next edition of the Symposium will take place in Spring 1978, while and Advanced Course on Clinical Enzymology will be organized in Rome instead of a Symposium in 1977.

PART 1

ENZYME PATHO-PHYSIOLOGY

THE CATABOLISM OF PLASMA ENZYMES

J.H. Wilkinson and A.R. Qureshi

Summary

The mechanisms whereby plasma enzymes are removed from the circulation have hitherto remained obscure. We injected radioactively-labelled enzymes into the rabbit and studied the rates of disappearance from the plasma of catalytic activity and radioactivity. With ^{125}I-labelled lactate dehydrogenase - 5, catalytic activity and radioactivity disappeared at similar rates during the first phase lasting 1-2 hours, when the fall in plasma activities is attributed to distribution throughout the extracellular fluid. Subsequently catalytic activity was found to disappear at a faster exponential rate ($t_{0.5}$ = 3.5 hours) than radioactivity ($t_{0.5}$ = 12.4 hours). This is interpreted as indicating intravascular inactivation of the injected enzyme. Studies with ^{125}I-labelled lactate dehydrogenase – 1 gave similar results. After the completion of the distributional phase, catalytic activity was removed at a faster rate ($t_{0.5}$ = 18 hours) than radioactivity ($t_{0.5}$ = 31 hours). In both series of experiments, appreciable radioactivity was detected in the intestinal contents, but little in the faeces.

It is concluded that the inactivated enzyme protein passively diffuses into the intestine where it is digested and the constituent amino-acids are reabsorbed.

The mechanisms whereby serum enzyme activities return to normal during recovery from an acute illness have intrigued clinical enzymolgists for more than 20 years. The marked elevations of the serum aspartate and alanine transaminases in acute viral hepatitis, or of the serum lactate dehydrogenase in severe megaloblastic anaemia, usually rapidly disappear during recovery over a period of a few weeks. In myocardial infarction the elevated serum creatine kinase activity returns to normal in about 2 days, while the less marked elevation in the serum aspartate transaminase usually takes 4-7 days to disappear. It is clear that enzymes are removed from the circulation at different rates, but hitherto little is known of the mechanisms involved. Most of our present knowledge derived from both clinical and experimental observations is of a rather negative nature.

Apart from those with relatively low molecular weights, e.g. amylase (48,000), uropepsinogen (42,500), most enzymes of current clinical interest

Department of Chemical Pathology, Charing Cross Hospital Medical School, London, W6 8RF, U.K.

cannot be filtered by the glomerulus in the absence of renal disease and do not appear in the urine in significant amounts (Rosalki and Wilkinson, 1959). For many years, the serum alkaline phosphatase was considered to originate in the bones and to be excreted in the bile but numerous studies during the past decade have shown this 'retention theory' to be untenable (Posen, 1970). The excretion of enzymes thus appears to be unlikely, and some alternative possibilities are listed in Table 1.

There is little evidence, however, in support of these suggestions and the theory that circulating enzymes undergo intravascular inactivation has been postulated (Posen, 1970; Wilkinson, 1970).

Table 1. *Possible clearance mechanisms for plasma enzymes*

1. Excretion in the urine or bile
2. Inhibition by small molecules in the circulating blood
3. Auto-immune reactions
4. Digestion of enzyme proteins by circulating proteins
5. Degradation of enzyme protein in the tissues, e.g. the reticuloendothelial system
6. Intravascular inactivation of enzymes, possibly by blood cells

Radio-iodination of lactate dehydrogenase isoenzymes

The work described in the present paper was designed to test this hypothesis, and for this purpose rabbit-muscle lactate dehydrogenase (EC 1.1.1. 27) isoenzymes, LD-1 and LD-5, were radio-iodinated with ^{125}I and injected intravenously into rabbits. LD-5 was labelled by direct iodination but the recovery of catalytically-active enzyme protein was only about 15% (Qureshi and Wilkinson, 1976*a*). This method failed completely with LD-1, but this isoenzyme was successfully radio-iodinated by the method of David (1972) using solid-state lactoperoxidase and hydrogen peroxide as oxidizing agent. The recovery of catalytic activity was 85% (Wilkinson and Qureshi, 1976). In both cases the products were purified by chromatography on Sephadex G-25 columns, and on subsequent electrophoresis on cellulose acetate catalytic activity detected by means of the tetrazolium reaction was found to coincide with the radioactivity detected by radioautography.

Methods involving the coupling of labelled esters or other reagents with enzyme proteins were considered unsuitable for catabolic studies because it was found that the amide linkages between a labelled ester and the enzyme protein were rapidly hydrolysed *in vivo* (Qureshi and Wilkinson, 1976*b*).

The labelled enzymes were injected intravenously into the marginal ear veins of rabbits, given Na ^{126}I in their drinking water to prevent thyroidal uptake of ^{125} iodide liberated during catabolism of the labelled enzymes. Blood was collected from the other ear at timed intervals into heparinized tubes, and the plasma was separated by centrifugation for measurement of the enzyme activity and radioactivity.

Disappearance rates of injected enzymes

The plasma lactate dehydrogenase activities at various times after the injection of unlabelled LD-1 and LD-5 are shown in Figure 1. In both cases catalytic activity disappeared according to a biphasic exponential system. The initial rapid phase appears to be due to distribution of the injected material throughout the extracellular fluid while the slower phase represents the breadkdown of the enzyme. As is well-known from clinical and experimental observations, LD-5 is catabolized at a faster rate than LD-1.

Figures 2 and 3 show the disappearance curves for the plasma radioactivity and catalytic activity after the administration of labelled LD-1 and LD-5 respectively. The catalytic activities disappeared at rates similar to those observed with the unlabelled enzymes. This suggests that the presence of the radioactive label does not interfere with the manner in which the enzyme is treated by the body. After the initial distributional phase, radioactivities were found to disappear from the plasma at slower rates than the catalytic

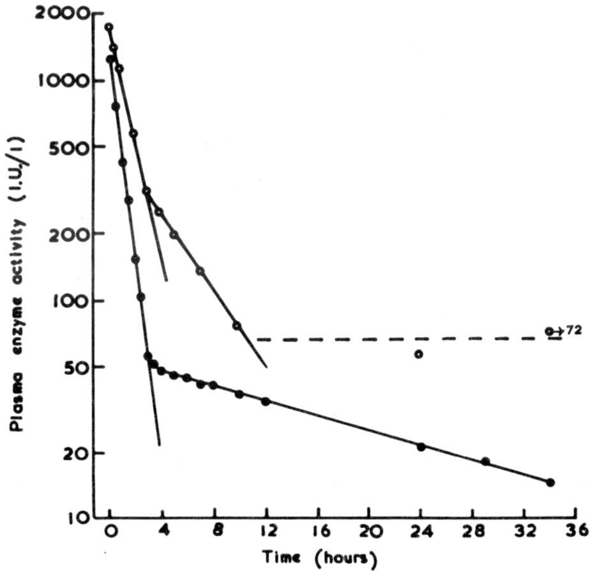

Fig. 1 – Disappearance curves for (●) the plasma LD-1 and (○) the plasma LD-5 after the injection of unlabelled lactate dehydrogenase isoenzymes in the rabbit. In this experiment the half-lives for LD-1 were 0.7 and 17.8 hours during the first and second phases, and the corresponding values for LD-5 were 0.9 and 3.5 hours respectively. LD-1 activity was determined with α-oxobutyrate as substrate (Rosalki and Wilkinson 1960).

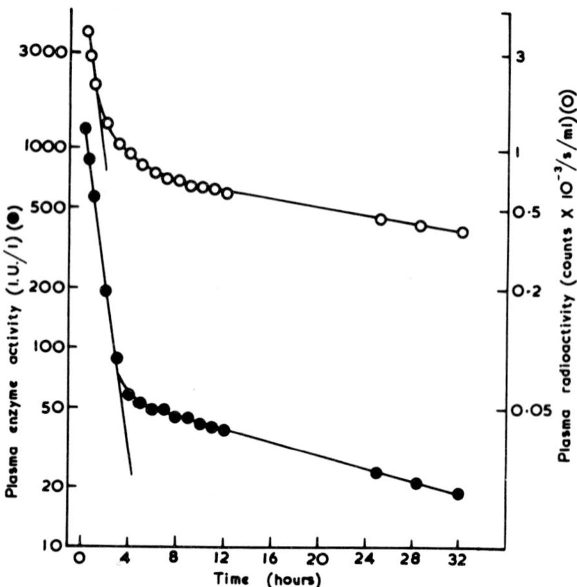

Fig. 2 – The plasma radioactivity (○) and catalytic activity (●) after the injection of ^{125}L-labelled lactate dehydrogenase-1 into the rabbit.

activities. The half-lives for the distributional and catabolic phases for both enzymes are listed in Table 2. In each case the difference between the disappearance rates for catalytic activity and radioactivity was statistically significant ($P<0.001$). After the return of the plasma enzyme activity to the normal range, plasma radioactivity continued to decline at a slow exponential rate.

These results suggest that both enzymes are inactivated in the circulating blood, and that the catalytically inactive, but radioactive, products remain in the blood for a considerable period. Our results contrast with those of earlier workers who reported similar rates for the disappearance of radioactivity and catalytic activity from the plasma after the injection of ^{131}I-labelled aldolase (Schapira, Dreyfus and Schapira, 1962), or ^{131}I-labelled aspartate and alanine transaminases (Massarat, 1965). This discrepancy might be due to the breakdown of enzymes labelled with ^{131}I, since Bär *et al.* (1972/1973) demonstrated the relative instability of ^{131}I-labelled LD-5. Enzymes labelled by coupling with radioactive agents, such as LD-5 labelled with ^{14}C-iodoacetate (Boyd, 1967) or ^{14}C-acetic anhydride (Bär et al., 1972/1973), have also been shown to have similar disappearance rates for radioactivity and catalytic activity. Our experience with LD-5 coupled with ^{125}I-labelled diiodophenylpropionate (Qureshi and Wilkinson, 1976*b*) suggests that amide linkages are readily hydrolysed *in vivo,* and similar instability

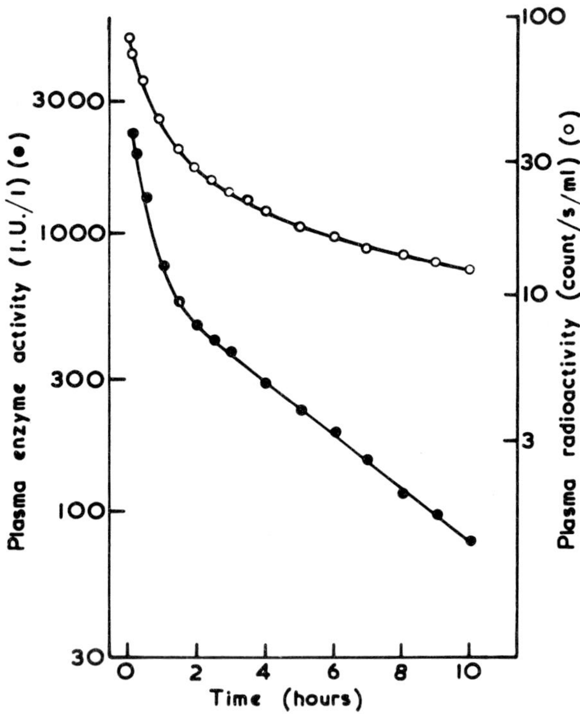

Fig. 3 – The plasma radioactivity (○) and catalytic activity (●) after the injection of ^{125}I-labelled lactate dehydrogenase-5 into the rabbit.

Table 2. *Half-lives for the disappearance of lactate dehydrogenase isoenzymes 1 and 5 from the plasma of the rabbit*

*Phase**	*Half-lives (hours ± S.D.)*			
	Lactate dehydrogenase-1		*Lactate dehydrogenase-5*	
Distributional	1.39 ± 0.13(4)†	0.62±0.07(4)	0.83±0.31(5)	0.70±0.24(5)
Catabolic	30.5 ± 1.7 (4)	18.4±0.5 (4)	12.4±2.3 (5)	3.5 ±0.5 (5)
Residual radioactivity	80	–	50	–

*The distributional phase persisted for about 9 hours and 4 hours for LD-1 and LD-5 respectively; the catabolic phase occupied the next 25 hours in the case of LD-1 and the next 6 hours for LD-5, after which the plasma enzyme activity had returned to normal. Residual radioactivity then disappeared at the slow rates shown.

†The figures in parentheses indicate the numbers of experiments performed.

might account for the differences between our results and those of investigators using enzymes labelled by coupling reactions.

Distribution of radioactivity in the tissues after the injection of radioactively-labelled enzymes

In order to study the disposal of the inactive products we measured the radioactivity in various tissues. Table 3 gives the percentage of the plasma

Table 3. *Percentage of plasma radioactivity in the tissues and intestinal contents of rabbits injected with ^{125}I-labelled lactate dehydrogenase isoenzymes 1 and 5*

Material	*Percentage of plasma activity*		
	LD-1	*LD-5*	
	6 hours	*2 hours*	*8 hours*
Spleen	16	100	150
Liver	15	55	32
Heart	12	17	18
Kidney	36	5	6
Skeletal muscle	3	3	6
Duodenum	17	50	30
Jejunum	7	75	55
Ileum	5	10	30
Gall bladder & contents	41	10	10
Duodenal contents	20	34	–
Jejunal contents	19	29	33
Ileal contents	29	26	–

Table 4. *Urinary and faecal radioactivity after the injection of ^{125}I-labelled lactate dehydrogenase isoenzymes 1 and 5 in the rabbit*

	Period of collection (hours)	*Percentage of dose*	
		LD-1	*LD-5*
Urine	0 – 24	34.2	42.5
	24 – 48	25.0	23.4
	48 – 72	11.6	7.5
	0 – 72	70.8	73.4
Faeces	0 – 72	2.5	5.1

radioactivity in a number of tissues at various times after the injection of labelled LD-1 and LD-5. With the possible exception of the spleen, no tissue exhibited significantly higher radioactivity than the plasma, hence we concluded that the removal of inactivated enzyme protein is a passive rather than an active process. It was interesting to note the appreciable amounts of radioactivity in the intestinal mucosa and the intestinal contents, but despite these relatively little appeared in the faeces. Most of the injected radioactivity was excreted in the urine (Table 4).

About 60% of the urinary radioactivity consisted of the radio-iodinated amino-acids and the remainder of inorganic iodide. The former consisted largely of mono- and di-iodotyrosines with smaller amounts of diiodohistidine and 3,5-diiodophenyl-lactate, a known metabolite of diiodotyrosine (Fletcher, 1957; Roche, Michel, Closon and Michel, 1958). The presence of free iodide in the urine is probably due to deiodination of iodinated amino-acids in the liver (Maclagan and Reid, 1957).

Our experiments with these two forms of lactate dehydrogenase suggest that after inactivation in the circulating blood, or in a compartment which releases the inactive products into the blood, the products are removed from the circulation via the small intestine. The presence of relatively high concentrations of radioactivity in the mucosal cells suggests a direct discharge into the intestinal lumen, but we cannot exclude the possibility that much of the labelled protein might reach the intestine via the bile and that the mucosal radioactivity might be due to re-absorption of the products of intestinal digestion. The plasma radioactivity was shown to be almost completely protein-bound, but in the tissues the fractions precipitable by trichloroacetic acid ranged from 20-41%. Only 10.6% of the radioactivity in the duodenal contents 6 hours after the injection of LD-1 was found to be protein-bound and the proportions in the jejunum and ileum were even less, but 80-90% of the ^{125}I in the intestinal contents was shown to be organically-bound (Wilkinson and Qureshi, 1976).

It seems therefore that after inactivation in the circulation and transport of the products to the intestine, the enzyme proteins are digested and the resulting aminoacids are reabsorbed to enter the amino-acid pool. In our experiments the labelled iodoamino-acids could not be used in the biosynthesis of protein and were excreted in the urine. Such a process has been suggested to account for the removal of plasma albumin (Glebert, Jarnum and Riemer, 1962; Jeejeebhoy, Jarnum, Singh, Nadkarni and Westergaard, 1968), and Fleisher and Wakim (1968) have shown that the small intestine plays a part in the clearance of injected alanine transaminase in the dog.

Enzyme inactivation in blood *in vitro*

If our theory is correct, it should be possible to demonstrate enzyme inactivation in blood in vitro. Experiments designed to test this hypothesis were carried out by incubating rabbit LD-5 under sterile conditions with heparinized rabbit whole blood at 37°C (Qureshi and Wilkinson, 1976*c*). Samples were removed at hourly intervals and centrifuged. The LD-5 activity in the plasma was determined either by deducting the control from the test value or by measuring activity in the presence and absence of 0.2 mM-oxalate to enable the proportions of LD-1 (from the erythrocytes) and LD-5 to be calculated (Emerson, Wilkinson and Withycombe, 1964). Both methods gave similar results.

After a short latent period lasting about 3 hours, LD-5 was found to be exponentially inactivated at a rate comparable with that observed *in vivo*. The mean half-life was 5.2 ± S.D. 1.0 hours. Since the loss of enzyme activity was negligible in parallel experiments in which LD-5 was incubated with plasma, it was concluded that the cellular components are responsible for enzyme inactivation.

Similar experiments in which the enzyme was incubated in rabbit plasma containing erythrocytes alone or leucocytes alone were carried out. The red cell had virtually no activating effect on the enzyme, while the leucocytes inactivated LD-5 at a rate (half-life, 4.9 ± S.D. 1.1 hours) similar to that found in whole blood. We cannot exclude the possibility that the platelets might also be involved, as our leucocyte preparations might have been contaminated with these cells.

These *in vitro* experiments provide support for the suggestion that circulating enzymes are inactivated in the blood, but it remains to be seen whether this conclusion applies to enzymes other than lactate dehydrogenase and species other than the rabbit.

References

1. Bär U., Friedel R., Heine H., Mayer D., Ohlendorf S., Schmidt F.W., and Trautschold I., Studeis on enzyme elimination. III. Distribution, transport and elimination of cell enzymes in the extracellular space. *Enzyme* (Basel), 14, 133, 1972-1973.
2. David G.S., Solid stata lactoperoxidase: a highly stable enzyme for simple, gentle iodination of proteins. *Biochem. Biophys. Res. Commun.*, 48, 464, 1972.
3. Emerson P.M., Wilkinson J.H., and Withycombe W.A., Effect of oxalate on the activity of lactate dehydrogenase isoenzymes. *Nature* (Lond.), 202, 1337, 1964.
4. Fleisher G.A., and Wakim K.G., The role of the small intestine in the disposal of enzymes. *Enzymol. Biol. Clin.*, 9, 81, 1968.
5. Fletcher K.: The fractionation of urinary iodine. 2. Metabolites excreted during treatment of carcinoma of the thyroid. *Biochem. J.*, 67, 140, 1957.
6. Glenert J., Jarnum S., and Reimer S.: The albumin transfer from blood to gastrointestinal tract in dogs. *Acta Chir. Scand.*, 124, 63 (1962).
7. Jeejeebhoy K.N., Jarnum S., Singh B., Nakdarni G.D., and Westergaard H.: *95*Nb-labelled albumin for the study of gastrointestinal albumin loss. *Scand. J. Gastroenterol.*, 3, 449 (1968).
8. Maclagan N.F., and Reid D.: The deiodination of thyroid hormones *in vitro*. In *Ciba Foundation Colloquia on Endocrinology*, vol. 10, *Regulation and Mode of Action of Thyroid Hormones*, p. 190. Ed. Wolstenholme G.E.W., and Millar E.C.P., Churchill, London, (1957).

9. Massarat S.: Enzyme kinetics, half-life, and immunological properties of iodine-131-labelled transaminases in pig blood. *Nature* (Lond.), 206, 508 (1965).
10. Posen S.: Turnover of circulating enzymes. *Clin. Chem.*, 16, 71 (1970).
11. Qureshi A.R., and Wilkinson J.H.: The fate of circulating lactate dehydrogenase-5 in the rabbit. *Clin. Sci. Molec. Med.*, 50, 1, (1976*a*).
12. Qureshi A.R., and Wilkinson J.H.: Rates of disappearance from the plasma of enzymes labelled by coupling with a radioactive iodo-ester. *Clin. Chem.*, in the press, (1976*b*).
13. Qureshi A.R., and Wilkinson J.H.: Inactivation *in vitro* of lactate dehydrogenase-5 in blood. *Ann. Clin. Biochem.*, in the press (1976*c*).
14. Roche J., Michel R., Closon J., and Michel O: Nature of the products eliminated in the urine after the administration of diiodo-*L*-tyrosine. *Rev. franç. Et. Clin. Biol.*, 3, 135, (1958).
15. Rosalki S.B., and Wilkinson J.H.: Urinary lactic dehydrogenase in renal disease. *Lancet,* ii, 327 (1959).
16. Rosalki S.B., and Wilkinson J.H.: Reduction of α-ketobutyrate by human serum. *Nature* (Lond.), 188, 1110 (1960).
17. Schapira F., Dreyfus J.C., and Schapira G.: La duree de séjour dans le plasma de l'aldolase chez el lapin: étude à l'aide d'une aldolase marquée à l'iode radioactif. *Rev. franç. Et. Clin. Biol.*, 7, 829 (1962).
18. Wilkinson J.H.: Clinical significance of enzyme activity measurements. *Clin. Chem.*, 16, 882 (1970).
19. Wilkinson J.H. and Qureshi A.R.: The catabolism of plasma enzymes studied with ^{125}I-labelled lactate dehydrogenase-1 in the rabbit. *Clin. Chem.*, in the press (1976).;

THYROID HORMONE ACTION ON MITOCHONDRIAL TRANSCRIPTION PROCESS AND FATTY ACID CHAIN ELONGATION SYNTHESIS IN RAT LIVER

Maria N. Gadaleta, C. Landriscina, C. Saccone, G.V. Gnoni and E. Quagliariello

Summary

The effects of thyroid hormones on two metabolic pathways of rat liver, namely 1) the *in vitro* mitochondrial RNA synthesis and 2) the fatty acid chain elongation synthesis will be discussed in this report.

1. *Mitocondrial RNA synthesis*

3,3',5-triiodo-L-thyronine administered subcutaneously enhances RNA polymerase activity in isolated mitochondria of rat liver either after chronic or after acute treatment. The increased activity of mitochondrial RNA polymerase is followed by a net increase in the RNA content of organelles.

Experiments will be presented showing that thyroid hormone affects mitochondrial transcription by acting at template as well as at enzyme level.

cAMP and its dibutyrryl derivative do not seem to be involved in this hormone action.

2. *Fatty acid chain elongation synthesis.*

As regards the effect of thyroid hormones on fatty acid chain elongation in rat liver, evidence has been found that microsomal fatty acid chain elongation synthesis and desaturation, as well as acetyl-CoA carboxylase and fatty acid synthetase, are strongly influenced by thyroid hormone levels. Conversely, the fatty acid chain elongation system in mitochondria does not seem significantly affected by the thyrotoxic state. Finally, a reduction in the hepatic cyclic AMP level of about 41% is reported in triiodothyronine-treated rats, after 19 days. A tentative interpretation is made of these observations concerning the possible role of the mitochondrial chain elongation synthesis in liver and cAMP alteration in the thyrotoxic state.

Although it is well known that thyroid hormones play in essential role in the fine regulation of cellular metabolism, the molecular basis of their action mechanism is still under investigation. The present study deals with the

Istituto di Chimica Biologica (Facoltà di Scienze) dell'Università di Bari - Bari

effect of thyroid hormones, at the molecular level, on hepatic cell mitochondrial and extramitochondrial metabolism. In particular, two different aspects of this problem in rat liver are discussed: the effect of thyroid hormones on the mitochondrial transcription process and the effect produced by the thyrotoxic state on fatty acid chain elongation synthesis.

A. Mitochondrial RNA synthesis

The most striking effect of the thyroid hormone on adult mammalian tissue is an increase in the basal metabolic rate accompanied by an increase in oxygen consumption[1]. This fact has induced many scientists to study the interrelationship between thyroid hormones and mitochondrial structure and function[2]. Initial reports in this field stressed the uncoupling effect of thyroxine on oxidative phosphorylation with consequent wasting of heat, loss of respiratory control and mitochondrial swelling[3]. Subsequently, however, Tata et al.[4] proved that the uncoupling effect on oxidative phosphorylation and the catabolic effect of thyroid hormones in general were to be considered the results of administration of toxic amounts of thyroid hormones to animals and that, on the contrary, the administration of physiological doses resulted in an anabolic effect and, in general, growth of the organism. Biochemical experiments[5] and electronic microphotographs of rat skeletal muscle[6] also showed that, at the mitochondrial level, besides an increase in the mitochondrial ratio of cristae to matrix, these physiological doses also promote an increase in the number of mitochondria per cell. In other words, Tata demonstrated that basal metabolism and hence respiration in mammalian tissues were regulated by thyroid hormones through the selective increase of the number of respiratory units, localized in the cristae of the inner mitochondrial membrane. Specifically, the action of thyroid hormones on mammalian tissues was shown to be initiated by induction processes occurring at the transcription level in nucleus and mitochondrion, followed by the increase of protein synthesis in both extramitochondrial and mitochondrial compartments[7]. These results have now assumed great significance in the light of more recent studies on mitochondrial biogenesis which have clearly shown the synthesis of new mitochondrial structures to be under the joint control of nuclear and mitochondrial genomes, thus requiring the cooperation of mitochondrial and extramitochondrial systems of macromolecular synthesis[8].

To examine more closely the action of thyroid hormones at the level of mitochondrial transcription process, RNA synthesis was studied in isolated rat liver mitochondria after administration of physiological doses of T_3 to thyroidectomized rats. Fig. 1 shows results obtained by injecting 3,3',5 triiodo-L-thyronine to 3 month old thyroidectomized rats, and 60 days after thyroidectomy either in a single dose or with chronic treatment. As is evident, both types of treatment increased ^{3}H-UTP incorporation activity in

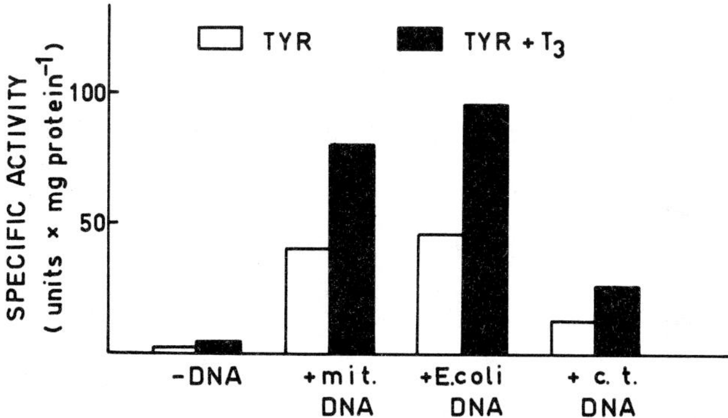

Fig. 1 – ^{3}H-UMP incorporation activity and RNA content in isolated rat liver mitochondria following acute or chronic treatment with 3,3',5-triiodo-L-thyronine. Hormone treatment was carried out by injecting 25 μg/100 gr. b.w. of T_3 in a single dose respectively, or, 5 μg/100 gr. b.w., daily. Operated animals were used as controls. For further details see ref. 10.

isolated mitochondria. However, mitochondrial RNA level rose only after seven days of chronic treatment[9].

The effect on *in vitro* mitochondrial RNA synthesis may be explained by direct action of the thyroid hormone on mitochondrial genome or mitochondrial RNA polymerase enzyme. This hormone may act on the DNA level by increasing the availability as template of the mitochondrial genome and on the enzymatic level, by increasing either the number of RNA polymerase molecules or their catalytic efficiency (see fig. 2). However, combined action on both components of the transcription system could also be considered.

In order to investigate hormone action at the mitochondrial genome level, experiments summarized in Table I were carried out. Hybridizing mitochondrial DNA with labeled RNAs newly synthesized by isolated mitochondria of rats respectively thyroidectomized (TYR), and thyroidectomized and T_3 treated (TYR + T_3), we found that, at the same DNA/RNA ratio, RNA isolated by mitochondria of hormone-treated animals hybridized more efficiently with mitochondrial DNA than RNA isolated from thyroidectomized animals. This could mean that RNA newly synthesized by isolated mitochondria of treated and untreated animals contain either different copies of mitochondrial genes or the same copies in different proportions[10].

When the above-reported hybridization experiments were repeated in the presence of competitive RNAs, that is, cold RNAs extracted respectively from the mitochondria of treated and untreated animals, different competition was found with the two types of RNA. In particular, the cold RNA from TYR animals completely inhibited the hybridization reaction of labeled RNA from TYR animals but only partially that of labeled RNA from

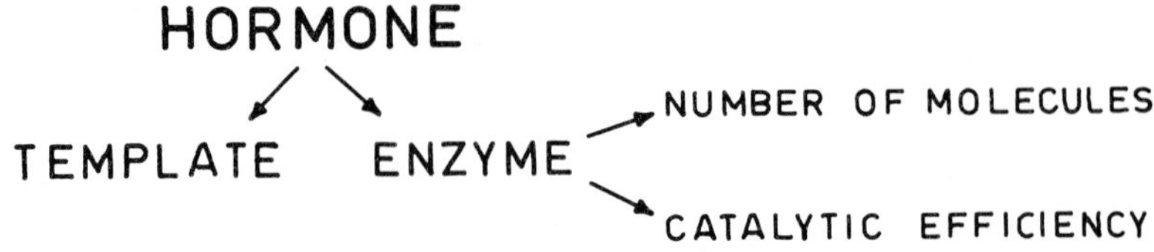

Fig. 2 – Possible T_3 action sites in the *in vitro* mitochondrial transcription process.

Table I. *Experiments performed to demonstrate the action of T_3 on the mitochondrial DNA transcription in rat liver*

a. Hybridization of newly synthesized mitochondrial RNA from TYR and TYR + T_3 rats with mitochondrial DNA[10].
b. Competition of cold mitochondrial RNA isolated from TYR and TYR + T_3 rat liver on the hybridization reaction reported in A)[10].
c. RNAase resistance of newly synthesized RNA in isolated mitochondria from TYR and TYR + T_3 rats[10].
d. Measurement of the percentage of newly synthesized stable RNA in mitochondria isolated from TYR and TYR + T_3 rats[12].

hormone-treated rats. In contrast, the cold RNA from treated rats competed equally well with labeled RNA from treated and untreated animals. The decreased capacity of cold mitochondrial RNA from untreated rats to compete with ^{3}H-RNA from hormone-treated rats may indicate that also *in vivo* the population of liver mitochondrial RNA of thyroidectomized rats differs somewhat from that of animals treated with the hormone[10].

Measurement of RNAase resistance of labeled RNA newly synthesized by isolated mitochondria of TYR and TYR + T_3 rats, either freshly isolated or after self-annealing, showed that RNA synthesized from mitochondria of TYR animals displayed a higher percentage of RNAase resistance[10]. In agreement with Attardi's proposal, the RNAase resistance of newly synthesized RNA could reflect a symmetric transcription of rat liver mitochondrial DNA. In fact, it has been demonstrated that in Hela cells both strands of mitochondrial DNA are copied but only the products of the H strand and of a small portion of the L strand remain undegraded and are therefore represented in the stable RNA species in vivo[11]. The higher level of symmetry found in RNA newly synthesized by mitochondria of thyroidectomized rats again suggests an impairment in these animals, probably in the processing of newly synthesized mitochondrial RNA.

Finally, the addition of rifampicin, a known inhibitor of the initial step in mitochondrial transcription *in vitro,* showed that a different percentage of easily degradable acid-insoluble radioactivity was synthesized in the two kinds of mitochondria after 10 min of incubation with labeled precursors of RNA synthesis. The percentage of solubilized radioactivity easily reflects that of the synthesis of unstable RNA, probably messenger RNA versus stable RNA. The higher content of easily degradable, acid-insoluble radioactivity in newly-synthesized RNA of mitochondria from hormone-treated rats could therefore mean that either a higher percentage of unstable messenger RNA chains have been synthesized in these organelles or a higher number of RNA chains have been initiated but not processed or terminated at the time of rifampicin addition in these mitochondria, thus resulting in a higher percentage of unstable molecules.

Results of all above-mentioned experiments strongly indicate that thyroid hormones exert their action at the mitochondrial level by specifically influencing gene activity.

However, since these results do not exclude action of the hormone at the enzyme level, another series of experiments was planned to purify RNA polymerase from liver mitochondria of TYR and TYR + T_3 rats to a stage at which the enzyme became dependent on endogenous template[12]. Enzyme purification was performed as reported by Gallerani et al.[13]. In order to measure enzyme activity, an exogenous source of template, either homologous DNA from normal rat mitochondria, E. Coli DNA or calf thymus DNA was used. Although the three DNAs were differently effective as templates, purified RNA polymerase from hormone-treated animals was in all cases 100% more active than purified RNA polymerase from thyroidectomized rats. This clearly demonstrated that regardless of template origin, the enzyme from animals treated with T_3 was more active, thus suggesting thyroid hormone action at the enzyme level as well. However, these experiments are insufficient to demonstrate whether the increased mitochondrial enzyme activity is due to enzyme molecule activation or to an increase of enzyme molecule content per mg of mitochondrial protein.

The addition of cAMP or its derivative dibutyrryl cAMP to mitochondria *in vitro* was also found to have no effect on RNA synthesis[14], thus excluding the possible role of this cyclic nucleotide in mediating the effect of 3,3',5-triiodo-L-thyronine.

In conclusion we wish to stress that in physiological conditions:

1. the induction of mitochondrial RNA synthesis as well as that of nuclear RNA seems to be an early action of thyroid hormone,
2. this induction, at least in mitochondria, seems to be carried out with action on the enzyme as well as on DNA, and finally
3. cAMP does not seem to mimic the action of thyroid hormones on mitochondrial RNA synthesis.

B. Fatty acid synthesis in subcellular fractions

It is generally accepted that fatty acid synthesis in liver proceeds according to two mechanisms, the *de novo* pathway via malonyl-CoA, followed by the chain elongation system operation. *De novo* synthesis takes place exclusively in cell sap, mediated essentially by acetyl-CoA carboxylase and by fatty acid synthetase, its most relevant product being palmitic acid. Meanwhile chain elongation synthesis takes place in microsomes as well as in mitochondria through specific multienzymatic systems, which elongate and/or desaturate palmitate formed *de novo,* producing a variety of higher saturated and unsaturated fatty acids[15].

Several studies have shown that abnormal levels of thyroid hormones influence various enzymatic activities of fatty acid synthesis in liver as well as in other tissues [16-18]. Following *in vivo* administration of triiodothyronine, a near doubling in activity has been observed in almost all lipogenic enzymes, both *in vivo* and *in vitro,* with recent studies[19,20] indicating an increase in acetyl-CoA carboxylase and fatty acid synthetase in rat liver as well.

Experiments described below were aimed at assessing alterations in fatty acid chain elongation synthesis of microsomes and mitochondria in rats during the thyrotoxic state.

The thyrotoxic state was confirmed by an increase in oxygen consumption by certain hepatic mitochondrial enzymes from triiodothyronine or thyroxine-treated rats and by a corresponding decrease in the case of propylthiouracil-induced hypothyroid rats.

Fig. 3 illustrates the activities of crude rat liver acetyl-CoA carboxylase (A), soluble fatty acid synthetase (B), fatty acid chain elongation and desaturation system of microsomes (C) and mitochondria (D) at different time intervals following administration of either propylthiouracil, triiodothyronine or thyroxine. As is evident from frame A of this figure, in propylthiouracil-treated animals a decrease of about 30% after three days of treatment is already apparent which reaches 55% within 6 days. At longer times, the decrease remains fairly constant. Differing behaviour is observed in the case of triiodothyronine and thyroxine-treated rats, with acetyl-CoA carboxylase activity remaining almost unchanged after the first 10 days of hormone administration, then increasing about 110 and 60 % respectively after 21 days of triiodothyronine and thyroxine administration. No difference was observed using either crude rat liver cell sap or ammonium sulphate precipitated cell sap (40 % stn). The same behaviour is shown by both soluble fatty acid synthetase (frame B), responsible for the *de novo* synthesis, and fatty acid chain elongation and desaturation systems of rat liver microsomes (C). In fact, in propylthiouracil-hypothyroidism, a near-50 % inhibition is already in evidence three days after drug administration, while in triiodothyronine and tyroxine-treated rats an increase of about 100 % and 50 % respectively was observed after twenty one days of treatment.

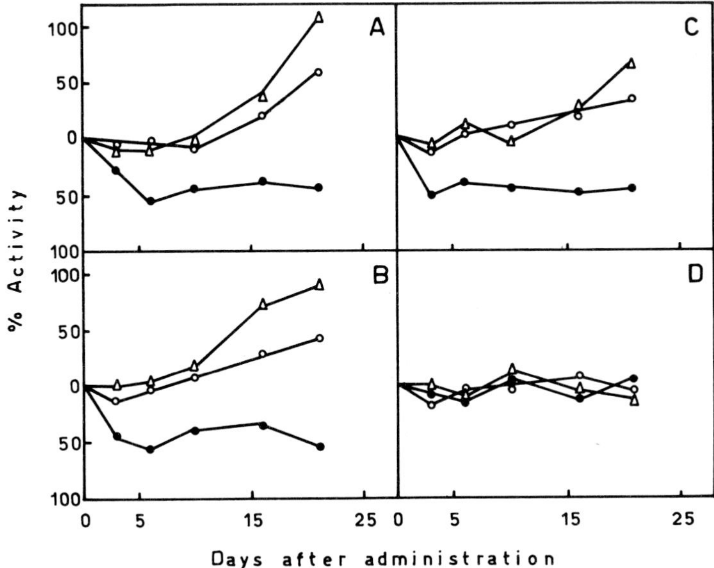

Fig. 3 – Fatty acid synthesis in rat liver subcellular fractions at different time intervals following thyrotoxic state induction.

Table II - *Effect of propylthiouracil, triiodothyronine or thyroxine administration on rat liver microsomal chain elongation and desaturation of fatty acids.*

Condition	*Specific Activity* nmoles [1,3-C14] malonyl-CoA incorporated into fatty acids/min/mg protein		*Relative rate*	
	Aerobiosis	*Anaerobiosis*	*Aerobiosis*	*Anaerobiosis*
Normal	0.35 ± 0.05	0.24 ± 0.02	1.00	1.00
Propylthiouracil	0.23 ± 0.03	0.17 ± 0.02	0.66	0.71
Triiodothyronine	0.57 ± 0.06	0.40 ± 0.03	1.63	1.67
Thyroxine	0.48 ± 0.04	0.37 ± 0.04	1.37	1.54

Experiments in anaerobiosis carried out under N_2 flow. Each result the mean of eight animals ± standard deviation (S.D.). Rats treated for 30 days used.

Finally, frame D shows that mitochondrial chain elongation synthesis, unlike microsomal and fatty acid synthetase activity, is almost unaffected by either propylthiouracil, triiodothyronine or thyroxine administration at all time intervals tested.

Table 2 shows that the change in hepatic microsomal fatty acid chain elongation and desaturation activities remains fairly constant even if the thyrotoxic period is extended from 21 to 30 days. Moreover, this table also indicates that microsomal incubation with labeled malonyl-CoA in anaerobiosis, when only chain elongation occurs, is altered during the thyrotoxic state similarly to that under aerobiotic conditions, when both elongation and desaturation reactions occur. These observations are in contrast with those of Faas[21], who observed that hyperthyroidism *in vitro* has no effect on hepatic microsomal chain elongation of fatty acids, thus indicating that both elongation and desaturation are effectively influenced by abnormal thyroid hormone levels.

The pattern of fatty acids synthesized in the thyrotoxic state by the rat-liver subcellular fractions under investigation was found to be similar to that obtained from untreated rats. On the contrary, radioactivity distribution in lipid classes isolated after incubation of microsomes with labelled malonyl-CoA, differed slightly in both propylthiouracil-treated and hyperthyroid rats. In fact, as indicated in Fig. 4, unlike cell sap, in which almost all radioactivity is associated with free fatty acids in all types of animals, microsomes isolated from hypothyroid rats show a higher percentage of labeled free fatty acid along with an equivalent decrease in di, triglyceride and phospholipid esterification. Moreover, the reverse seems to occur in microsomes from triiodothyronine and thyroxine-treated animals; i.e., an increase in di, triglyceride and phospholipid esterification coinciding with a decrease in free fatty acids. As expected, in no case were variations observed in mitochondria.

Finally, Table III reports a sensitive modification of the hepatic cyclic AMP level in rats by the thyrotoxic state. In particular, while in PTU or T_4-treated animals no significant variation was observed after 19 days of treatment, in triiodothyronine-induced hyperthyroidism a reduction of 41 % occurs. As this cyclic nucleotide has recently been reported to be an inhibitor of lipid synthesis at various enzymatic steps, the present liver experiments indicating its reduced amount along with an increase in protein enzyme content or by some unknown direct thyroid hormone action, could account for enhanced fatty acid synthesis in hyperthyroid animals.

The results reported on fatty acid synthetic activity clearly show that, besides a strong increase in both acetyl-CoA carboxylase and fatty acid synthetase in rat liver from hyperthyroid rats, the microsomal system is strongly influenced by an abnormal thyroid hormone level while mitochondrial chain elongation synthesis is in no way influenced. We can affirm that this is not due to permeability problems, since several other mitochon-

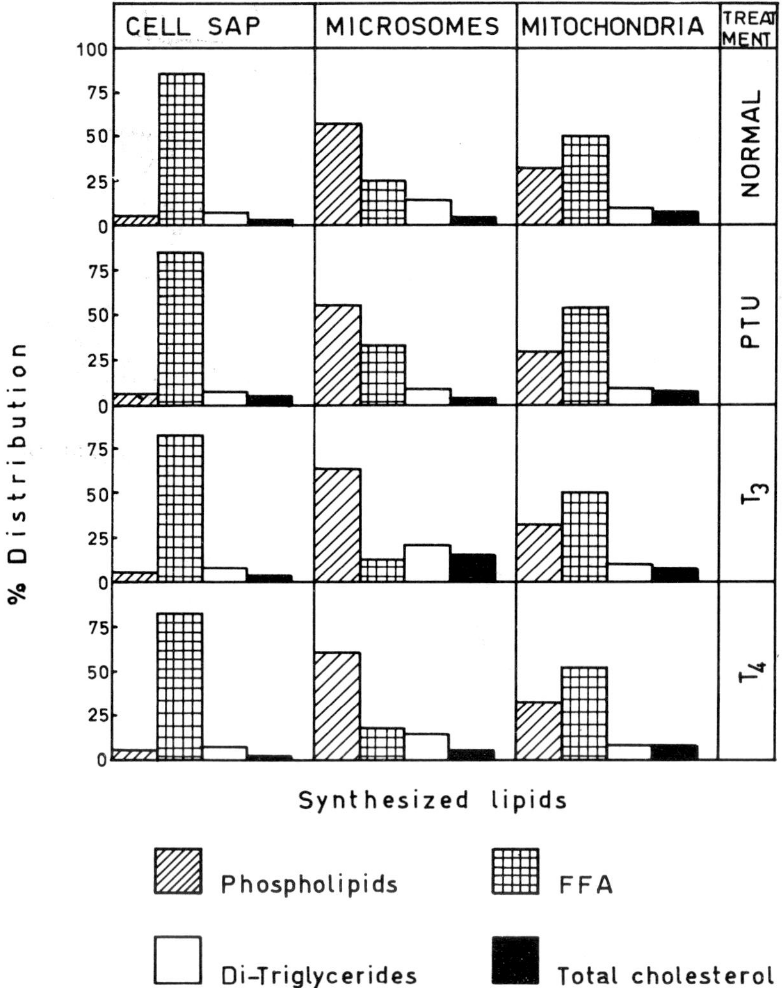

Fig. 4 – Pattern of lipids synthetized by rat liver subcellular fractions in the thyrotoxic state.

drial enzymatic activities have been seen to be influenced by the thyrotoxic state, thus excluding the possibility that the hormonal signal cannot go beyond the mitochondrial membrane. Previous evidence from our laboratory, namely the insensitivity of this system both to such hypolipidemic drugs as clofibrate[22] and to other fatty acid synthesis inhibitors, in addition to the lack of any response to the thyroid hormones, led us to conclude that in all probability mitochondria have an entirely secondary or non-existent role in hepatic fatty acid synthesis.

Table III. *Hepatic cyclic adenosine 3',5'-monophosphate level in thyrotoxic rats.*

Condition	*Cyclic AMP content (pmoles/mg tissue)*	*Relative amount*
Normal	0.96 ± 0.07	1.00
Propylthiouracil	0.88 ± 0.17	0.92
Triiodothyronine	0.57 ± 0.13	0.59
Thyroxine	0.84 ± 0.12	0.88

Another interesting result which emerges from the present study is the significantly lower cyclic AMP level observed in liver following 19 days of triiodothyronine administration to rats. These data may prove that, whereas in adipose tissue hyperthyroidism due to lypolisis stimulation raises the total long chain fatty acyl-CoA level, thus reducing fatty acid synthesis, in liver these inhibitory compounds are quickly removed by enhanced oxidation with a simultaneous decrease in cyclic nucleotide levels as well. Therefore, the possible increase in some enzyme-protein contents caused by thyroid hormones and the reduced cAMP amounts in liver could account for the increase in fatty acid synthesis observed in the hyperthyroid state. This last point, however, deserves further investigation.

References

1. Barker S.B.: Physiol. Rev., *31,* 205 (1951).
2. Tata J.R.: In Lutwack G. and Kritchevsky D.: Actions of hormones on Molecular Processes. Wiley, New York, p. 58 (1964).
3. Martius C. and Hess B.: Arch. Biochem. Biophys., *33,* 486 (1951).
4. Pitt-Rives R. and Tata J.R.: The Thyroid Hormones, Pergamon, London p. 115, 1959.
5. Tata J.R., Ernster L., Lindberg O., Arrhenius E., Pedersen S., and Hedman R.: Biochem. J., *86,* 408 (1963).
6. Gustafsson R., Tata J.R., Lindberg O. and Ernster L.: J. Cell. Biol., *26,* 555 (1965).
7. Tata J.R.: In: Tager J., Papa S., Quagliariello E. and Slater E.C.: Regulation of metabolic processes in mitochondria. Elsevier, Amsterdam, p. 489 (1966).
8. Kroon, A.M. and Saccone C.: The Biogenesis of Mitochondria. Academic Press, New York (1974).
9. Gadaleta M.N., Barletta A., Saccone C. and De Leo T.: Rass. Med. Sper., *5-6,* 146 (1970).
10. Gadaleta M.N., Barletta A., Caldarazzo M., De Leo T. and Saccone C.: Eur. J. Biochem., *30,* 376 (1972).
11. Aloni Y. and Attardi G.: Proc. Nat. Acad. Sci. U.S.A., *68,* 1757 (1971).
12. Gadaleta M.N., Di Reda N., Bove G. and Saccone C.: Eur. J. Biochem., *51,* 495 (1975).
13. Gallerani R., Saccone C., Cantatore P. and Gadaleta M.N.: FEBS Letters, *22,* 37 (1972).

14. Gadaleta M.N., Caldarazzo M., and Saccone C.: Boll. Soc. It. Biol. Sper., *49,* 247 (1973).
15. Landriscina C., Gnoni G.V. and Quagliariello E.: Eur. J. Biochem., *29,* 188 (1972).
16. Hamburgh M. and Flexner L.B.: J. Neurochem., *1,* 279 (1956).
17. Eayrs J.T.: In Michael, R.P.: Endocrinology and Human behaviour. Oxford University Press, London, p. 239, (1968).
18. Colton D.G., Mehlman M.A. and Ruegamer W.R.: Endocrinology, *90,* 1521 (1972).
19. Volpe J.J. and Kishimoto Y.: J. Neurochem., *19,* 737 (1972).
20. Diamant S., Gorin E. and Shafrir E.: Eur. J. Biochem., *26,* 553 (1972).
21. Faas F.H., Carter W.J. and Wynn J.: Endocrinology, *91,* 1481 (1972).
22. Landriscina C., Gnoni G.V. and Quagliariello E.: Biochem. Med., *12,* 356 (1975).

IDENTIFICATION OF THE ORIGINS OF CIRCULATING ENZYMES

D.W. Moss

Summary

The diagnostic value of measurements of enzyme activity in serum is greatly enhanced if the observed changes can be attributed to disease of a particular organ or tissue. Often circumstantial evidence allows this to be done with sufficient certainty. However, additional objective criteria are always useful and are frequently essential. The most direct evidence is derived from comparison of enzymes obtained from tissues with the corresponding enzymes of serum. This depends on the existence of differences between homologous enzymes from various tissues which are preserved after the enzymes are released. It is essential to base these comparisons on enzyme preparations of equivalent purity, since the properties of some enzymes (e.g., alkaline phosphatase) are affected by the matrix in which they are dissolved. Enzymes which appear similar when extracted *in vitro* from various tissues may show differences after their release into the circulation *in vivo:* for example, aspartate transaminase entering serum from skeletal muscle appears to be less saturated with pyridoxal phosphate than the enzyme released from liver. The origin of this interorgan difference is not yet known.

The established approach of measuring an enzyme in serum which is also present in high concentrations in a particular organ can be refined by studying the distribution of the enzyme amongst the individual cell-types which together constitute the organ. We have measured alkaline phosphatase, 5'-nucleotidase and γ-glutamyl transferase in constituent cells of rat liver separated by perfusion of the organ with collagenase solution. These enzymes are differently distributed amongst the various cell types and also respond differently to ligation of the bile duct. Investigations of this nature may provide additional insights into the interpretation of enzyme levels in serum.

The diagnostic value of measurements of enzyme activity in serum is greatly enhanced if the changes observed can be attributed to disease of a particular organ or tissue. Often, circumstantial evidence allows this to be done with sufficient certainty. However, additional objective criteria are always useful and are frequently essential. Several approaches have been

Royal Postgraduate Medical School London W12 OHS, England

developed by which an element of organ-specificity can be added to the data provided by the clinical enzymologist. These approaches include the comparison of patterns of relative activity of several enzymes in the serum in disease with those of various tissues; the measurement of activities in serum of enzymes which are present in exceptionally high concentrations in particular tissues, and the direct comparison of the properties of an enzyme in serum with those of the homologous enzyme in tissue.

All these approaches retain their usefulness and continue to undergo refinement. The purpose of this communication is to review some recent developments in the field of organ-specificity in diagnostic enzymology.

Patterns of relative activity of several enzymes in serum

Changes in enzyme activity in serum in disease commonly affect not one but several enzymes. This is especially the case when the changes originate from the leakage of enzymes from damaged or necrotic cells. The relative activities of the affected enzymes in serum may be matched to the corresponding pattern of tissues in which damage is suspected, to indicate the probable location of the lesion[1]. In some cases, comparison of the serum enzyme patterns with the enzymic composition of a whole tissue is inappropriate: in the well-known application of this principle to the elucidation of relative activities of aspartate and alanine aminotransferases in acute infective hepatitis[2], for example, the appropriate comparison is with the enzymic composition of hepatic cytoplasm, rather than with that of whole liver[3].

The study of relative enzyme activities in serum becomes easier, and more appropriate to present-day practice in clinical enzymology, with the introduction of multichannel analysis in which several enzyme activities are routinely determined on each serum sample. However, an incomplete enzyme "profile" may be misleading in the absence of definite clinical data. For example, both aminotransferases are frequently included in multichannel profiles, but other enzymes such as creatine kinase less commonly so. Consequently, slight or moderate comparable elevations of aspartate and alanine aminotransferases may be found, suggesting at first sight a minimal parenchymal lesion: but, if creatine kinase is also determined, a marked elevation of this enzyme may reveal a myopathy, perhaps due to metabolic or toxic causes.

A recent contribution to the interpretation of relative enzyme activities has come from the observation that the total aspartate aminotransferase of cardiac and skeletal muscle apparently contains a greater ratio of apoenzyme to holoenzyme than is the case for the corresponding enzyme of liver, and that this difference persists after the enzymes are released into the circulation[4,5,6]. Thus, the magnitude of the effect of supplementation with pyridoxal phosphate on the activity of aspartate aminotransferase in serum may provide an additional criterion by which liver and muscle may be distinguished as potential sources of the enzyme (Fig. 1).

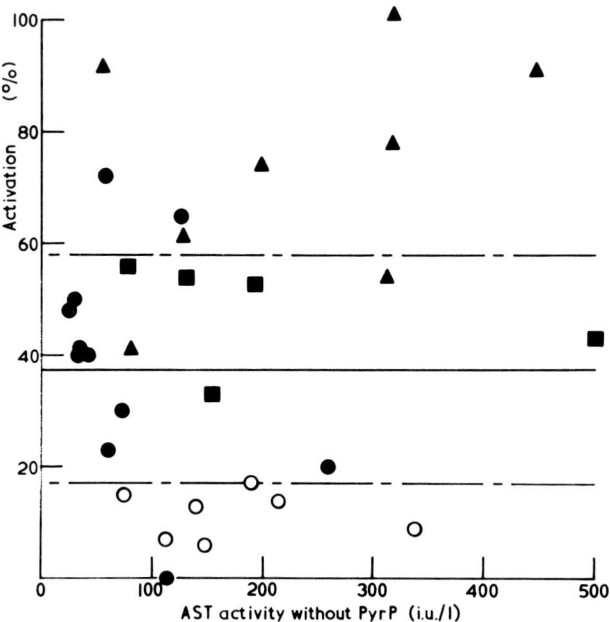

Fig. 1 – Percentage activation by pyridoxal phosphate (PyrP) of aspartate aminotransferase (AST) activity in serum in various conditions as a function of the activity without added pyridoxal phosphate: ○, liver disease; ●, congenital or metabolic myopathies; ■, after cardiac surgery with extracorporeal circulation; ▲, myocardial infarction. The solid horizontal line represents the average percentage activation in sera with normal AST activities and the broken lines one standard deviation above and below this. (Data from Moss[6] and S. Ratnaike & D.W. Moss; unpublished).

Association of certain enzymes with particular cells or organs

The ability of cells in general to carry out the major metabolic transformations essential to life implies that many of the enzymes of interest to the clinical enzymologist are widely distributed and, therefore, that changes in their activity in serum do not by themselves convey organ-specific information. Nevertheless, several enzymes are essentially organ-specific, notably the tartrate- inhibited form of acid phosphatase derived from prostatic cells, and creatine kinase, which essentially reflects only changes affecting striated (skeletal and cardiac) muscle.

All organs of the body are composed of cells of more than one type. A further refinement of organ-specific diagnosis is therefore possible if changes in enzymic activity originating from a particular organ can be localized to specific cells within that organ. The liver consists of cells of various morphologies and functions which may be selectively affected in disease. We have

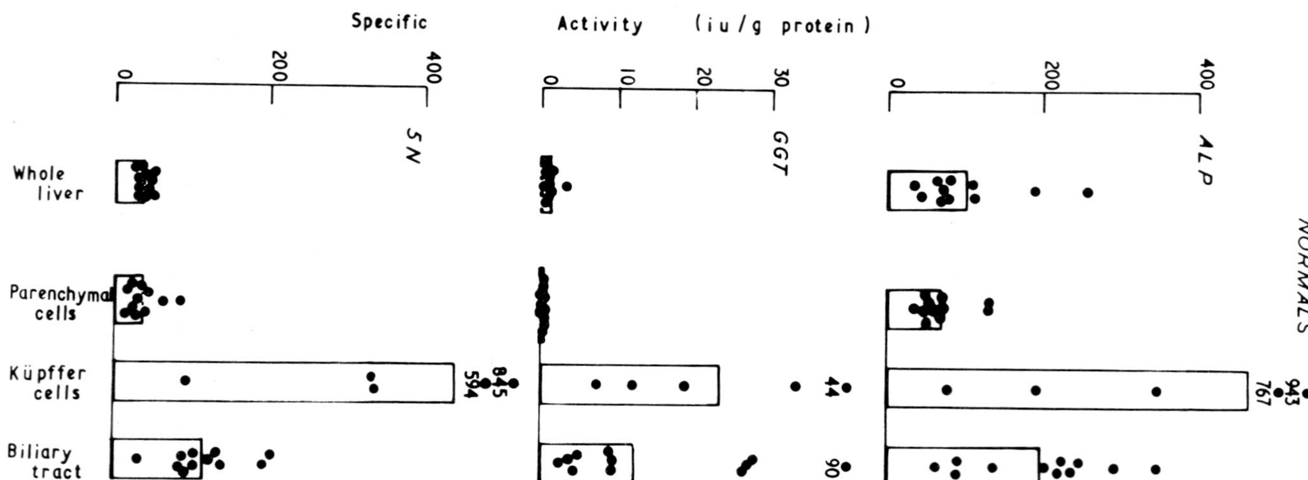

Fig. 2 – Specific activities (in./g protein) of alkaline phosphatase (ALP), γ-glutamyl transferase (GGT) and 5,-nucleotidase (5N) in normal whole rat liver, and after perfusion with collegenase solution and separation into parenchymal-cell, Küpffer-cell and biliary-tract fractions. The average specific activites are shown by the vertical bars. (A.M. Wootton, G. Neale and D.W.Moss; unpublished).

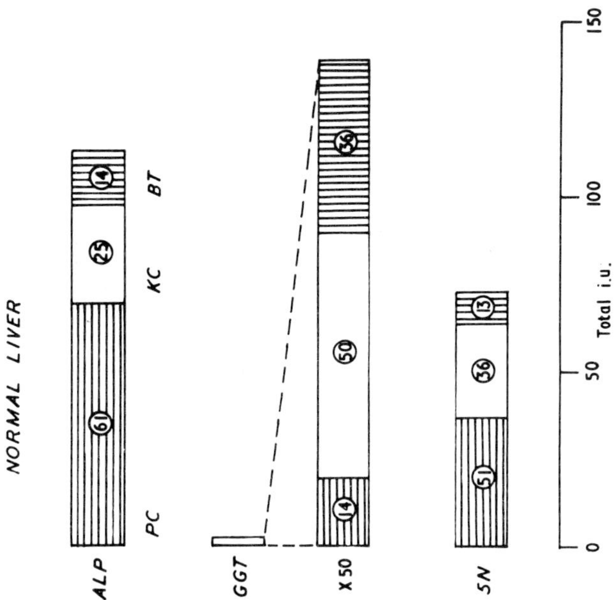

Fig. 3 – Total activities of alkaline phosphatase (ALP), γ-glutamyl transferase (GGT) and 5'-nucleotidase (5N) in normal rat liver. The total activity (represented by the length of each horizontal bar) is divided up into sections representing the contribution of parenchymal (PC), Küpffer (KC) and biliary-tract (BT) cells for each enzyme. The circles numbers indicate the percentage of the total enzyme contributed by each cell type. The bar for GGT has been expanded 50-fold for clarity. (A.M. Wootton, G. Neale and D.W. Moss; unpublished).

recently been engaged in isolating cells of various types from rat liver by perfusion with collagenase solution, in order to study the activities of several enzymes of diagnostic interest in them[7].

Three fractions are obtained: parenchymal cells; a "Küpffer cell" fraction (obtained by digestion of the supernatant from the parenchymal-cell fraction with Pronase to remove contaminant parenchymal cells), and a residual fraction consisting of the larger elements of the biliary tree and connective tissue. The "Küpffer cell" fraction is probably heterogenous, containing epithelial cells from the biliary tract as well as true Küpffer cells.

The specific activities of three enzymes (alkaline phosphatase, γ-glutamyl transferase and 5'-nucleotidase) in each of these fractions from normal rat livers are shown in Fig. 2. When the estimated relative masses of the various types of liver cells in the whole organ and their respective protein contents are taken into account, the contribution of each cell type to the total enzyme activities in the liver is as shown in Fig. 3.

These Results show that, although the specific activities of alkaline phosphatase and 5'-nucleotidase in the Kupffer-cell and biliary tract fractions are higher than in parenchymal cells, the latter contribute more than half of the total activity of these enzymes in whole liver, because of the greater mass of these cells in the organ. The γ-glutamyl transferase activity of whole, normal rat liver is low, as previously reported[8,9], but again, the highest concentrations of this enzyme are found in the biliary-tract and Küpffer-cell fractions. However, the concentration of this enzyme in parenchymal cells is particularly low, so that these cells make a correspondingly small contribution to total liver γ-glutamyl transferase activity, in spite of their greater mass.

The activity of alkaline phosphatase in parenchymal cells rises rapidly after ligation of the bile duct and it is presumably from these cells that the raised activity of this enzyme in the serum originates. Some increase in alkaline phosphatase activity in non-parenchymal cells and of γ-glutamyl transferase in parenchymal and biliary-tract cells is noticeable 7 days after occlusion. Changes in the specific activity of 5'-nucleotidase are not marked in any cell fraction following bile-duct ligation. Thus, these results do not provide a ready explanation, such as increased production of these enzymes in the liver, for the rapid rise of serum 5'-nucleotidase and γ-glutamyl transferase seen after biliary obstruction.

Further extensions of the specificity of enzyme measurements will probably result from characterization of enzymes by their subcellular, as well as their cellular, origins. The utility of this approach has already been shown by measurement of mitochondrial enzymes such as glutamate dehydrogenase in serum. Zonal ultracentrifugation of biopsy specimens has been used to compare the intracellular distribution of several enzymes in normal and pathological human jejunum[10] and to examine changes in enzyme patterns in human liver after iron-overload[11]. Furthermore, preliminary work indicates that combination of the techniques of isolation of specific cell types and zonal ultracentrifugation promises to provide additonal insights into enzyme changes in liver disease. When cells isolated from normal rat liver by collagenase perfusion are examined in the ultracentrifuge, alkaline phosphatase, γ-glutamyl transferase and γ-nucleotidase are recovered in the membrane fraction of biliary-tract cells. However, the small amount of γ-glutamyl transferase activity of parenchymal cells is found in the cytoplasmic fraction, in contrast to the predominantly membrane-bound location of the other two enzymes in these cells. (T.J. Peters and A.M. Wootton; unpublished).

Comparison of properties of enzymes in serum and tissues

The demonstration of organ-specific characteristics after the release of enzymes into the circulation has extended our understanding of the chemical pathology of several enzymes, as well as adding to the diagnostic value of routine investigations. A notable example of this approach is the characterization of alkaline phosphatase in serum[12]. Comparisons of enzymes derived from tissues and from serum may be invalidated if the degrees of purity of the respective samples are dissimilar. For example, comparisons of kinetic properties of tissue and serum alkaline phosphatases were found to be valid only if the enzymes from each source were submitted to similar purification procedures[13].

Apparent dissimilarities between enzymes in serum and corresponding tissue enzymes may arise if the former become aggregated or attached to serum proteins or tissue débris. The formation of complexes between alkaline phosphatase and other proteins, probably lipoproteins, was noted early in the study of the electrophoretic heterogeneity of this enzyme[14], and the association of alkaline phosphatase with immunoglobulin/IgG has been held to account for the appearance of electrophoretically-slow zones of the enzyme in some sera[15]. Several enzymes have been found to be associated with particulate elements in serum (koinozymes), derived apparently from the fragmentation of the plasma membrane of damaged parenchymal cells[16]. Effects such as these, which modify the observed properties of enzymes in serum, must be taken into account when attempting to identify the origins of circulating enzymes by a study of their properties. However, if correctly interpreted, they may also provide new information on the ways by which enzymes are released from cells in disease.

Conclusions

Considerable progress has been made and will continue to be made towards fulfilling the aim of the clinical enzymologist to provide organ-specific diagnostic information. These practical advances have been based on developments in fundamental understanding, particularly of the nature and origins of structural diversity amongst functionally-similar enzymes. In this latter respect, enzymology in general owes a considerable debt to the problem-orientated approach of the clinical enzymologist.

References

1. Henley K.S., Schmidt E. and Schmidt F.W.: Enzymes in Serum (Charles C. Thomas, Springfield, Ill., 1966).
2. De Ritis F., Coltorti M. and Giusti G.: Clin. Chim. Acta, *2,* 70 (1957).
3. Schmidt F.W.: Verh. Deutsch. Ges. Inn. Med., *70,* 612 (1964).
4. Cheung T. and Briggs M.H.: Clin. Chim. Acta, *54,* 127 (1974).
5. Rosalki S.B. and Bayoumi R.A.: Clin. Chim. Acta, *59,* 357 (1975).
6. Moss D.W.: Clin. Chim. Acta, *67,* 169 (1976).
7. Wootton A.M., Neale G. and Moss D.W.: Clin. Chim. Acta, *61,* 183 (1975).
8. Ideo G., Morganti A. and Dioguardi N.: Digestion, *5,* 326 (1972).
9. Kryszewski A.J., Neale G., Whitfield J.B. and Moss D.W.: Clin. Chim. Acta, *47,* 175 (1973).
10. Peters T.J., Heath J.R., Jones P.E., and Peacham A.D.: Gut, *16,* 826 (1975).

11. Peters T.J., and Seymour C.A.: Clin. Sci. Mol. Med., *50,* 75 (1976).
12. Moss D.W.: Enzyme, *20*, 20 (1975).
13. Moss D.W., Campbell D.M., Anagnostou-Kakaras E., and King E.J.: Biochem. J., *81,* 441 (1961).
14. Moss D.W. and King E.J.: Biochem. J., *84,* 192 (1962).
15. Nagamine M., and Ohkuma S.: Clin. Chim. Acta, *65,* 39 (1975).
16. De Broe M.E., Borgers M. and Wieme R.J.: Clin. Chim. Acta, *59,* 369 (1975).

ANTI-CHOLINESTERASE ACTION OF NON-DEPOLARISING MUSCLE RELAXANTS

J. King

Summary

The anti-acetylcholinesterases achieve a muscle-relaxant effect through the accumulation of acetylcholine depolarising the motor end plate, while the non-depolarising drugs act post-synaptically by occupying the receptor sites for acetylcholine. Tubocurarine, gallamine and pancuronium are three non-depolarising muscle relaxants producing typical curariform blockade. These have been examined for inhibition of human serum cholinesterase hydrolysis of 50 μM benzoylcholine and of 5 mM acetylthiocholine and human erythrocyte acetylcholinesterase hydrolysis of 500 μM acetylthiocholine.

All three relaxants were found to inhibit benzoylcholine hydrolysis and the degree of inhibition varied with the cholinesterase phenotype. This was particularly evident with pancuronium which, like the depolarising drug succinyldicholine, could be used to differentiate the phenotypes. Furthermore, and unlike succinyldicholine, pancuronium at in vivo concentrations had a potent inhibitory effect on the hydrolysis of benzoylcholine by the usual cholinesterase isoenzyme. Neither tubocurarine nor gallamine inhibited benzoylcholine hydrolysis at in vivo concentrations.

Pancuronium inhibition of acetylthiocholine hydrolysis was less phenotype dependent and occurred at concentrations higher than those occurring in vivo.

Hydrolysis of acetylthiocholine by haemolysates was also inhibited by pancuronium and by tubocurarine, the inhibition curves closely following those of atypical homozygote serum. This last observation is particularly interesting in view of the substrate specificities of the E_1^a E_1^a enzyme.

In the light of modern theories of neuro-muscular transmission, the classification of muscle–relaxant drugs becomes increasingly complex. In the classical picture which satisfied simple stereochemical, pharmacological and physiological precepts, such drugs were separated into two categories, depolarising and non-depolarising. The former were long thin molecules which achieved their effect by depolarising the motor end plate in the same way as acetylcholine. Indeed high levels of acetylcholine itself produce blockade and it is through this action that the antiacetylcholinesterases achieve their effect.

Department of Biochemistry, Royal Infirmary, Glasgow, U.K.

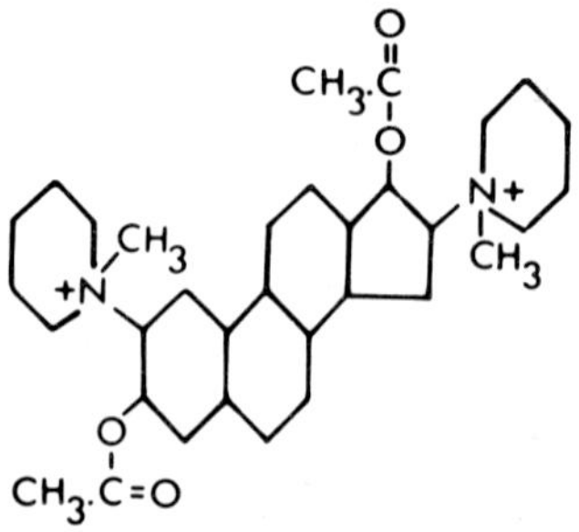

Pancuronium, Pavulon.[R]

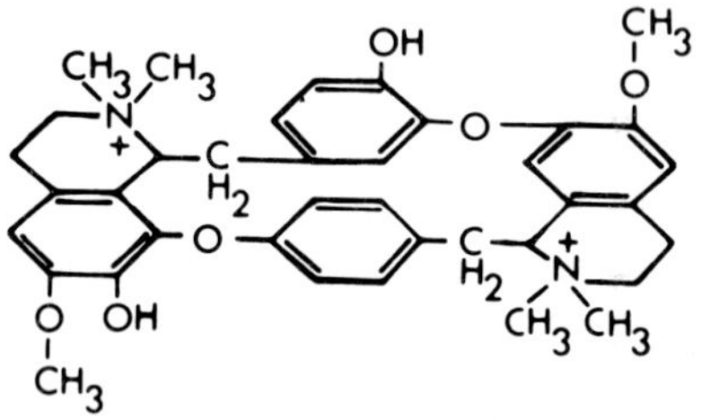

Tubocurarine, Tubarine.[R]

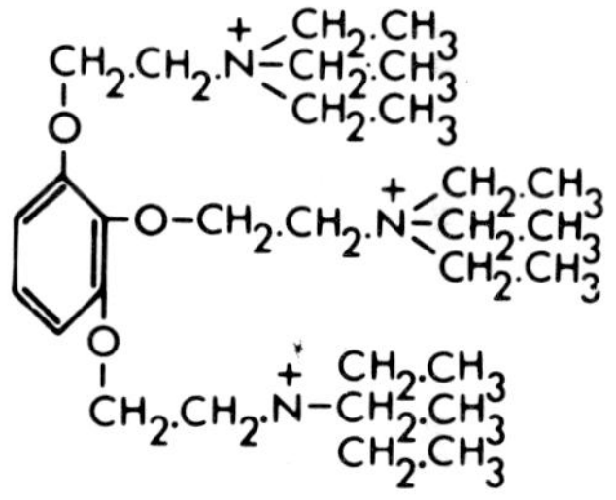

Gallamine, Flaxedil.[R]

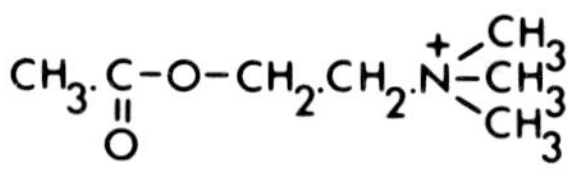

Acetylcholine.

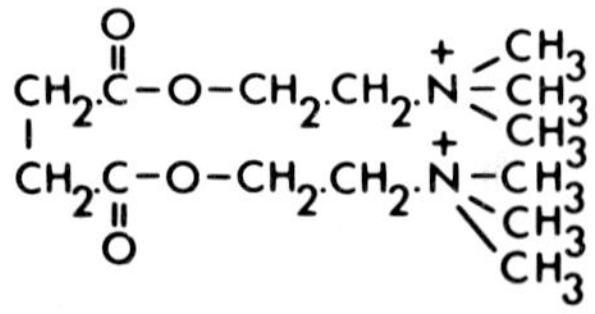

Succinyldicholine, Scoline.[R]
Suxamethonium.

Fig. 1 – Formulae of some cholinesterase inhibitors.

These are therefore classed as depolarising agents. The non-depolarising relaxants were bulky molecules which acted post-synaptically by occupying the receptor sites for acetylcholine. Succinyldicholine and tubocurarine are typical examples of depolarising and non-depolarising muscle relaxants respectively and their structural formulae are shown in Figure 1.

Pancuronium bromide is a recently-introduced non-depolarising muscle relaxant. The structural formula (Figure 1) shows it to be steroidal with two methyl pyridine and two acetyl residues attached. It is interesting to note that in both instances the ester oxygens are separated by two carbon atoms from the quaternary nitrogens, similar to the spatial arrangement of acetylcholine and other cholinesterase substrates. Pharmacologically the neuromuscular block produced by pancuronium is of curariform type and stereochemically the molecule is typically non-depolarising. Nevertheless the presence of the methylated quaternary nitrogens and acetyl groups suggested possible anticholinesterase potential and this was accordingly investigated. Unlike succinyldicholine, pancuronium is reputedly not hydrolysed by cholinesterase and therefore any inhibition observed would not be an index of its rate of enzymic hydrolysis but of a simple blocking of the active site. Tubocurarine and gallamine, two other non-depolarising drugs were also examined for comparison.

The inhibition of the serum cholinesterases was studied employing a 50μM benzoylcholine substrate (Kalow and Lindsay, 1955) at 25°C on a Unicam SP8000 recording spectrophotometer or using 5mM acetylthiocholi-

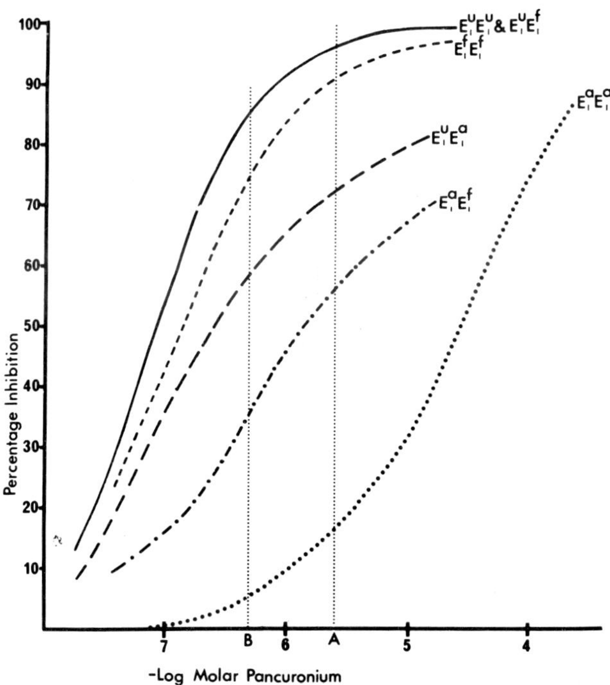

Fig. 2 – Pancuronium inhibition of benzoylcholine hydrolysis by serum cholinesterases.

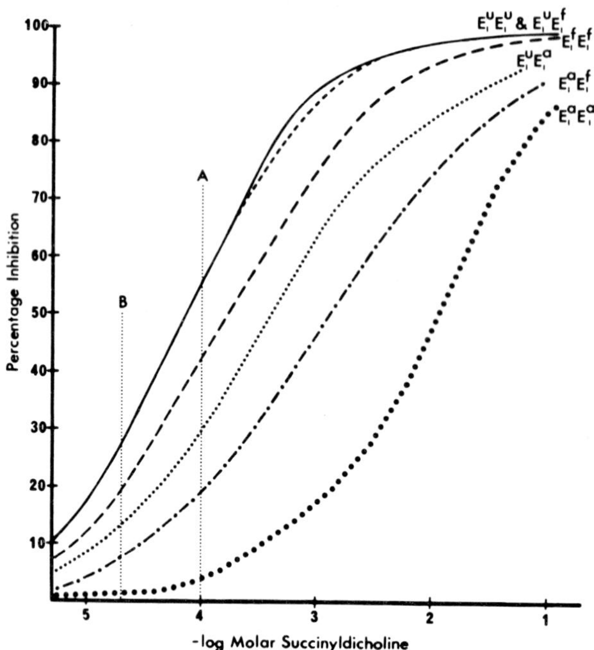

Fig. 3 – Succinyldicholine inhibition of benzoylcholine hydrolysis by serum cholinesterases. (King and Griffin, 1973).

ne (King and Ritchie, 1973) at 37°C on an LKB8600 reaction rate analyser. Erythrocyte acetylcholinesterase experiments were also conducted at 37°C on an LKB8600 but a 500μM acetylthicoline concentration. Pancuronium bromide (Pavulon) was obtained from Organon Laboratories, tubocurarine chloride (Tubarine) from Burroughs Wellcome, gallamine triethiodide (Flaxedil) from May and Baker Ltd. and succinyldicholine chloride from BDH Chemicals.

Figure 2 illustrates the inhibition of the serum cholinesterase hydrolysis of benzyoylcholine by pancuronium. In general this is very similar to that of succinyldicholine in Figure 3 (King and Griffin, 1973) and indicates that pancuronium could readily be employed for differentiating the allelic variants but for the lack of contrast in effect between the usual and fluoride-resistant enzymes.

The plasma levels of most injected substances show a biphasic reduction, the first rapid, followed by a more gradual decline. The half life of the first phase is generally of the order 4-7 minutes. Five circulation times or about $7\frac{1}{2}$ minutes is usually accepted as necessary for even distribution throughout the plasma of injected substances and therefore it is inferred that the first ra-

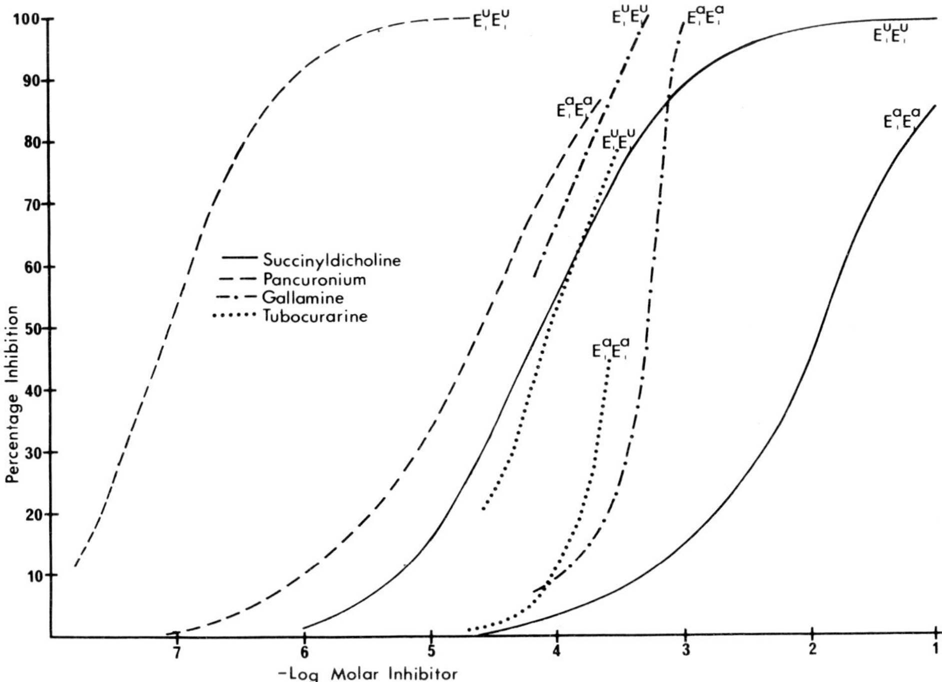

Fig. 4 – Comparative inhibition of serum cholinesterase by pancuronium, succinyldicholine, tubocurarine and gallamine.

pid phase represents a transfer from the plasma to the interstitial fluid (Kalow, 1959). Taking the circulating plasma as 5% of body weight and extracellular fluid as 20-25% or 3.5 litres and 14-18 litres respectively in a 70 Kg person, it can generally be assumed that 20-30 minutes after injection only 20 % of the calculated initial concentration will remain in the plasma.

This discloses a difference between the inhibitions by succinyldicholine and pancuronium. The maximal dose of 100 mg succinyldicholine, assuming immediate even distribution, would give a plasma level of 100 μM and after 20-30 minutes, assuming no hydrolysis, the concentration would be 20 μM. Neither assumption of course is justified but even these maximum possible levels of 10^{-4} M and 2 x 10^{-5} M (A and B in Figure 3) barely come into the range of concentrations at which succinildycholine inhibits benzoylcholine hydrolysis in vitro. On the other hand a full dose of pancuronium of 7 mg would result in an initial plasma level of about 2.5 μM and after 20-30 minutes, 0.5 μM. Both of these in vivo levels (A and B respectively, Figure 2) are well within the effective range of in vitro Pavulon inhibition of benzoylcholine hydrolysis the initial level representing near maximum inhibition of the usual homozygote. Neither gallamine nor tubocurarine exhibit inhibitions at such concentrations (Figure 4). A dose of 15 mg Tubarine gives an

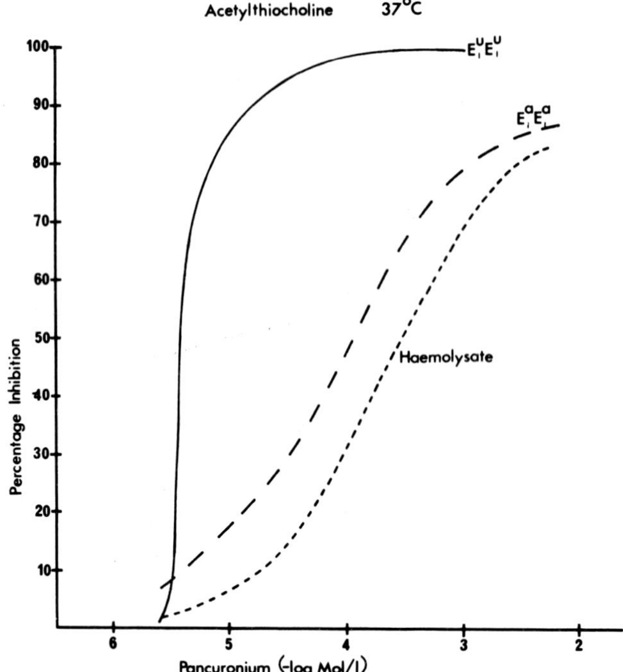

Fig. 5 – Pancuronium inhibition of acetylthiocholine hydrolysis by serum cholinesterases (5mM) and erythrocyte acetylcholinesterase (500 μM).

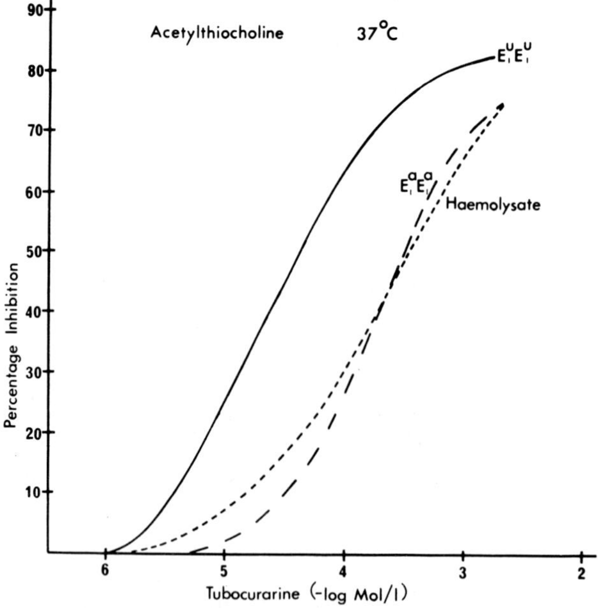

Fig. 6 – Tubocurarine inhibition of acetylthiocholine hydrolysis by serum cholinesterases (5mM) and erythrocyte acetylcholinesterase (500 μM).

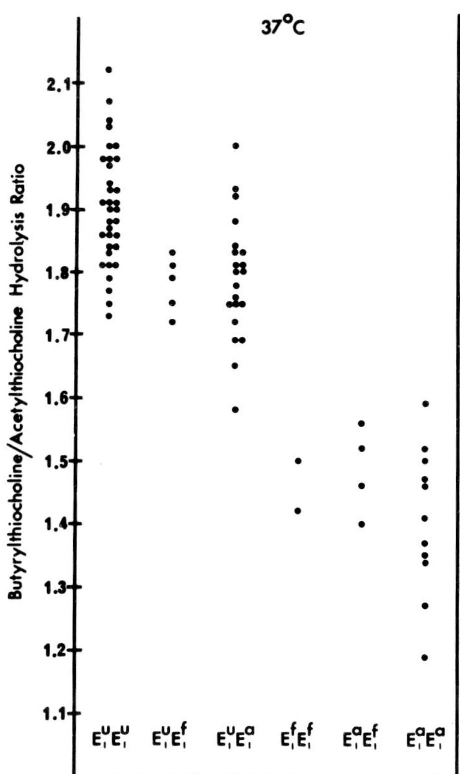

Fig. 7 – Relative hydrolysis rates at 37°C of Butyrylthiocholine (3mM) and acetylthiocholine (5mM) by serum cholinesterase variants.

initial plasma concentration of 5 μM and 120 mg gallamine about 40 μM.

The experiments into the effect of the non-depolarising relaxants on the hydrolysis of acetylthiocholine by usual and atypical homozygous sera confirm the inhibitory action of pancuronium and tubocurarine. There is however a relative difference in that an initial plasma level of 2.5 μM pancuronium is too low to have any effect on either variant (Figure 5) while an initial plasma level of 5 μM tubocurarine would only slightly inhibit the usual enzyme (Figure 6). The inhibition of erythrocyte acetylcholinesterase which with both pancuronium and tubocurarine closely resembles that of the atypical cholinesterase is particularly interesting in view of the substrate specificities of this variant (Figure 7) which approach most closely to that of acetylcholinesterase (King and Ritchie, 1973).

Acetylcholinesterase does not hydrolyse succinyldicholine and it is a similar lack of affinity in the atypical cholinesterase (Figure 8) which results in Suxamethonium apnoea in individuals homozygous for this variant. The re-

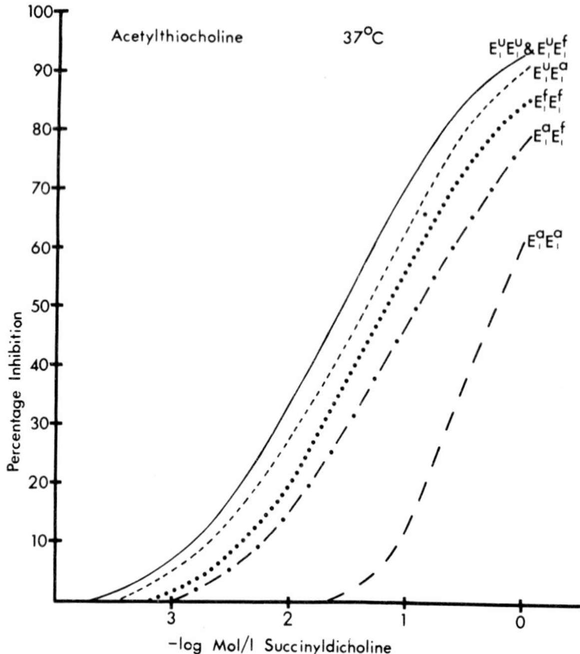

Fig. 8 – Succinyldicholine inhibition at 37°C of 5mM acetylthiocholine by serum cholinesterases.

sults with benzoycholine hydrolysis would tend to suggest that a combination of succinyldicholine and pancuronium would in all cases result in apnoeic responsens, certainly if pancuronium was administered first. Under these circumstances the phenotypes which readily hydrolyse succinyldicholine thereby limiting the duration of its action would appear to be largely inhibited by pancuronium while those almost unaffected by pancuronium are unable to hydrolyse succinyldicholine. On the other hand the results from the acetylthiocholine studies indicate that pancuronium does not inhibit any cholinesterase variant at concentrations achieved in vivo. These latter observations are borne out in anaesthetic practice. This in turn highlights the dangers in extrapolating from results of in vitro experiments employing nonphysiological substrates and reaction temperatures.

References

1. Kalow W.: Canad. J. Physiol. Pharmacol., *42,* 161-168 (1959).
2. Kalow W. and Lindsay H.A.: Canad. J. Biochem. Physiol., *33,* 568-580 (1955).

3. King J. and Griffin D.: Brit. J. Anaesth., *43,* 450-454 (1973).
4. King J. and Ritchie G.: 5th International Congress on Clinical Enzymology, Venice (1973).

HUMAN HYPOXANTINE-GUANINE PHOSPHORIBOSYLTRANSFERASE: PURIFICATION AND PROPERTIES

C. Salerno and A. Giacomello

Summary

Hypoxanthine guaine phosphoribosyltransferase (EC 2.4.2.8) was purified to apparent homogeneity from erythrocytes. In the presence of the purified enzyme, pyrophosphate and magnesium ions, IMP pyrophosphorolysis was demonstrated. The possible role of this reaction in purine metabolism is discussed.

Hypoxanthine guanine phosphoribosyltransferase (HGPRT)* catalyzes the reaction between phosphoribosylpyrophosphate (PRPP) and hypoxanthine or guanine to form inosine 5'-phosphate (IMP) or guanosine 5'-phosphate (GMP) respectively.

The absence or partial deficiency of this enzyme, observed in the Lesch-Nyahn syndrome and in some adult gouty subjects, has been shown to be associated with purine overproduction and elevated cellular concentration of PRPP[1]. These findings have focused interest on this enzyme.

In the present paper we describe the purification of HGPRT from human erythrocytes and some of its properties.

Experimental procedure

Materials: IMP, hypoxanthine and xanthine oxidase were purchased from Boeringer A.G. . PRPP tetrasodium salt was obtained from Sigma Inc. . DEAE cellulose from Whatman Inc. Dowex 1-X2 50-100 Mesh J.T. Baker Chemical Co., and Sephadex CM 50 and G 150 were obtained from Pharma-

Istituto di Chimica Biologica dell'Università di Roma e Centro di Biologia Molecolare del C.N.R. - Roma; Istituto di Reumatologia dell'Università di Roma.

*Abbreviations: HGPRT = Hypoxanthine guanine phosphoribosyltransferase; APRT = Adenine phosphoribosyltransferase; XOD = Xanthine oxidase; PRPP = Phosphoribosylpyrophosphate; IMP = Inosine 5'-phosphate; GMP = Guanosine 5'-phosphate.

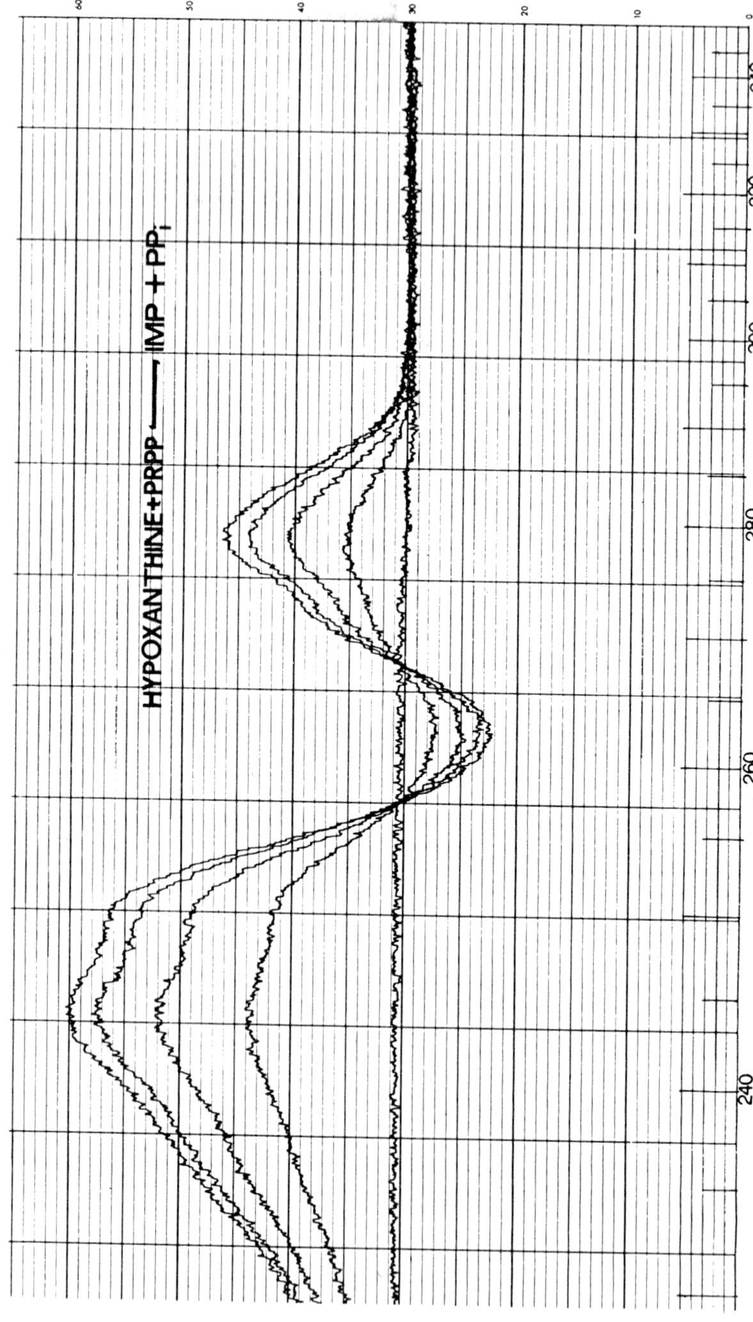

Fig. 1 – Differential absorption spectra recorded at 15 min. intervals between a solution containing 0.1 M Tris HCl pH 7.4, 4×10^{-2}M $MgSO_4$, 1.2×10^{-4}M hypoxanthine, 7×10^{-4}M PRPP and 1 ml HGPRT in a final volume of 2 ml and a solution containing the same substances without PRPP.

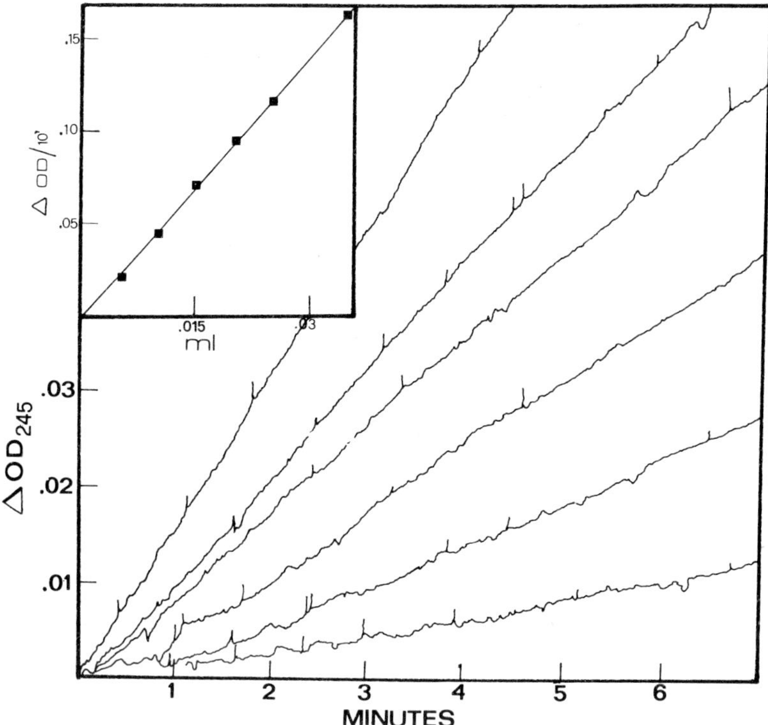

Fig. 2 – Hypoxanthine phosphoribosyltransferase assay. Optical density variation at 245 nm as a function of time. Inset: initial optical density change (referred to 10 min) as a function of enzyme concentration. Incubation mixtures contained 7 x 10^{-4} M PRPP, 1.2 x 10^{-4} M hypoxanthine, 4 x 10^{-2} M $MgSO_4$, 0.1 M Tris HCl buffer pH 7.4 and HGPRT (0.005 to 0.035 ml) in a final volume of 2 ml. The experiments were performed at 37°C.

cia. All other reagents were high purity commercial samples from Merck A.G. and Fluka A.G.

Enzyme assay: The HGPRT catalyzed reaction may be followed spectrophotometrically. Fig. 1 shows differential absorption spectra recorded at 15 min. intervals between a solution containing 0.1 M Tris-HCL, pH 7.4, 4 x 10^{-2} M $MgSO_4$, 1.2 x 10^{-4} M hypoxanthine, 7 x 10^{-4} M PRPP and 1 ml HGPRT in a final volume of 2 ml and a solution containing the same substances without PRPP. Fig. 2 shows that the initial velocity of HGPRT catalyzed reaction (as monitored from the increase in absorbance at 245 nm) was linear and proportional both to time and enzyme concentration.

Enzyme purification: The enzyme was purified from hemolysed human

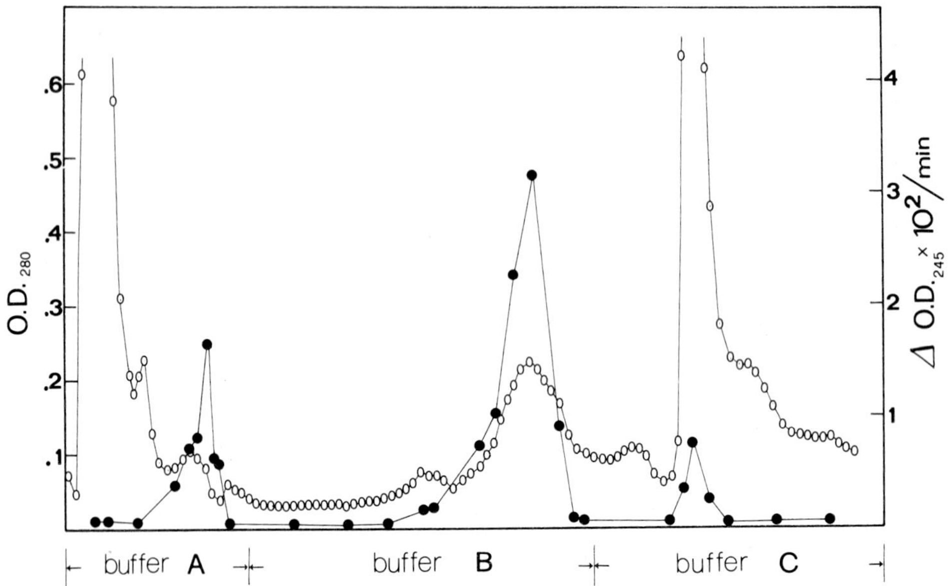

Fig. 3 – CM 50 Sephadex elution profile of HGPRT. ooo optical density at 280 nm; ●●● enzyme activity. The column (2 x 60 cm) was eluted as described in the text.

erythrocytes. After fractionation on DEAE cellulose according to Arnold and Kelley[2], the hemolysate was dialyzed against 10 mM K^+ – citrate buffer, pH 5.8, containing 1 mM mercaptoethanol and 5 mM $MgSO_4$ and loaded on CM Sephadex C50 column (2 x 60 cm) equilibrated with the same buffer. A stepwise elution was carried out with the following buffers:

a. 10 mM citrate, 1 mM mercaptoethanol, 5 mM $MgSO_4$, pH 5.8 (1700 ml)
b. 10 mM citrate, 1 mM mercaptoethanol, 5 mM $MgSO_4$, 25 mM KCl, pH 6.2 (3800 ml)
c. 50 mM Tris HCl, 1 mM mercaptoethanol, 5 mM $MgSO_4$, 25 mM KCl, pH 7.4 (3500 ml).

The elution profile is presented in fig. 3. After heat treatment according to Olsen and Milman[3], the enzyme from peak 2 from CM Sephadex C 50 was loaded on a Sephadex G 150 column (3 x 70 cm). The elution was carried out with 5 mM Tris HCl, 1 mM $MgSO_4$, pH 7.4.

Polyacrylamide gel electrophoresis: A discontinuous Tris glycine buffer, pH 8.9, was used as described by Davis[4]. The gels were stained for protein or assayed for HGPRT activity; for the latter purpose gels were sliced into 2 mm sections and allowed to stand for 48 hr in 0.2 ml of 5 mM Tris HCl

Table I. *Purification of hypoxanthine guanine phosphoribosyltransferase*

Step	*Specific activity nmoles/mg protein x hr*	*Total protein mg*	*Recovery initial activity*	*Purification % – fold*
DEAE eluate	8.9	4,600	100	1
CM eluate I	23.6	94	5	2.6
II	121.6	200	59.4	13.6
III	2.0	911	4.4	0.2
Heat treatment	4560	4	44.5	512
G150 eluate	4560	3.9	43.4	512

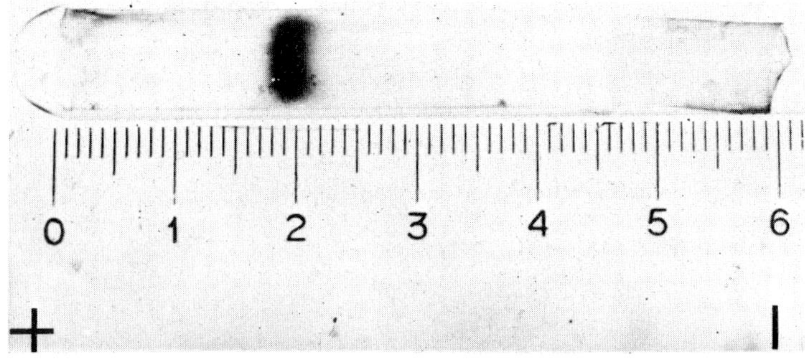

Fig. 4 – Polyacrylamide gel electrophoresis of purified HGPRT. Gel electrophoresis was performed following the procedure of Davis[4]. The gel was stained overnight in 0.1% Coomassie bleu (in methanol-acetic acid-water, 5 : 1 : 5) and destained electrophoretically. The cathode is on the right hand side.

buffer, pH 7.4, containing 1 mM $MgSO_4$. The eluate was assayed for HGPRT activity as described.

Results and Discussion

Enzyme purification. The results of enzyme purification are summarized in table I. As fig. 3 shows, three peaks of HGPRT activity were found in CM Sephadex C50 chromatography; this is in keeping with the observations by other investigators of three electrophoretic variants of HGPRT in normal human hemolysate.

Samples of 10 μl of the enzyme obtained after chromatography on G 150 were run on polyacrylamide gel electrophroesis. The gel displayed only one major band (fig. 4). A single peak of HGPRT activity located by assaying gel slices coincided with the protein band.

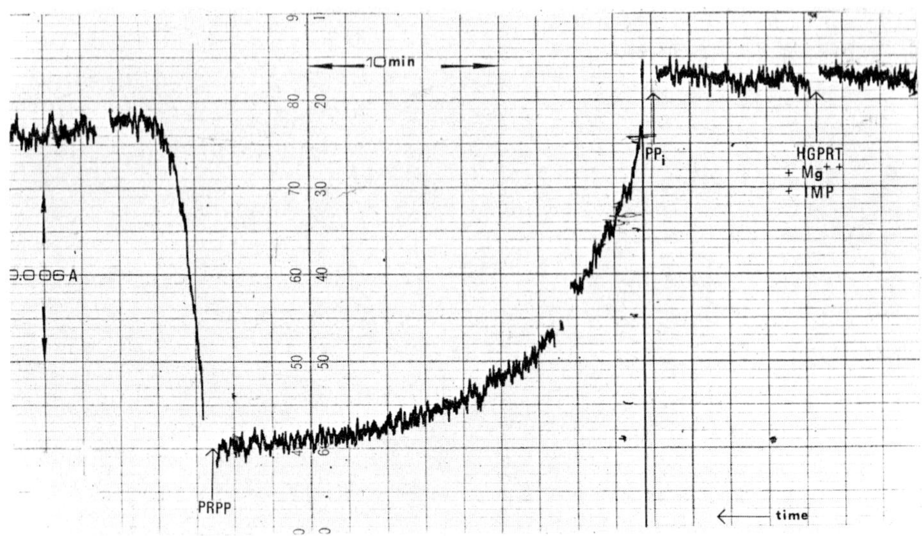

Fig. 5 – Direct demonstration of HGPRT catalyzed IMP pyrophosphorolysis. Optical density variation at 245 nm relative to the absorption at 257 nm. The experiment was performed at 37°C as described in the text.

IMP pyrophosphorolysis. The reversibility of the HGPRT catalyzed reaction was studied by direct spectrophotometric measurement of IMP cleavage in the presence of Mg pyrophosphate. These measurements are complicated by the low solubility of the pyrophosphate. Nevertheless the decrease in optical density at 245 nm relative to that at 257 nm (isosbestic point for conversion of IMP to hypoxanthine) was observed when 10 μl of 0.15 M pyrophosphate were added to 2.5 ml of a solution containing 0.1 M Tris HCl pH 7.4, 1.66 x 10^{-4} IMP, 6.5 x 10^{-4} $MgCl_2$ and 1.5 ml of the purified enzyme (fig. 5). An optical density variation in the direction of a return to the initial absorption value was observed after the additon of 5 μl of 2 x 10^{-2} M PRPP.

Moreover, hypoxanthine formation could be demonstrated when a solution containing 1 mM IMP, 1 mM pyrophosphate, 10 mM $MgSO_4$ and 20 mM Tris HCl, pH 7.4, was incubated at 37°C for 8 hr with 1 ml of the G 150 enzyme in a final volume of 10 ml. After purification by chromatography on a Dowex 1 column chloride form (fig. 6), the reaction product was identified as hypoxanthine by thin layer chromatography according to Randerath[5] and by its ultraviolet absorption spectra at two pH values (1.2 N HCl and 1 N NaOH).

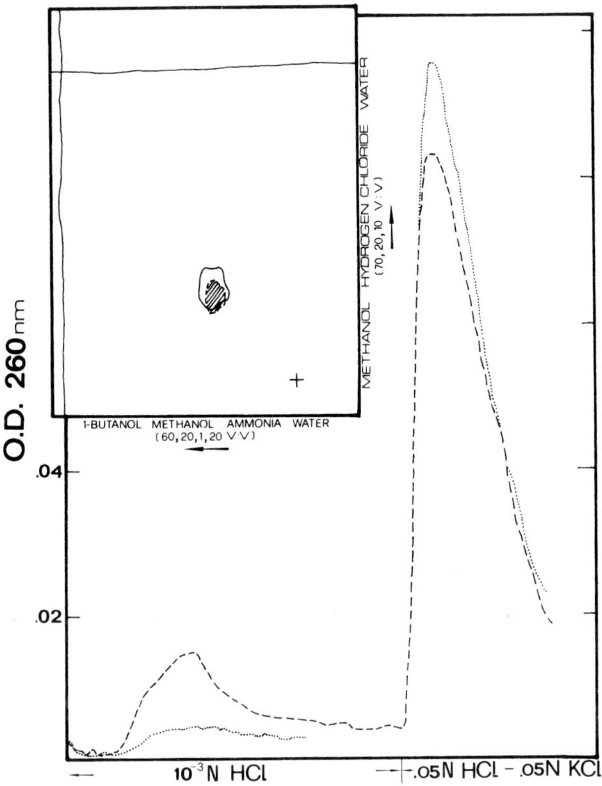

Fig. 6 – Purification of the reaction product of IMP pyrophosphorolysis on Dowex 1. The column (1 x 4 cm) was first eluted with 70 ml of 0.001 N HCl (sufficient to elute the reaction product hypoxanthine) and then with 0.05 N HCl – 0.05 M KCl which promptly eluted IMP. ———————— test run; . . . . control run with preheated enzyme. Inset: identification of the reaction product as hypoxanthine by thin layer chromatography according to Randerath[5].

□ hypoxanthine; ▨ reaction product of IMP pyrophosphorolysis.

PRPP formation was demonstrated by its quantitative determination according to Kornberg[6].

In the presence of an excess of xanthine oxidase (XOD) activity, IMP pyrophosphorolysis can be followed by the continuous spectrophotometric determination of the final product uric acid at 293 nm. As fig. 7 shows, the initial reaction progress was linear in time and its rate proportional to HGPRT concentration.

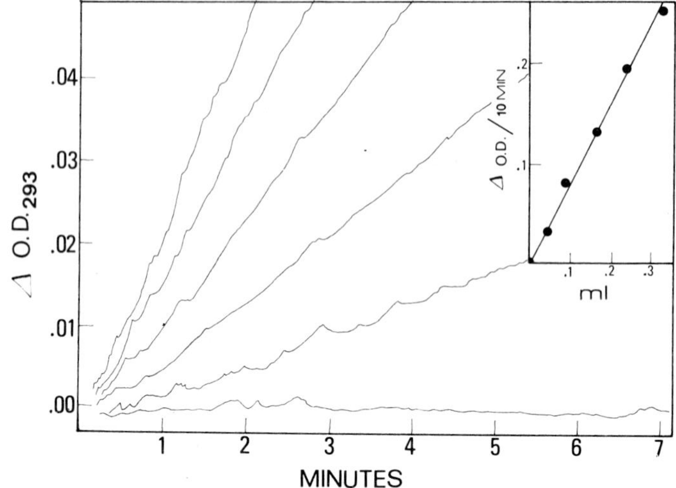

Fig. 7 – IMP pyrophosphorolysis assay at 37°C. Optical density variation at 293 nm as a function of time. Inset: initial optical density variation referred to 10 min a a function of enzyme concentration. Incubation mixtures contained: 1 mM pyrophosphate, 0 1 mM IMP, 40 mM $MgSO_4$, 100 mM Tris-HCl, pH 7.4, 0.08 U/ml xanthine oxidase and HGPRT (0 to 0.3 ml) in a final volume of 1 ml.

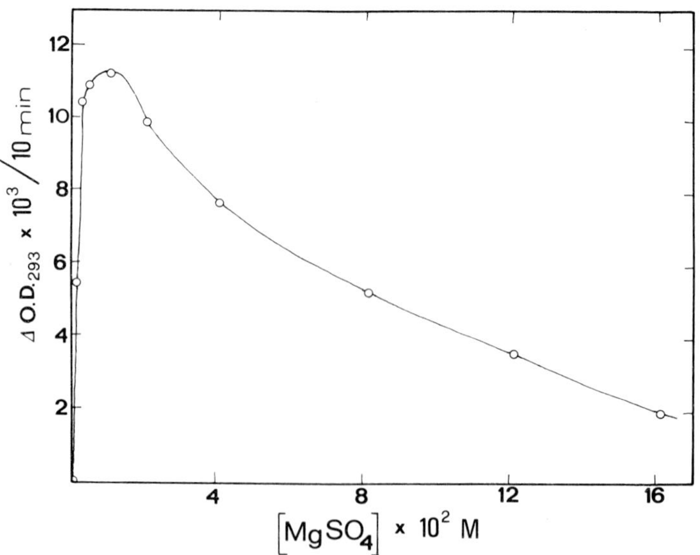

Fig. 8 – The effect of Mg^{2+} on the rate of IMP pyrophosphorolysis. Abscissa: $MgSO_4$ concentration; ordinate: IMP pyrophosphorolysis rate.

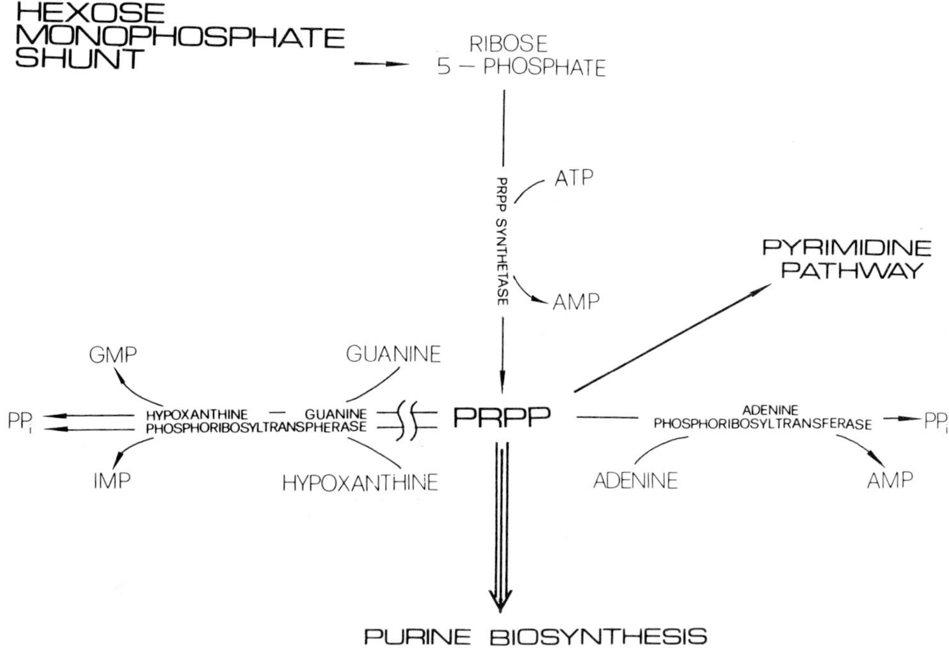

Fig. 9 – Proposed mechanism of "de novo" purine overproduction associated with HGPRT deficiency, assuming a reduced consumption of PRPP in the phosphoribosyltransferase reaction and a resultant surplus of PRPP for "de novo" biosynthesis[1].

The reaction has an absolute requirement for magnesium ions. As fig. 8 shows, the optimum magnesium ion concentration was 10^{-2} M when pyrophosphate and IMP levels were respectively 10^{-3} M and 10^{-4} M. Marked inhibition was observed at higher concentrations of magnesium ions. Similar results were obtained by other investigators studying the magnesium ion dependance of HGPRT activity at fixed hypoxanthine and PRPP concentrations[7].

Although the absence of HGPRT in humans is associated with accelerated "de novo" purine biosynthesis and elevated cellular concentration of PRPP, the physiological role of this enzyme remains still unclear. The mechanism of "de novo" purine overproduction in the Lesch-Nyhan syndrome and in adult gouty subjects with partial HGPRT deficiency is generally considered to involve reduced consumption of PRPP in the phosphoribosyltransferase reaction and a resultant surplus of PRPP for the "de novo" synthesis[1] of phosphoribosylamine and purines (fig. 9). However, a partial deficiency of the closely related enzyme adenine phosphoribosyltransferase, which also operates in the "salvage pathway", is not associated with purine overpro-

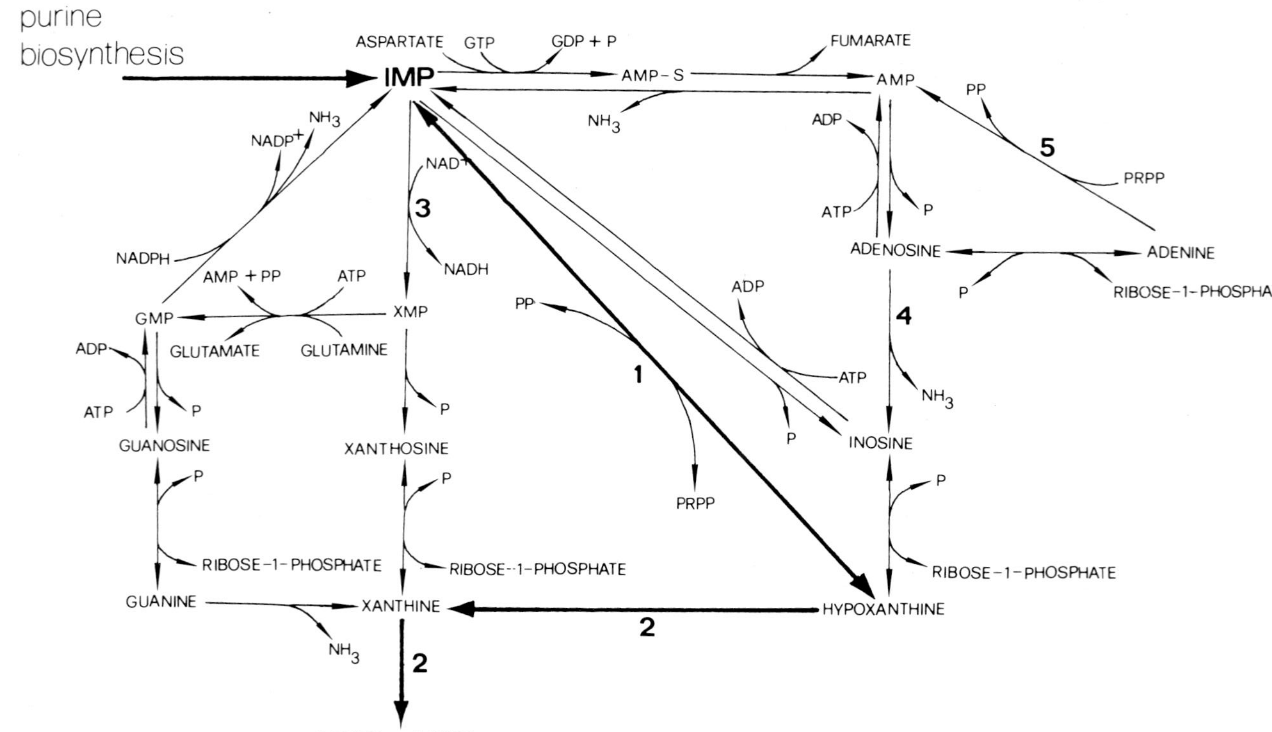

Fig. 10 – Purine reutilization, interconversion and catabolic pathways assuming an "in vivo" HGPRT catalyzed IMP pyrophosphorolysis. 1 hypoxanthine guanine phosphoribosyltransferase; 2 xanthine oxidase; 3 IMP dehydrogenase; 4 adenosine deaminase; 5 adenine phosphoribosyltransferase.

duction and elevated cellular concentration of PRPP[10], although the Michaelis constant for PRPP reported for this enzyme is considerably lower than that for HGPRT[8,9]. Recently the case of a patient with complete APRT deficiency and a serum urate level of only 6 mg/100 ml has been reported[11]. Moreover, Reen[12] demonstrated an increase in PRPP-synthetase activity in cultured lymphocytes from Lesch-Nyhan patients and suggested that this biochemical abnormality in PRPP metabolism may play an important role in the genesis of purine overproduction.

In the present paper it was shown that HGPRT catalyzes IMP pyrophosphorolysis. The possibility that this reaction may have a role in the regulation of purine metabolism should be considered, especially in the liver, where XOD activity is great. This problem cannot be resolved at the present time, since the intracellular concentration of all substrates and products of the reaction, their compartmentalization and the equilibrium constant of the HGPRT catalyzed reaction are not known.

Assuming that a HGPRT-catalyzed IMP pyrophosphorolysis occurs "in vivo", IMP would be directly transformed into hypoxanthine, and, in the presence of XOD, eventually into uric acid (fig. 10). In the absence of HGPRT, IMP could be catabolized only through the reactions catalyzed by 5'-nucleotidases and by purine nucleoside phosphorylases, either directly or after being converted to other nucleotides. The IMP-dehydrogenase increase, found in the erythrocytes from subjects with HGPRT deficiency[13], could be taken as supporting evidence for the activation of these pathways. The finding of a significant increase of the adenosine deaminase activity in the erythrocytes from subjects with primary gout and normal HGPRT activity[14], suggests the existence of a correlation between the activation of catabolic pathways alternative to the IMP pyrophosphorolysis and purine overproduction. Therefore, it seems possible to justify the accelerated "de novo" purine biosynthesis associated with the HGPRT deficiency, assuming an IMP pyrophosphorolysis catalyzed by this enzyme "in vivo". Moreover, this hypothesis is in good agreement with the lack of purine overproduction associated with an APRT deficiency.

References

1. Kelley W.N., Wyngaarden J.B.: in Advances in Enzymology, A. Meister Ed., vol. 41, John Wiley & Sons, New York, p. 1 (1974).
2. Arnold W.J., Kelley W.N.: J. Biol. Chem., *246,* 7398 (1971).
3. Olsen A.S., Milman G.: J. Biol. Chem., *249,* 4030 (1974).
4. Davis B.J.: Ann. N.Y. Acad. Sci., *121,* 404 (1964).
5. Randerath K.: Nature *205,* 908 (1965).
6. Kornberg A., Lieberman I., Simms E.S.: J. Biol. Chem., *215,* 389 (1955).
7. Krenitsky T.A., Papaioannou R., Elion G.B.: J. Biol. Chem., *244,* 1263 (1969).
8. Henderson J.F., Brox L.W., Kelley W.N., Rosenbloon F.M., Seegmiller J.E.: J. Biol. Chem., *243,* 2514 (1968).

9. Hori M., Henderson J.F.: J. Biol. Chem., *241,* 3404 (1966).
10. Fox I.H., Meade J.C., Kelley W.N.: Am. J. Med., *55,* 614 (1973).
11. Cartier P., Hamet M.: C.R. Acad. Sci. Paris, *279*, 883 (1974).
12. Reem G.H.: Science *190,* 1098 (1975).
13. Pehlke D.M., Mc Donald J.A., Holmes E.W., Kelley W.N.: J. Clin. Invest., *51,* 1398 (1972).
14. Nishizawa T., Nishida Y., Akaoka I., Yoshimura T.: Chim. Clin. Acta, *58,* 277 (1975).

ENZYMES OF LYSOSOMAL ORIGIN IN SERUM: ASSAY AND SIGNIFICANCE

G. Tettamanti, A. Lombardo, L. Caimi and V. Foà

Summary

The activity in the plasma (or serum) of the following enzymes of lysosomal origin was investigated: alpha-galactosidase; beta-galactosidase; beta-glucosidase; beta-glucuronidase; beta-N-acetylglucosaminidase; alfa-mannosidase; alfa-fucosidase. In all cases a fluorimetric procedure, employing the corresponding 4-methylumbellyferyl-glycoside as the substrate, was used. The optimum assay conditions were established: optimum pH, ionic strength and substrate concentration; time course of the reaction; protein: activity relationship; maintenance of enzyme activity with time at different temperature and upon freeezing and thawing. A reliable and convenient automatic assay method was also set up for each enzyme.

The mean normal values (expressed in nanomoles hydrolyzed substrate/hour/ml) of enzyme activity were the following: alpha-galactosidase, 12; beta-galactosidase, 17; beta-glucosidase, 10; beta-glucuronidase, 120; beta-N-acetylglucosaminidase, 1000; alpha-mannosidase, 22; alpha-fucosidase, 380.

A systematic study on the influence of environmental factors (Hg vapours) and of pathological conditions on the plasma levels of these enzymes has been initiated. Evidence has been provided that lysosomal enzymes play an interesting role in clinical investigation.

1. Biology of lysosomal enzymes

a. *Lysosomes and lysosomal enzymes*

Lysosomes are organelles virtually present in all cells. They are limited by a single unit membrane and contain many hydrolytic enzymes, having optimum pH in the range 3-6[1].

Lysosomal enzymes are glycoproteins[2]. The protein portion of their molecule, after being biosynthesized in the endoplasmic reticulum, moves to the Golgi apparatus, where the saccharide portion is attached[3,4]. The final products leave the Golgi apparatus, wrapped up in a membrane, so that they form vesicle-like particles, named primary lysosomes[3,4].

Istituto di Chimica Biologica, Facoltà di Medicina e Chirurgia, Università di Milano e Istituto di Medicina del Lavoro, Facoltà di Medicina e Chirurgia, Università di Milano.

Extracellular material of different nature, when interacting with the cell, may be incorporated, giving rise to the so-called endocytic vacuoles (pinocytosis or phagocytosis)[1, 5, 6]. Various intracellular components or macromolecules may be involved as well in the formation of intracellular vacuoles (autophagocytosis)[1, 5, 6]. Endocytic or autophagocytic vacuoles fuse with primary lysosomes and form secondary lysosomes[1, 5, 6]. At this moment the hydrolysis of the various molecules takes place. An active proton pump, or/and the presence of acid glycosaminoglycans and glycolipids have been postulated as responsible for lowering pH to acidic values: a condition which is essential for lysosomal enzymes action[7]. Once the material contained in secondary lysosomes has been completely degraded, the catabolites obtained diffuse through the membrane into the cytosol. Now lysosomes are again ready to fuse with the next substrate containing vacuole and a new degradation cycle takes place. This scheme emphasizes the role played by lysosomal enzymes in the intracellular digestion of material of both cellular and exogenous origin.

b. *Release and uptake of lysosomal enzymes by the cell*

Primary or secondary lysosomes can fuse with the cell membrane and release their content of enzymes and metabolites into the extracellular space (exocytosis)[8]. Thus, lysosomal enzymes are actually excreted by the cell and may enter the blood stream. Of course the presence of lysosomal enzymes in blood could also be caused by a process of cell and organelle lysis.

It has been shown that cultured cells easily uptake, by phagocytosis, lysosomal enzymes added to the culture medium[9, 10]. Interestingly, the uptaken enzymes are present in secondary lysosomes and are functionally active on their physiological substrates. The following example stresses the importance of this phenomenon[11]. Fibroblasts from patients suffering from metachromatic leukodistrophy are genetically deficient in arylsulphatase A. Consequently, they accumulate the natural substrate of this enzyme, the sulphatide. When arylsulphatase A (the missing enzyme) is added to the culture medium of these fibroblasts an increased activity of intracellular arylsulphatase A and a decreased sulphatide content are promptly observed.

According to recent lines of investigation[10, 11], the uptake of lysosomal enzymes by the cell requires a specific binding between the enzyme and a membrane receptor. The enzyme should contain a proper "marker", very likely a terminal carbohydrate unit, which is recognized by the receptor[12]. An interesting hypothesis[11] postulates that the "marker" glycosyl unit is introduced by a specific glycosyltransferase lying on the external surface of the plasma membrane[13].

Very likely both the release and uptake of lysosomal enzymes occur also in vivo, causing a sort of flow of enzymes between the intracellular and the extracellular compartments.

The release and uptake of lysosomal enzymes by the cell constitutes a new and impressive chapter of cell biology. Further studies are needed for understanding both the mechanism and the physiological implications of this phenomenon. Anyway the very presence of lysosomal enzymes in body fluids raises some questions of great interest. Are there any functions of lysosomal enzymes in body fluids? Molecules (glycosaminoglycans, glycolipids, glycoproteins, proteins etc.) which may act as substrate for lysosomal enzymes, are present in the extracellular space and in blood. These enzymes surely act in the metabolism and chemical modulation of the above molecules. Thus the hypothesis that lysosomal enzymes (besides their involvement in intracellular digestion) play a role in the processes of cell "social" behaviour and of ion exchange between cellular and extracellular compartments appears more and more likely.

Of course researches aiming at establishing the actual possibilities for these enzymes to act in their physiological environment (the various body fluids) would greatly contribute to further knowledge of cell function.

c. *The I-cell disease: leakage of lysosomal enzymes into extracellular fluids*

A unique occasion for studying the relationship between lysosomes and extracellular fluids has been the I-cell disease. Inclusion cell (I-cell) disease (or mucolipidosis II)[14, 15] is a fatal degenerative disease of childhood, transmitted by an autosomal recessive inheritance. It is clinically similar to the mucopolysaccharidoses but usually shows normal urinary excretion of mucopolysaccharides[14, 15]. I-cell disease is characterized by the presence of several cytoplasmic inclusions in most cells and in cultured fibroblasts. These inclusion particles probably originate from lysosomes and are the sites for abnormal accumulation of lipids, glycolipids, glycoproteins and mucopolysaccharides[16, 17, 18].

Many hydrolases (β-glucuronidase, β-galactosidase, β-fucosidase, α-mannosidase, α-iduronidase, arylsulphatases) were shown to be deficient in cultured fibroblasts derived from I-cell disease-affected patients[14-18]. Some enzymes, namely β-galactosidase and β-N-acetylglucosaminidase, were found to be lowered in tissues and leukocytes too[14-18]. In contrast, increased activity of several lysosomal hydrolases has been ascertained in the plasma of all I-cell disease-affected patients[14-18]. . This situation is shown in Table I, which refers to a patient suffering from I-cell disease we had the opportunity to study in our laboratory. The leukocyte content (that is the intracellular content) of several hydrolases is in the normal range (β-N-acetylhexosaminidase, α-mannosidase) or moderately lower (α-fucosidase, β-galactosidase, arylsulphatase A). The level of the same hydrolases in serum (that is in the extracellular fluids) is, in contrast, extraordinarily increased (from 10 to 100-fold).

The exact pathogenetic mechanism of I-cell disease in still unknown. On the basis of the above-mentioned observations it has been suggested that

Table I. *Content of some hydrolases of lysosomal origin in leukocytes and serum of normal subjects and of a patient affected from "I-Cell disease". The activity of each enzyme is expressed as nMoles hydrolyzed substrate/hour/ml serum, for serum enzymes; as nMoles hydrolyzed substrate/hour/mg protein for leukocyte enzymes*

Enzyme	*Normal range*		*I-Cell disease*	
	Leukocyte	*Serum*	*Leukocyte*	*Serum*
β-N-acetylglucosaminidase	2090-5350	710-3580	2690	41000
β-galactosidase	136-330	13-76	145	3000
α-mannosidase	170-350	9-64	300	1050
α-fucosidase	90-300	40-1200	90	6500
Arylsulphatase A*	76-238	1-2	53	221

*All enzymes assayed by fluorimetric procedures except Arylsulphatase A, determined colorimetrically (substrate: p-nitrophenylsulphate).

there is excessive leakage of lysosomal enzymes in I-cell disease[19]. A reason for that may be excessive exocytosis due to some anomaly on the lysosomal membrane. However, and this is consistent with the most recent data[10, 12], the increased excretion of enzymes could also derive from decreases or lack of capacity of the cell to uptake the enzyme, when in the extracellular medium. Assuming that the uptake requires recognition by the cell membrane surface of a specific "marker" on the enzyme, in the I-cells a mutation alters the recognition site so that these enzymes are unable to enter adjacent cells. An interpretation at the molecular level of the uptake impairment has been given[20]. In fact it has been observed that the isoenzymes of β-N-acetylglucosaminidase, excreted by cultured fibroblasts from I-cell disease-affected patients, differ electrophoretically from the isoenzymes excreted by fibroblasts from normal subjects and from the intracellular isoenzymes of both cell types (I-cell disease and normal). Better, in the I-cells the intracellular β-N-acetylglucosaminidase isoenzymes are different from the extracellular ones, in contrast with normal cells where the two patterns are identical. The difference was proved to consist of higher amounts of sialic acid present in the excreted enzymes. After treatment with sialidase the excreted enzymes are converted to the intracellular (and normal) forms and are easily uptaken. The explanation for the excessive sialic acid residues present in the enzymes excreted from I-cell disease fibroblasts is, at present, hypothetical. One possibility could be an abnormal exocytosis allowing oversialylation catalyzed by sialyltransferases located on the Golgi apparatus[20]. A second possibility is deficient desialylation by a neuraminidase located on the cell plasma membrane and responsible for adjusting the recognition site of the enzyme molecule.

2. Assay of several hydrolases of lysosomal origin in serum

a. *Optimum conditions for the assay of several lysosomal hydrolases in serum*

Reports on the assay of lysosomal hydrolases in serum and other body fluids have been given by different laboratories[21-38], with special reference to variations of the enzyme activities under various pathological conditions. However the problem of reproducible and optimized enzyme tests for routine practice was only occasionally raised[29, 30, 34].

We performed a careful investigation in this direction in our laboratory[39]. The following enzymes, all glycohydrolases, were studied: β-galactosidase, α-galactosidase, β-glucosidase, β-N-acetylglucosaminidase, β-N-acetylgalactosaminidase, β-glucuronidase, α-mannosidase, α-fucosidase. All of them were determined by fluorimetric techniques using the correspondent 4-methylumbelliferyl-glycosides as the substrate (see Table II). This approach (see Table III) was chosen for two reason: the high degree of sensitiveness; the occurrence of the same reaction product (4-methylumbelliferone) in all assays.

Table II. *Substrates for the various glycohydrolases of lysosomal origin assayed in serum by fluorimetric procedures*

Enzyme	*Substrate**
β-galactosidase	MU-β-D-galactoside
α-galactosidase	MU- -D-galactoside
β-glucosidase	MU-β-D-glucoside
β-N-acetylglucosaminidase	MU-β-D-Nac. glucosaminide
β-N-acetylgalactosaminidase	MU-β-D-Nac. galactosaminide
β-glucuronidase	MU-β-D-glucuronide
α-mannosidase	MU-α-D-mannoside
α-fucosidase	MU-α-L-fucoside

*MU = 4-methylumbelliferyl.

The optimum values of the principal parameters for enzyme assays (pH, type and molarity of buffers, K_m, linear time course, protein range for linear V/E response), which were obtained, are reported in Table IV. Optimum buffers were : 0.1 M sodium acetate-acetic acid for β-galactosidase, α-galactosidase, β-glucuronidase, α-glucosidase; 0.1 M sodium citrate-citric acid for β-N-acetylglucosaminidase, β-N-acetylgalactosaminidase, α-mannosidase, and α-fucosidase. pH optima ranged from a minimum of 3.3 (β-galactosidase) to a maximum of 5.0 (α-fucosidase) and were very close to those recorded, for

Table III. *Principle of the fluorimetric assay for glycohydrolases of lysosomal origin in serum*

MU-glycoside $\xrightarrow{\text{Enzyme}}$ glycoside + 4-methylumbellyferone

4-methylumbellyferone : fluorescent
excitation wavelength : 460 nm
– used filter with Turner Model 110 fluorimeter: 7-60 (365 nm)

emission wave length : 448 nm
– used filter with Turner Model 110 fluorimeter: 2A (415 nm)

Table IV. *Optimum conditions for the fluorimetric assay in the serum of various glycohydrolases of lysosomal origin.*

Enzyme	*Optimum pH*	*Optimum buffer*	*Km**	*Time course***	*Protein (mg) range****
β-galactosidase	3.3	acetate	3.8	3	0.5 – 5
α-galactosidase	4.5	acetate	22.2	4	0.5 – 5
β-glucosidase	4.5	acetate	14.1	3	1 – 7
β-glucuronidase	3.6	acetate	4.0	6	0.025 – 3
β-N-acetylgluco-saminidase	4.2	citrate	12.0	6	0.025 – 1
β-N-acetylgalac-tosaminidase	4.2	citrate	0.46	6	0.025 – 3
α-mannosidase	4.0	citrate	6.1	2	0.25 – 1
α-fucosidase	5.0	acetate	0.23	5	0.1 – 2

*x 10^{-4} M
**linear up to .... hours
***for linear V/E response.

the same enzymes, in tissues, leukocytes and cultured fibroblasts. The course of the enzyme reaction was linear with time from a minimum of 2 hours (α-mannosidase) to at least 6 hours (β-N-acetylglucosaminidase, -N-acetyl-galactosaminidase, α-glucosidase). Anyway 15-20 minutes incubation provided sufficiently high readings for accurate and reproducible assays in all cases. The protein range for a linear V/E relationship was rather narrow for α-mannosidase (0.25 - 1 mg) and β-glucosidase (1-7 mg) but acceptably large for all other enzymes (from. 0.5-5 mg to 0.025-3 mg). The K_m values varied from a minimum of 0.23 x 10^{-4} M to a maximum of 22.2 x 10^{-4} M (α-galactosidase). In other words the enzymes tested displayed affinities toward the correspondent 4-methylumbelliferyl- glycosides which differed in the 100-fold range. Consequently such high substrate concentrations should be taken for saturating some enzymes (α-galactosidase, β-glucosidase, β-Nacetylglucosaminidase) that adequate precautions should be taken for maintaining the substrate in solution (for instance addition of albumin, detergent, etc., see reference 39).

The analytical reproducibility of the fluorimetric assays was high (see Table V). Thus the assays should be considered fairly reliable and suitable for routine practice.

Table V. *Fluorimetric assay of various lysosomal glycohydrolases in the serum: analytical reproducibility. 40 determinations on a pool of freshly prepared sera were carried out. Each enzyme was assayed using different quantities of protein, chosen within the range of linear relationship between enzyme activity and amount of protein. The enzyme activity is expressed as nMoles hydrolyzed substrate/hour/ml of serum.*

Enzyme	*Enzyme activity ± S.E.*
β-galactosidase	30 ± 0.4
α-galactosidase	20 ± 0.3
β-glucosidase	3 ± 0.02
β-N-acetylglucosaminidase	1200 ± 14
β-N-acetylgalactosaminidase	150 ± 1.8
β-glucuronidase	205 ± 2.2
α-mannosidase	30 ± 0.5
α-fucosidase	510 ± 3.2

b. *Determination of serum levels of several hydrolases of lysosomal origin by automatic procedures*

The optimized assay conditions reported above were applied to automation[40]. The general indications of Lowden et al.[29], who set up an automatic procedure for the assay of serum β-N-acetylglucosaminidase, were followed. A.C. Erba model S. Autoanalyzer, supplied with a Turner model 111 fluorimeter and a recorder, was employed. No substantial modifications to the manual procedure were introduced. The substrate solution was maintained under constant stirring; the incubation time was 20 min (for details see reference 40).

An example of repetitive automatic determinations of one of the tested enzymes, α-fucosidase, is given in Figure 1. The response in several determinations is practically constant, which means that the degree of reproducibility is very high.

The results of the automatic determinations of the same α-fucosidase in

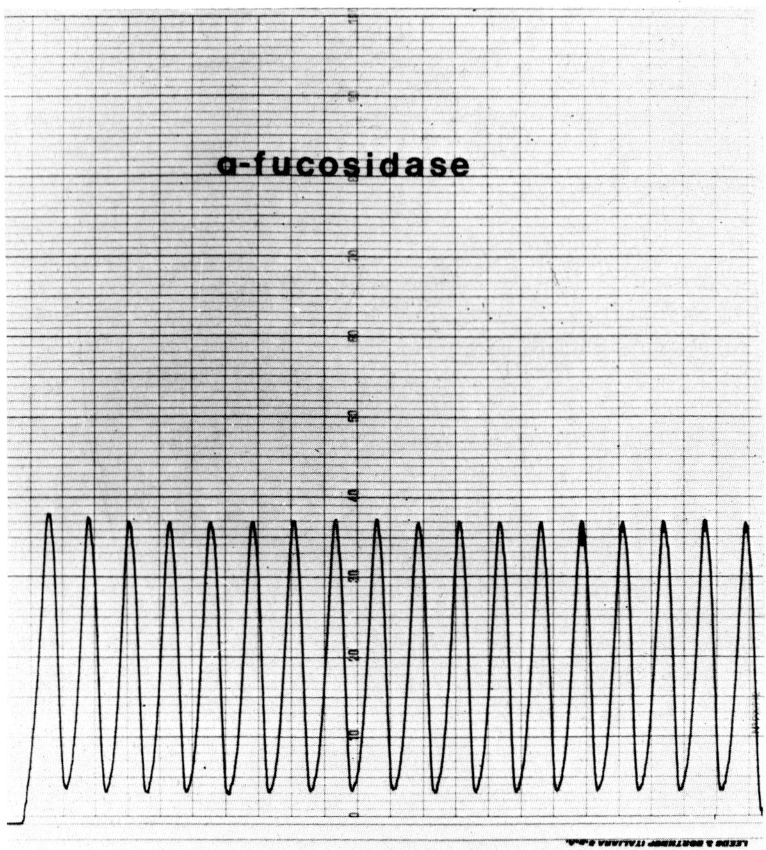

Fig. 1 – Repetitive determinations of α-fucosidase activity in a pool of fresh sera obtained from normal subjects.

the serum of different individuals is reported in Figure 2. It is worth stressing that sample 374 (the first of the series) gave the same value of enzyme activity, when repeated (the last of the series) on a 1:6 diluted serum.

The levels of enzyme activity determined by the manual and the automatic techniques were practically identical in the same serum (see Table VI).

c. *Normal activity values of several hydrolases of lysosomal origin in serum*

The data collected thus far indicate that the serum activity values of the examined lysosomal enzymes undergo marked variations in the different individuals. The mean, minimum and maximum values of activity of six lyso-

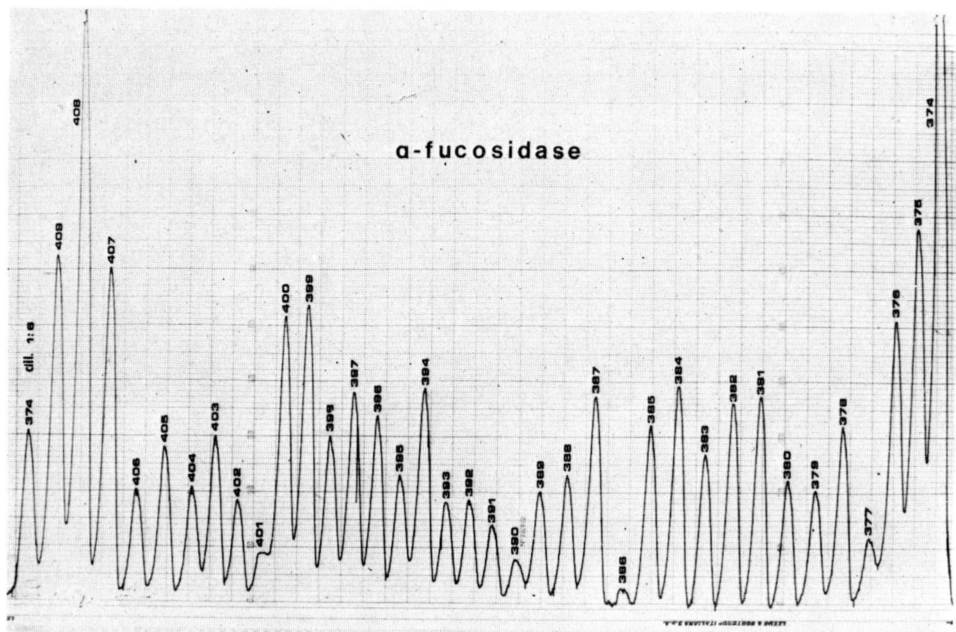

Fig. 2 – Automatic determinations of serum α-fucosidase activity in patients suffering from different diseases. Note patient 374 (the first and the last of the series); the two determinations were carried out on an undiluted sample (the first one) and on a 1/6 diluted sample (the last one).

Table VI. *Fluorimetric determination of various glycohydrolases of lysosomal origin in the serum: comparison between manual and automatic procedures. A pool of sera obtained from healthy subjects was employed. 15 determinations for each enzyme were performed. The enzyme activity is expressed, in all cases, as nMoles hydrolyzed substrate/hour/ml serum.*

Enzyme	*Enzyme activity ± S.E.*	
	Manual procedure	*Automatic procedure*
β-galactosidase	33 ± 0.5	32.6 ± 0.4
β-glucosidase	2.6 ± 0.03	2.7 ± 0.03
β-N-acetylglucosaminidase	1250 ± 15	1280 ± 18
β-glucuronidase	195 ± 2.5	191 ± 2.7
α-mannosidase	31 ± 0.6	31.3 ± 0.5
α-fucosidase	505 ± 3	502 ± 4

Table VII. *Levels of various lysosomal glycohydrolases in the serum of normal adult subjects (= 208). The mean, minimum and maximum recorded values are given. The activity of each enzyme determined fluorimetrically is expressed as nMoles/hour/ml serum.*

Enzyme	*Normal range*		
	Mean	*Minimum*	*Maximum*
β-N-acetylglucosaminidase	1357	711	3579
β-galactosidase	37.1	13	76
β-glucosidase	1.7	0.4	4.8
β-glucuronidase	184.5	116	476
α-mannosidase	32.2	9	64
α-fucosidase	519	40	1200

somal hydrolases recorded in 208 healthy individuals are reported in Table VII. The frequency distribution of the same enzymes is given, by nomograms, in Figure 3. It can be observed that the so-called "normal range" is rather wide for all the enzymes.

So far no indications concerning possible variations of serum levels of lysosomal hydrolases in relation to sex, age, circadian rhythms etc., have been reported.

The occurrence of a wide range of enzyme levels in normal subjects is not surprising. Similar behaviour is shown by all the serum enzymes used in the routine practice of clinical enzymology. Of course this situation must be kept in mind when giving a diagnostic significance to a "high" recorded value of enzyme activity.

3. Behaviour of some hydrolases of lysosomal origin in serum under different pathological conditions

a. *Serum levels of some hydrolases of lysosomal origin in patients suffering from different diseases*

An inspection on the behaviour of serum lysosomal hydrolases in various pathological conditions was conducted, 216 patients who entered one of the Milan hospitals (Ospedale S. Raffaele) being examined. These patients were suffering from different diseases: chronic inflammatory processes (n = 40); subacute inflammatory processes (n = 61); cancer (n = 30); diabetes (n = 25); hypertension with altered blood lipid levels (n = 20); various cardiovascular diseases (n = 15); liver diseases (n = 25). The following enzymes were assayed automatically: β-galactosidase, β-glucoronidase, β-N-acetylglucosaminidase, α-fucosidase, α-mannosidase. The results of this preliminary investigation are reported, by nomograms in Figures 3 and 4.

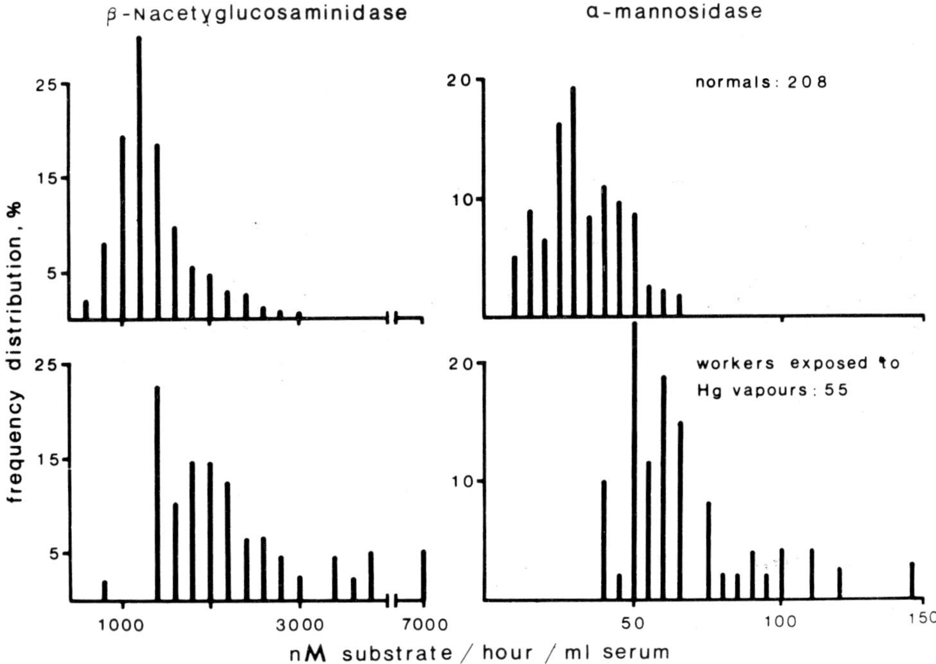

Fig. 3 – Frequency distribution of serum levels of β-galactosidase, β-glucuronidase and α-fucosidase in normal subjects (= 208) and in patients (= 216) suffering from various diseases.

Each enzyme was determined fluorimetrically by automatica procedures. The activity of each enzyme is expressed as nMoles hydrolyzed substrate/hour/ml serum.

Three situations emerge from a comparison of the serum levels of the various enzymes in the group of patients and in the control group of healthy people:

1. the frequency distribution of serum activity levels of β-galactosidase, β-glucuronidase, and α-fucosidase in the patient group overlaps that of the healthy group (see Figure 3); in other words no significant alterations of these enzymes levels derive from the pathological conditions;
2. the frequency distribution for serum activity levels of β-N-acetylglucosaminidase and α-mannosidase is slightly but significantly shifted toward higher values (see Figure 4), thus indicating an implication of these enzymes in the pathological process;
3. very high serum values of some enzymes (α-fucosidase, β-N-acetyl-glucosaminidase, α-mannosidase) occur occasionally but actually among the patients, which means that some enzymes mobilize from the cell more easily under pathological conditions.

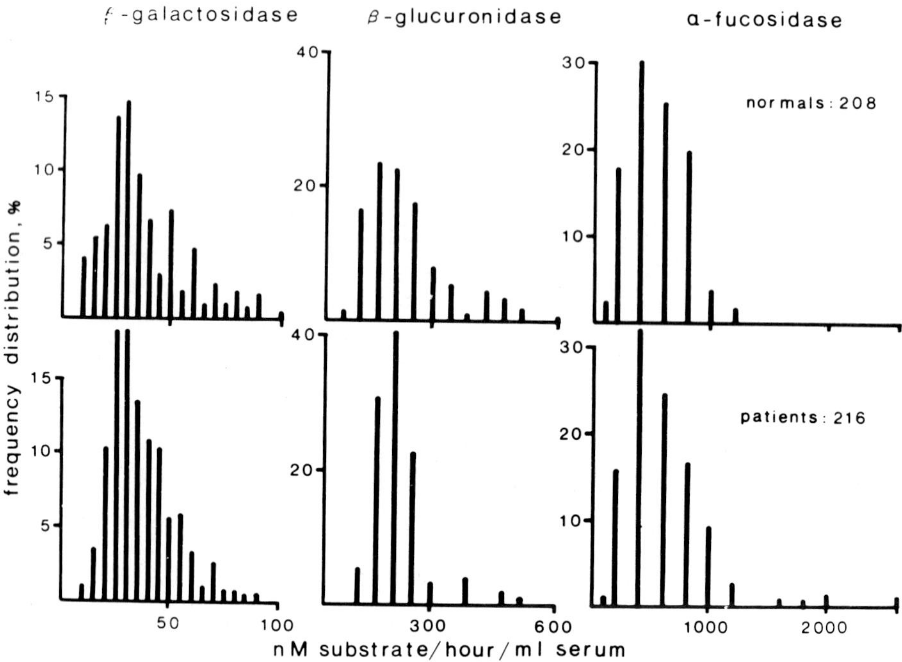

Fig. 4 — Frequency distribution of serum levels of β-N-acetylglucosaminidase and α-mannosidase in normal subjects (= 208) and in patients (= 216) suffering from various diseases.

Each enzyme was determined fluorimetrically by automatic procedures. The actvity of each enzyme is expressed as nMoles hydrolyzed substrate/hour/ml serum.

A deeper analysis within the different classes of diseases showed that:

1. in all observed cases of chronic and subacute inflammatory processes none of the tested enzymes showed increased serum levels;
2. in hypertensive diseases, liver diseases, diabetes and cancer a significant increase of α-fucosidase, α-mannosidase, β-N-acetylglucosaminidase was recorded, specially marked for α-mannosidase. This evidence confirms previous, more limited, findings[26, 31, 34];
3. for some diseases, for instance liver diseases, increased serum levels of α-fucosidase and α-mannosidase were somewhat dependent on the seriousness of the disease;
4. in the few (four) examined cases of varicose veins the increase of serum levels of lysosomal enzymes was very marked, in accordance with previous evidence[41].

On the whole, the results of the pilot investigation reported here seem promising. Lysosomal enzymes, like α-fucosidase, α-mannosidase, β-N-acetylglucosaminidase, are surely involved in a large number of pathological conditions: some of theses enzymes are subjected to easier mobilization toward extracellular fluids, either all together or individually.

b. *Serum levels of some hydrolases of lysosomal origin in workers exposed to inorganic mercury vapours*

The serum levels of β-N-acetylglucosaminidase, β-galactosidase, β-glucuronidase, α-mannosidase were determined in a gropu of a plant producing sodium hydroxyde[42]. These workers (55 subjects) were exposed to airborne inorganic mercury at concentrations between 0.06 to 0.3 mg/m^3; none of them showed any preclinical or clinical signs of mercury poisoning. The nomograms reproduced in figures 4 and 5 show that the serum levels of all four enzymes in the examined workers underwent a marked shift toward higher values. The same data are presented in table VIII, provided with statistical evaluation. The mean values of enzyme activity in the subjects exposed to

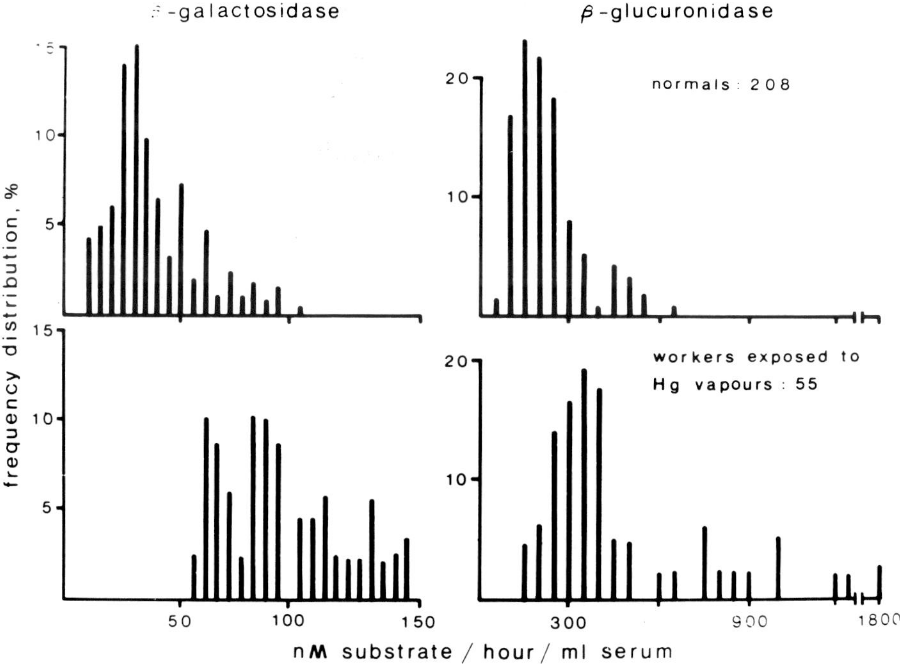

Fig. 5 – Frequency distribution of serum levels of β-galactosidase and β-glucuronidase in normal subjects (= 208) and in workers exposed to mercury vapours (= 55). Each enzyme was determined fluorimetrically by automatic procedures. The activity of each enzyme is expressed as nMoles hydrolyzed substrate/hour/ml serum.

Table VIII. *Serum levels of some hydrolases of lysosomal origin in normal subjects (= 208) and in workers exposed to mercury vapours (= 55). The activity of each enzyme, determined fluorimetrically, is expressed in nMoles hydrolyzed substrate/hour/ml serum. The mean level of enzyme activity ± S.D. is reported.*

Enzyme	*Exposed subjects*		*Normal subjects*
β-N-acetylglucosaminidase	2295.2 ± 1339.6		1356.8 ± 472.1
		$t = 6.47$ $p < 0.001$	
α-mannosidase	63.5 ± 20.3		32.2 ± 13.4
		$t = 8.79$ $p < 0.001$	
β-glucuronidase	479.6 ± 327.6		184.4 ± 74.9
		$t = 4.06$ $p < 0.001$	
β-galactosidase	106.9 ± 43.4		37.1 ± 13.0
		$t = 15$ $p < 0.001$	

mercury vapours were from 2 to 3 times higher than those recorded in the group of healthy, unexposed individuals, with an extremely high degree of significance.

A more detailed study of the working conditions of the same individuals showed that the rise of the enzyme activities in serum was dependent on the length of exposure and the degree of the exposure at their place of work.

The increased serum content of lysosomal hydrolases indicates some damaging action of mercury on lysosomes or on the process of release and uptake of lysosomal enzymes. The mechanism of this damage, which does not appear to be related to cell lysis, is unknown. Mercury could labilize lysosome membrane allowing the leakage of lysosomal enzymes into extracellular fluids[43]. Another possibility is that mercury blocks the membrane receptor(s) specific for the uptake of the lysosomal enzymes, thus determining their accumulation in the extracellular fluids.

The interest of these serum enzyme determinations in preventive medicine is evident. It should be remembered that the increase in lysosomal enzyme activities has been observed in people without any complaints or symptoms related to mercurial intoxication. Thus it is possible that these biochemical measurements could be very useful in monitoring workers exposed to the metal and detecting any undue mercury absorption very early.

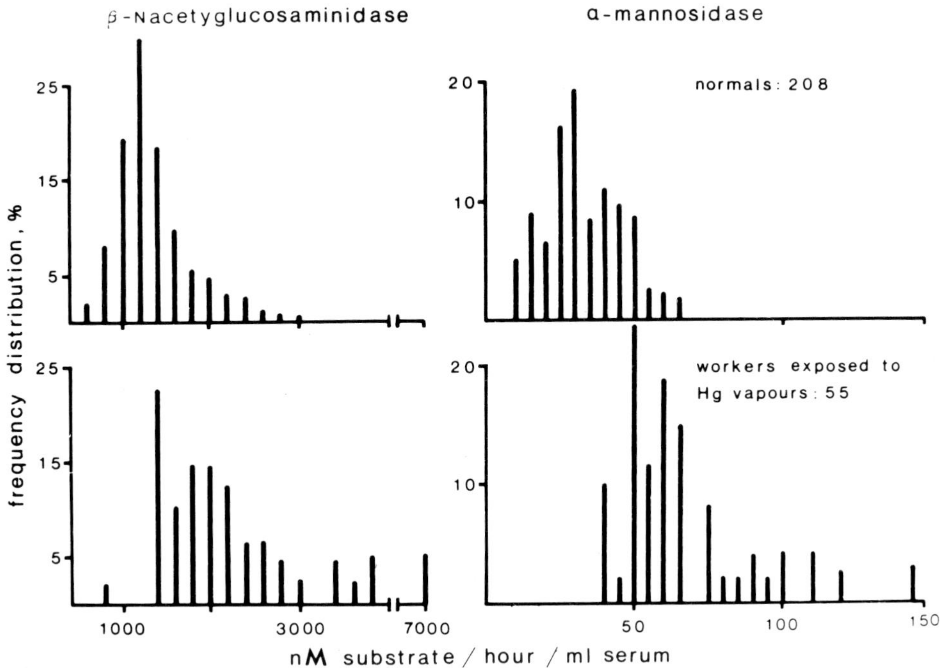

Fig. 6 — Frequency distribution of serum levels of β-N-acetylglucosaminidase and α-mannosidase in normal subjects (= 208) and in workers exposed to mercury vapours (= 55).
Each enzyme was determined fluorimetrically by automatic procedures. The activity of each enzyme is expressed as nMoles hydrolyzed substrate/hour/ml serum.

4. Perspectives

Reliable, reproducible and automatic methods are now available for determining in serum the activity of a number of hydrolases of lysosomal origin. Therefore the implications of lysosomes and lysosomal enzymes in pathological conditions may be inspected also in the routine practice of clinical enzymology.

The assay of lysosomal enzymes has been used so far in the diagnosis of inherited storage diseases: it became the elective tool not only for detecting patients suffering from these rare diseases, but also for screening heterozygous carriers.

At present the diagnostic potential of the determination of lysosomal enzymes in serum and other body fluids must also be taken into very serious consideration for acquired diseases. In fact an increased mobilization of these enzymes from the cell into the extracellular fluids may derive from two quite different types of cell damage: dramatic damage, leading to cell lysis; or mild, specific lesion at the level of the process of release and uptake of lysosomal enzymes involving primarily cell plasma membrane. In other words lysosomal enzymes may be "marker signals" either of cell lysis or of damage of cell membrane.

It is thus to be expected that in the near future the assay of lysosomal enzymes will enter the diagnostic potential of clinical enzymology.

References

1, De Duve C.: in "Ciba Foundation Symposium Lysosomes" (Eds. de Reuck A.V.S., and Cameron M.P.), Little and Brown, Boston, pp. 1-35 (1963).
2. Goldstone A., Koenig H.: Life Sciences. *9*, 1341-1354 (1970).
3. Daudwalder M., Gordon Wheley W., Kephard J.E.: Sub. Cell. Biochem., *1*, 225-275 (1972).
4. Novikoff A.B.: in "Lysosomes and storage diseases" (Eds. Hers H.G., Van Hoof F.), Academic Press, New York, pp. 1-41 (1973).
5. De Duve C.: J. Theor. Biol., 6, 33- (1964).
6. De Duve C: Harvey Lectures, Ser. 59, 49- (1965).
7. De Duve C., De Barsey Th., Poole B., Trouet A., Tulkens B., Van Hoof F.: Biochem. Pharmacol., 2595-2631 (1974).
8. Dingle J.T.: in "Lysosomes in Biology and Pathology" (Eds. Dingle J.T. and Fell H.B.), North Holland, Amsterdam, Vol. II, pp. 421-434 (1969).
9. Wiesmann U.N.: in "Enzyme therapy in lysosomal storage diseases (Eds. Tager J.M. and Hooghwinkel J.M., and Daems W.Th.), North Holland, Amsterdam, pp. 85-94 (1974).
10. Hickman S., Neufeld E.F.: Biochem. Biophys. Res. Comm., *49,* 992-999 (1972).
11. Wiesmann U.N., Di Donato S.: in "Lipid storage diseases: Genetics, Enzymology and Biochemical diagnosis" (Eds. Tettamanti G., Di Donato S. and Zambotti V.), Biochem. Exptl. Biol., in press (1976).
12. Hickman S., Shapiro L.J., Neufeld E.T.: Biochem. Biophys., Res. Comm., *57,* 55-61 (1974).
13. Wiesmann U.N., Herschkowitz N.N.: Pediat. Res., *8,* 865-870 (1974).
14. Leroy J.G., Spranger J.W., Feingold M., Opitz J.M., Crocker A.C.: J. Pediat., *79,* 360-365 (1971).
15. Leroy J.G., De Mars R.I., Opitz J.M.: in "The clinical delineation of birth defects", Vol. V, part IV, The March of Dimes (1969).
16. Martin J.J., Leroy J.G., Farriaux J.P., Fontaine G., Desnick R.J., Cabello P.: Acta Neuropath., *33,* 285-305 (1975).
17. Gilbert E.F., Dawson G., Zu Rhein G.M., Opitz J.M., Spranger J.W.: Z. Kinderheilk., *114,* 259-292 (1973).
18. Schmickel R.D., Distler J.J., Jourdian G.W.: J. Lab. Clin. Med., *86,* 672-682 (1975).
19. Wiesmann U.N., Vassella D., Herschkowitz N.N.: New Engl. J. Med., *285,* 1090-1091 (1971).
20. Vladutiu G.D., Rattazzi M.C.: Biochem. Biophys. Res. Comm., *67,* 956-964 (1975).
21. Jacox R.F., Feldman A.: J. Clin. Invest., *34,* 263- (1955).
22. Allen N., Reagan F.: Arch. Neurol., *11,* 144- (1964).
23. Shuttleworth E.C., Allen N.: Neurology, *18,* 534- (1968).
24. Dott. H., Dingle J.T.: Exptl. Cell. Res., *52,* 523 (1968).

25. Leroy J.G., Van Elsen A.F.: Humangenetick, *20,* 119-123 (1973).
26. Belfiore F., Lo Vecchio L., Napoli E, Borzi V.: Clin. Chem., *20,* 1229- (1974).
27. Kaback M.M., Zeiger R.S., Reynolds S.L.W., Sonneborn M.: Progress in Medical Genetics, Vol. X, Grune and Stratton, pp. 103-134.
28. Woollen J.M., Turner P.: Clin. Chim. Acta, *12,* 671 (1965).
29. Lowden J.A., Skomorowski M.A., Henderson F., Kaback M.M.: Clin. Chem., *19,* 1345- (1973).
30. Baum H., Dodgson K.S., Spencer B.: Clin. Chim. Acta, *4,* 453-455 (1959).
31. Belfiore F., Vecchio L.L., Napoli E.: Amer. J. Med. Sci., *264,* 457-466 (1972).
32. Boyland E., Wallace D.M., Williams D.C.: Brit. J. Cancer, *9,* 62-79 (1955).
33. Dzialoszynski L.M.: Clin. Chim. Acta, *2,* 542-547 (1957).
34. Goldbarg J.A., Pineda E.P., Banks B.M., Rutenburg A.M.: Gastroenterology, *36,* 193-201 (1959).
35. Kar N.C., Pearson C.M.: Proc. Soc. Explt. Biol. Med., *142,* 398-400 (1973).
36. Lawson J.G.: Obstet. Gynaec. Brit. Emp., *67,* 305-308 (1960).
37. Pineda E.P., Goldbarg J.A., Levitan R., Rutenburg A.M.: Amer. J. dig. Dis., *7,* 797-803.
38. Pineda E.P., Goldbarg J.A., Banks B.M., Rotenburg A.M.: Gastroenterology *36,* 202-213 (1959).
39. Lombardo A., Caimi L., Giuliani A., Tettamanti G; in preparation.
40. Caimi L., Lombardo A., Corti G., Giuliani A., Tettamanti G.: in preparation.
41. Niebes P.: Clin. Chim. Acta, *42,* 399-408 (1972).
42. Foà V., Caimi L., Amante L., Antonini C., Gattinoni A., Tettamanti G., Lombardo A., Giuliani A.: Int. Arch. Occup. Environ., *37,* 115-124 (1976).
43. Lauwerys R., Buchet J.P.: Eur. J. Biochem., *26,* 535-542 (1972).

COLLAGENASE PRESENCE IN REMODELING PATHOLOGICAL PROCESSES

M.A. Coletti-Previero*, A. Previero and B. Vallet

Summary

λ and k light chains of immunoglobulins behave as specific substrates for bacterial collagenase. This proteolytic cleavage takes place even when immunoglobulins are in the native state. Immunoglobulin seem therefore to be another molecule susceptible to collagenase in addition to the natural substrate of the enzyme, collagen. This fact is discussed in view of an alternative patho-physiological role of collagenase.

The enzyme and the substrate: collagenase and collagen

Collagenase is a hydrolytic enzyme which cleaves collagen in its native conformation, attacking the domains of this substrate when part of an helical conformation. The necessity for the existence of specific collagenases is created by the resistance of native collagen to general tissue proteases. Collagen was considered[1] the unique susceptible substrate of collagenase and no effect could be noticed on any other fibrous or globular proteins investigated: studies[2] on the nature of collagenase activity using synthetic polypeptides resembling collagen have shown a high specificity of the enzyme limited to the sequence –Pro–X–Y–Pro–, with cleavage between X and Y.

It is a practically unique feature in the mode of action of proteolytic enzymes to attack "native" better than "denaturated" substrate. This implies that the specificity requirements for the enzyme attack must be more complex than those of other proteolytic enzymes, which hydrolyse definite peptide bonds; in other words the local conformation around the cleavage site is essential for this attack. Theoretically there is little probability for this situation to exist in proteins other than collagen, but it does when proline content is relatively high.

Unité de Recherches INSERM 147, Avenue des Moulins 34000 Montpellier, Inst. de Biologie, Faculté de Médicine, Montpellier (France)* Chargee de Recherche C.N.R.S.

The new substrate: immunoglobulin light chains

Immunoglobulins are proteins with specific antibody activity, composed of two light and two heavy chains linked to one another by disulfide bridges: free light chains are excreted into the urine of patients suffering from multiple myeloma and are called Bence-Jones proteins. A large body of chemical data on the molecular structure of these proteins has accumulated in recent years[3] because of the favorable conditions of isolation and subsequent purification: within the antigenic class, they all show the same structural situation, a C-terminal half which is relatively constant, and an N-terminal half showing a marked structural variability. Each half folds in a distinct region ("domain"), as shown by crystallographic studies[4], connected by a "switch" region where the peptidic chain is extended.

Human immunoglobulin light chains of λ and k type behave as substrates for collagenase as reported in Table 1.

Table I

Carboxymethylated protein	*N terminal (residues/moles of protein)*	*Liberated N-terminal (residues/moles of protein)*	
		30 min	*60 min*
Lysozyme	Lys (1)	0	0
Myoglobin	Val (1)	0	0
β_2 microglobulin	Ile (1)	0	0
Lambda chain Ta	Ser (1)	Leu (0.7) Ser (0.5)	Leu (1.1) Ser (0.8) Ala & Val (tr.)
Lambda chain Ve	Tyr (1)	Leu (0.7) Phe (0.9) Ser (0.4)	Leu (1) Phe (1) Ser (0.8) Asp (tr.)
Kappa chain Ge	Glu (1)	Leu (1) Phe (0.8)	Leu (1) Phe (1) Ala & Val (tr.)
Kappa chain Bo	Asp (1)	Leu (1) Phe (0.7)	Leu (1.1) Phe (1) Lys & Val (tr.)

Light chains were isolated from urine of patients suffering from multiple myeloma[5] and β_2 microglobulin from patients with renal tubular disease[6]: the proteolytic cleavage was followed by NH_2-terminal determination[7,8,9] since so few points of enzymatic attack could hardly be evidentiated with other methods.

From the results presented in Table 1, it appears that only immunoglobulin light chains are substrates for bacterial collagenase (Clostridium collagenase, Worthington Corp. Freehold, N.J.); the action of the enzyme is fairly rapid ([E]/[S] = 1/100; 30°C; 30 min. pH 8) and this together with the negative results obtained on proteins other than light chains, allows us to consider the commercial preparation of collagenase as virtually free from spurious proteolytic activity. It is noteworthy that β_2 microglobulin, whose structure is considered related to the constant region of immunoglobulins[10], does not show the same susceptibility to collagenase attack.

Localization of the enzyme attack

We have investigated the exact nature of the sequences involved in the collagenase cleavage of light chain immunoglobulins to establish, if possible, whether or not sequences of the collagen type (–Pro–X–Y–Pro–) are concerned. Peptide separation and classical chemical work[11] on two k chains (Ge and Bo) gave the results summarized in Table II.

Table II. *Collagenase proteolytic cleavages on kappa chain Ge (capital letters) and Bo as substrates*

k Ge GLU–ILE–VAL–LEU–THR–GLX–SER–(PRO, GLY, THR, LEU, SER, LEU, SER, PRO, GLY, GLX)–ARG–ALA–THR–

k Bo asp–ile–gln–met–thr–gln–ser–pro–(ser, ser,

LEU–(SER, CYS)–ARG-ALA-SER–(GLX,...ALA)–PRO–ARG–LEU–↓–LEU–ILE–TYR–VAL–ALA–(SER, SER, ARG,... –leu-ile-tyr-

47

...ASX)↓–PHE–TYR–PRO–ARG–GLU–ALA–LYS–VAL–(GLN, TRP, LYS, ...GLU)–CYS

...asx)↓–phe–tyr–pro–arg–glu–ala–lys–val–(gln, trp, lys., ... glu)–cys

139 215

↓ site of collagenase cleavage.

minor cleavage.

Surprisingly, only a very faint action resembling the collagen-type can be detected, in the sense that there always is a proline near the susceptible bond, but no evidence can be presented as to why those particular sequences have specific affinity for collagenase. Probably a better knowledge of the microenvironment around the susceptible site wold clarify this problem.

As one can easily forsee, this very limited proteolytic cleavage will be a helpful tool for sequence studies. The use of proteolytic enzymes in structural studies of proteins is well established: actually the progress in automatic procedures[12, 13] requires a strategy involving the production of a small number of large fragments from the sequensable protein and therefore the use of "unconventional" enzymes may, as in this case, give encouraging results.

Incidence of collagenase activity during pathological events

Beside the immediate application for structural studies, further questions arise from the evidence of a collagenase action on immunoglobulin light chains. The first obvious one is to know if light chains are still susceptible to collagenase action when inserted in their native immunoglobulin molecule. We submitted a native immunoglobulin prepared from the serum of a patient suffering from myeloma and possessing a light chain of the k type to the enzyme action. In analogous experimental conditions a comparable attack of collagenase takes place, leading to the appearance of new NH_2 terminals. The fragments were also evidentiated by polyacrylamide gel electrophoresis after disulfide bridges disruption.

A second point is the incidence of the collagenase presence and eventually its physiological significance. *Bacterial* collagenases (microorganism) are assumed to have an invasive and, probably, nutritional role, while for *tissue* collagenases (multicellular organisms) the function seems to be distinct and certainly involved in collagen resorption[1, 14]. But an enhanced collagenase activity has been found during particular physiological events, such as normal (postpartum uterus[15]) or pathological remodeling procedures (tumoral tissues[16]). The tumor associated collagenase activity has been interpreted[1, 14, 16] in terms of invasiveness (in association with other disrupting enzymes) but it remains to be ascertained whether the presence of such large amounts of collagenase *in vivo* represents an unusual activation of the inactive zymogen (pro-collagenase[17]) or a real overproduction by tumoral cells or, finally, an example of saturation of the available serum inhibitors[18].

Whatever the reason may be, in the light of the results on immunoglobulins presented here, a new proposal on the function of collagenase in pathological conditions may be advanced. This second hypothesis may be visualized as complementary to the invasive one and should be called immunodisruptive; *in vivo* it could protect the tumor against the immunological defences of the organism. Even if the immunoglobulin molecule is not completely destroyed by the collagenase attack, since the disulfide bridges keep the fragments together, the immunological function may be considered to be, if not suppressed, at least diminished. More probably the molecule becomes more sensitive to other proteases (always present in tumors with invasive characteristics[19]), which probably were ineffective on the native immunoglobulin, leading to its complete destruction; this event may be ofbiological importance.

Acknowledgements

This work was partially supported by a grant of the Fondation pour la Recherche Médicale Française. The immunoglobulins were the generous gift of Dr. Yves Manuel from the Institut Pasteur, Lyon.

References

1. S. Seifer & E. Harper: The Enzymes, *3,* 645-697 (1971).
2. E. Harper: Collagenase, I. Mandl ed., Gordon & Breach, New York, 15-24 (1972).
3. M. Dayhoff: Atlas of Protein Sequence & Structure, Nat. Biomedical Res. Found., Washington D.C., D375-378 (1972).
4. A.B. Edmunson, M. Shiffer, K.R. Ely & M.K. Wood: Biochemistry, *11*, 1822-1831 (1972).
5. M.A. Coletti-Previero, J.C. Cavadore & C. Tonnelle: Immunochem., *12,* 93-95 (1975)
6. I. Beggård & A.C. Bearn: J. Biol. Chem., *243,* 4055 (1968).
7. J.C. Cavadore, G. Nota, G. Prota & A. Previero: Analyt. Biochem., *60,* 608-616 (1974).
8. B. Blomback, M. Blomback, P. Edman & B. Hessel: Biochim. Biophys. Acta, *115,* 371-378 (1966).
9. A. Previero, A. Gourdol, J. Derancourt & M.A. Coletti-Previero: FEBS Lett., *51,* 68-72 (1975).
10. P.A. Peterson, B.A. Cunningham, I. Beggård & G.M. Edelman: Proc. Nat. Acad. Sci. U.S.A., *69,* 1967-72 (1972).
11. B. Vallet, C. Tonnelle & M.A. Coletti-Previero: (in press).
12. P. Edman & G. Begg: Eur. J. Biochem., *1,* 80-91 (1967).
13. R.A. Laursen: Eur. J. Biochem., *20,* 89-102 (1971).
14. E.D. Harris jr. & S.M. Krane: New Engl. J. of Medicine, 537-563, 605-612; 652-661 (1974).
15. J.J. Jeffrey & J. Gross: Biochemistry *9,* 268-273 (1970).
16. E.D. Harris jr., C.S. Faulkner & S. Wood jr.: Biochem. Biophys. Res. Com., *48,* 1247-1253. M.N. Dresden, S.A. Helmen & J.D. Schmidt: Cancer Res., *32,* 993-996 (1972).
17. E. Harper & J. Gross: Biochem. Biophys. Res. Com., *48,* 1147-1152 (1972).
18. A.Z. Eisen, K.J Bloch & T. Sakai: J. Lab. Cli. Med., *75,* 258-263 (1970).
19. G.G. Glenner, M.S. Burstone & D.B. Meyer: J. Natl. Cancer Inst., *23,* 857-873 (1959).

METHODS OF ESTERASE DETERMINATION USEFUL IN CLINICAL LABORATORIES

Klas-Bertil Augustinsson

Summary

The principles of various assay methods for esterases, particularly for cholinesterases and arylesterases, will be introduced; the methods useful in clinical laboratories in more detail. The advantages of these methods will be examined, particularly the methods based upon a change in pH, titrimetric methods, the use of thiolesters as substrates. methods based upon chemical determination after disappearance of substrate, radiometric assay methods, and methods using various aromatic and heterocyclic ester substrates. The principle of the indicator methods will be discussed in connection with the so-called Acholest test and automated procedures. The normal variation of human plasma cholinesterase activity and its clinical significance will be presented.

Blood contains several esterases of various types, the most important of which are cholinesterases present in human blood plasma as a *butyrylcholinesterase* (BuChE) and the membrane bound *acetylcholinesterase* of human blood red cells (Fig. 1). These enzymes are also of interest in clinical medicine. The second esterase present in human plasma is *arylesterase* (ArE), which probably exists in several forms and is still very little investigated from a physiological or clinical viewpoint. The presentation of the various methods for esterase determination useful in clinical laboratories will therefore be restricted in the first place to those being successfully tested for cholinesterases.

Clinical significance of esterase determination

The physiological functions and biological substrates for blood esterases are unknown, even for BuChE which has been studied in detail in almost every conceivable physiological and pathological state (Fig. 2). Many attempts have been made to correlate serum (plasma) BuChE with neurological disorders, skeletal muscle dystrophies, anemias, and with hepatic diseases.

Biochemical Department, Arrhenius Laboratory, University of Stockholm, S-104 05 Stockholm, Sweden

Fig. 1 – Esterases present in normal human blood.

Plasma

Arylesterase ArE (EC 3.1.1.2)
  two types: I and II
  albumin-esterase
Acetylesterase AcE (EC 3.1.1.6)?
Butyrylcholinesterase BuChE (EC 3.1.1.8)
  normal (including variants)
  atypical

Erythrocytes

Membrane

Acetylcholinesterase AChE (EC 3.1.1.7)
Arylesterase?

Cytoplasm

Butyrylesterase?

Thrombocytes

Acetylcholinesterase AChE (EC 3.1.1.7)

Fig. 2 – Cases in which the estimation of serum butyrylcholinesterase has been found to be of clinical value.

1. Liver function and differential diagnosis of jaundice; malnutrition resulting from starvation, anorexia, etc.
2. Excessive exposure to irreversible ChE-inibitors, such as organophosphates and carbamates
3. Predicting susceptibility to prolonged apnea after administration of the muscle relaxant succinylcholine, due to qualitative or quantitative abnormalities in BuChE
4. Investigation of the inheritance of variants of BuChE, of which at least 5 are determined by specific genes

There is, however, almost no universal agreement on the value of estimating serum BuChE in most of these conditions, except for assessing liver function and in differential diagnosis of jaundice. In case of the erythrocyte AChE, this activity has been reported to be reduced significantly in Marchiafava-Micheli disease (paroxysmal nocturnal hemoglobinuria). The subnormal levels of AChE in hematologic disorders have been discussed recently by Tanaka (1975).

If the significance of ChE-estimation in clinical medicine is limited, it is of great value as a diagnostic tool in poisonings caused by irreversible ChE-inhibitors, such as organophosphates and carbamates, frequently used as insecticides, and for monitoring the health-safety aspects of applicator operations with these compounds. Serum BuChE levels may be reduced by more than 50 % before signs of poisoning are evident.

The utility of measurement of BuChE activity is also important when predicting susceptibility to prolonged apnea after administration of the muscle relaxant succinylcholine. This susceptibility is in most cases due to qualitative or quantitative abnormalities in serum BuChE. The assay of serum BuChE activity is also used in studies of the inheritance of variants of this enzyme, of which at least 5 isozymes are determined by specific genes.

Determination of blood cholinesterases

There are several methods described for measuring the ChE activity, including simple screening tests (Augustinsson 1957, 1971). I will restrict may discussion to methods used with blood samples. The differential estimation of AChE and BuChE activity of whole blood is made by the use of substrates hydrolysed by only one enzyme or by the use of selective inhibitors. Thus, acetyl-ß-methylcholine (MeCh) is hydrolysed only by AChE present in the erythrocytes and not by plasma BuChE which rapidly splits butyrylcholine not split by AChE. However, since in human blood the two enzymes are already separated, acetylcholine or some other ester can be used as substrate for both enzymes if their activities are measured separately. In the latter case the choice of substrate will be very important and critical.

Choice of substrate

It is generally accepted that no ChE so far investigated has an absolute specificity for choline esters. All these enzymes also split ordinate esters, the various enzymes having distinct specificity patterns. For example, AChE splits acetic acid esters more rapidly than propionic or butyric acid esters,

Fig. 3 – Substrates recommended in the determination of esterase activity. choline and thiocholine esters.

Substrate	*Active esterases (E)*	*Choice of method*
Choline esters	All eserine sensitive E, i.e. all types of known ChE	Great variety of choice of method: gasometric, pH methods, etc.
Thiocholine esters	The same	Thiocholine produced determined by the Ellman method

Fig. 4 – Substrates recommended in the determination of esterase activity, phenyl and thiophenyl esters.

Phenyl esters	ArE, but also all other known types of E at lower rates	Particularly methods based on the determination of liberated Ph
Thiophenyl esters	The same	Thiophenol produced determined by the Ellman method
Nitrophenyl esters	All types of E and proteinases, acylase, carbonic anhydrase	Production of yellow NO_2-pehnolate ion assayed at 400 nm

Fig. 5 – Substrates recommended in the determination of esterase activity naphthyl and indoxyl esters.

Naphthyl esters	Particulary carboxyl-E and ChE; poor substrates for ArE	Naphthol produced determined after diazotation or fluorimetrically
Indoxyl esters, N-methyl	ChE and also certain other E	Indigo produced determined fluorimetrically

whereas human plasma BuChE catalyses the hydrolysis of butyric acid esters at a higher rate than the esters of the lower homologous acid. It is important to realize that a great many hydrolases can be responsible for the reaction in a crude enzyme preparation when using a noncholine ester. The main problem in the use of a nonspecific substrate, e.g., noncholine esters in ChE studies, is to determine whether the ester is actually split by a ChE alone and/or by another esterase as well. We can make the generalization that any ester can be used as substrate for assaying the activity of a specific esterase, e.g., a ChE, assuming that the preparation studied contains only the specific esterase in question.

As long as choline or thiocholine esters (Fig. 3) with more or less selective specificity for various ChE are available, such esters are preferable to the less specific noncholine esters (Figs 4 and 5). The thiocholine esters are getting more and more popular, because of the excellent methods available for measuring free thiocholine. The same explanation holds for the use of thiophenyl esters for measuring arylesterase activity, but this reaction is far less specific.

Principles of various methods for esterase determination

The principles of the methods used in esterase determination, particularly ChE determination, are summarised in Fig. 6. These methods are based upon

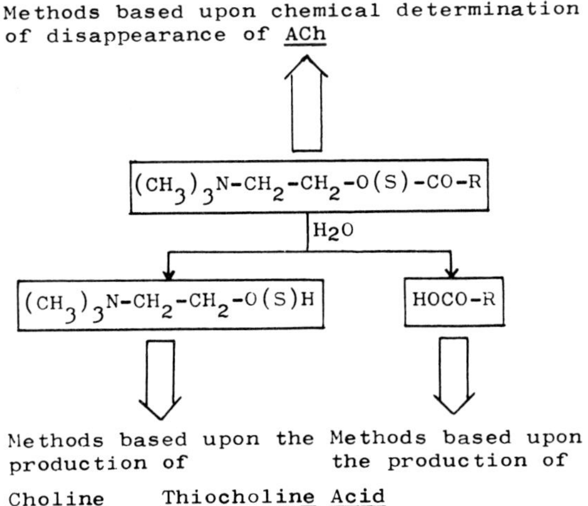

Fig. 6 – Principles of various assay methods for cholinesterases using a choline or thiocholine ester.

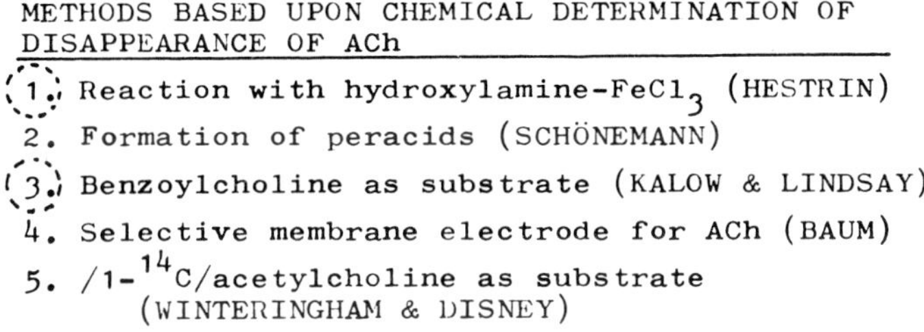

Fig. 7 – Methods of cholinesterase assay based upon the chemical determination of disappearance of acetylcholine.

the determination of the three different components in the enzymatic system:

1. disappearance of *substrate* (a choline ester) (Fig. 7);
2. production of *acid* (Fig. 8); and
3. production of free *alcohol* (Fig. 9), *thioalcohol* (Fig. 10), *phenol* or *thiophenol* (e.g., choline or thiocholine).

The various techniques described in each case are summarized in Figs. 7-10. The techniques most useful in clinical laboratories are the:

1. Gasometric method;

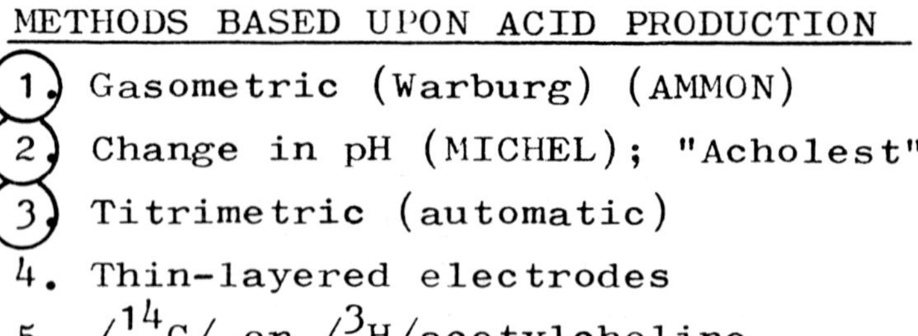

Fig. 8 – Methods of cholinesterase assay based upon acid production.

Fig. 9 – Methods of cholinesterase assay based upon the production of choline.

1. Microbiological
2. *Neurospora*
3. Selective electrodes
4. Oxidation linked to the reduction of cytochrome *c*

METHODS BASED UPON THE PRODUCTION OF THIOCHOLINE

1. Reaction with dithiobis-(2-nitrobenzoate) (ELLMAN)
2. Polarographic (GUILBAULT; FISEROVA-BERGEROVA)
3. Iodometric; nitroprusside-Na; and others

Fig. 10 – Methods of cholinesterase assay based upon the production of thiocholine.

2. Michel electrometric method;
3. Change in colour of an indicator ("Acholest" method);
4. pH-state reference method; and
5. Ellman thiolester method.

Gasometric method

This method has been carefully investigated and used successfully for more than 25 years in our laboratory (Fig. 11) (Augustinsson 1957). The enzymatic hydrolysis of the ester proceeds in a bicarbonate-CO_2 buffer in a

GASOMETRIC METHOD

Sample (Each spot cut in small pieces into the Warburg vessel containing a CO_2-HCO_3^--buffer)		0.05 ml whole blood on paper	
Substrate		Butyryl-choline	Acetyl-β-Me-choline
Activity measured		BuChE in plasma	AChE in erythrocytes
Activity in µkat l^{-1} (µmoles $s^{-1}l^{-1}$ blood):			
mean normal values	males	38 ± 7	18 ± 3
	females	34 ± 9	16 ± 3
range	males	22 - 65	12 - 45
	females	21 - 60	11 - 43

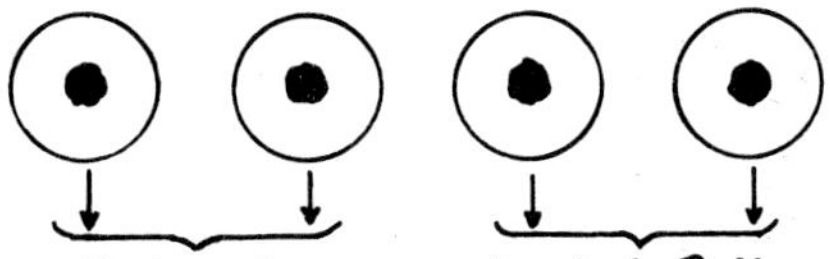

Special equipment: Warburg apparatus

Fig. 11 – Gasometric method in esterase determinations.

closed vessel attached to a manometer, usually a Warburg apparatus. The CO_2 concentration is fixed by the pressure of CO_2 in an atmosphere of N_2 and CO_2. The acid liberated reacts with the bicarbonate ion to form CO_2 given off in equivalent amounts to the acid produced and estimated manometrically, usually at constant volume.

We still use this technique for routine determinations of human blood ChE in 0.05 ml samples absorbed on filter paper and dried at room temperature. Altogether, more than 3.000 samples have been tested, not only those from Swedish and Scandinavian hospitals, but also samples collected under tropical conditions (278 samples from Senegal and Nigeria 1963, and 100 samples from Kenya 1975) and mailed to our laboratory in Stockholm for analysis (Augustinsson & Holmstedt 1965). The filter paper method, being simple and practical, is ideally suited for the control of spraymen and exposed populations when insecticides with ChE inhibiting properties are used.

The Warburg technique is rather complicated, however, and needs trained technicians. Few clinical laboratories therefore use this method in their routine analyses, but other simple techniques can be used as well with the filter-paper samples, according to our own experience.

As an example of the results obtained with this method I will present two cases of suicides with organophosphates (Fig. 12). In one case, the person tried parathion of one high dose, in the other bromophos was chosen. Pa-

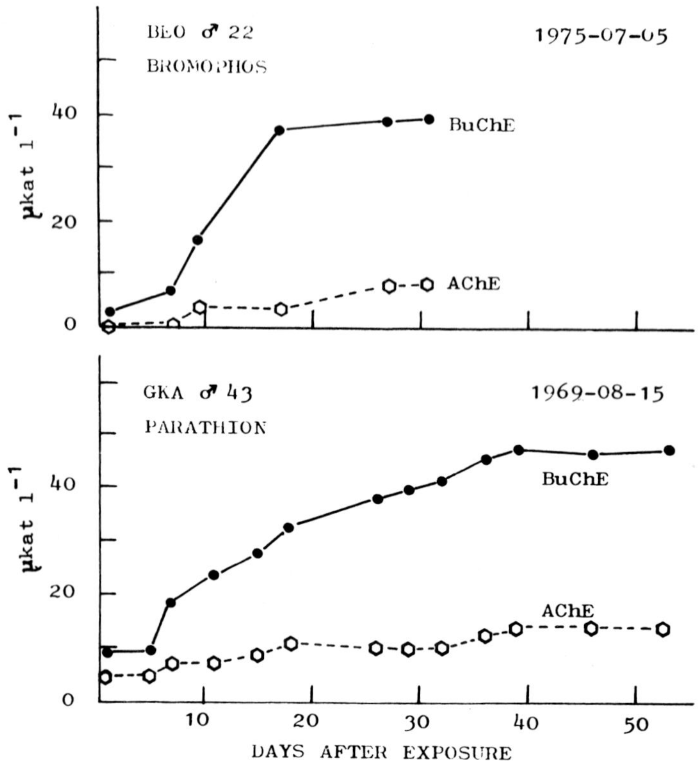

Fig. 12 – Blood cholinesterase activity in two cases of suicides with organophosphate insecticides. Serum butyrylcholinesterase, BuChE; erythrocyte acetylcholinesterase; AChE.

rathion is a stronger inhibitor of serum BuChE of erythrocyte AChE, for bromophos the inhibiting effect is the reverse. This difference could be observed when testing the velocity with which the activities returned to normal.

Electrometric method according to Michel

In this method (Fig. 13), also based upon acid production, the pH of the mixture of enzyme and substrate in a standard buffer solution is measured at the beginning and the end of a fixed interval (normally 1-2 hrs). The rate of change of pH (usually ΔpH hr^{-1}) is a measure of the enzyme activity. The control of temperature is necessary to obtain good precision and to avoid inter- and intralaboratory variations in values (Ellin *et al.* 1975).

Fig. 13 – Electrometric esterase method according to Michel (1949).

Sample, 20 ul	Whole blood		Plasma in 1.0 ml H_2O	Erythrocytes in 1.0 ml saponin soln
Buffer, pH 8.0, 1.0 ml	Phosphate		Phosphate low concentr.	Phosphate higher conc.
Substrate, 0.20 ml	BuCh	MeCh	ACh or BuCh	ACh
Activity measured	BuChE	AChE	BuChE	AChE
Activity in pH per hr			0.92 ± 0.20	0.77 ± 0.09

Special equipment: pH meter.

Michel's method, 25 years after its first description, is still one of the most popular methods for routine determination of blood ChE. Its absolute accuracy and sensitivity are not as great as those of the titrimetric, Warburg and Ellman methods. This is due primarily to the fact that one measures the pH, which is a logarithmic function of the acid concentration, rather than determining the acid production itself. In contradistinction to other more accurate methods, the activity is not measured at several time intervals. An improvement of the technique in this respect is the description of a recording pH meter, calibrated to relate the pH of the solution to the rate of addition of acetic acid.

Change in colour of an indicator; Acholest

The change in pH from the production of acid can be estimated by the change in colour of an indicator rather than with a pH meter. This change in colour will approximate to a linear function of the decrease in pH if the pK of the buffer solution (e.g., phosphate-barbital) is not more than 0.1 unit from that of the indicator used. Most of these procedures have been devised for serum or plasma; only a few are available for whole blood or erythrocytes. It must be confirmed that the indicator (e.g., phenol red or bromothymol blue) does not combine with protein, particularly serum albumin, and that it has no inhibiting effect on the enzyme. Bromothymol blue is the indicator of choice and a number of techniques have been described with it, including special test kits, and automated devices (Wilhelm *et al.* 1973). The use of an Auto-Analyzer has been described not only in clinical work with blood samples, but also for the determination of water-soluble organophosphorus insecticides.

The principle of the indicator methods has been used for the development of screening tests. A most popular and useful variant is to use a filter paper impregnated with buffer, a choline ester and bromothymol blue and subsequently dried. These test-papers are commercially available as "Acholest",

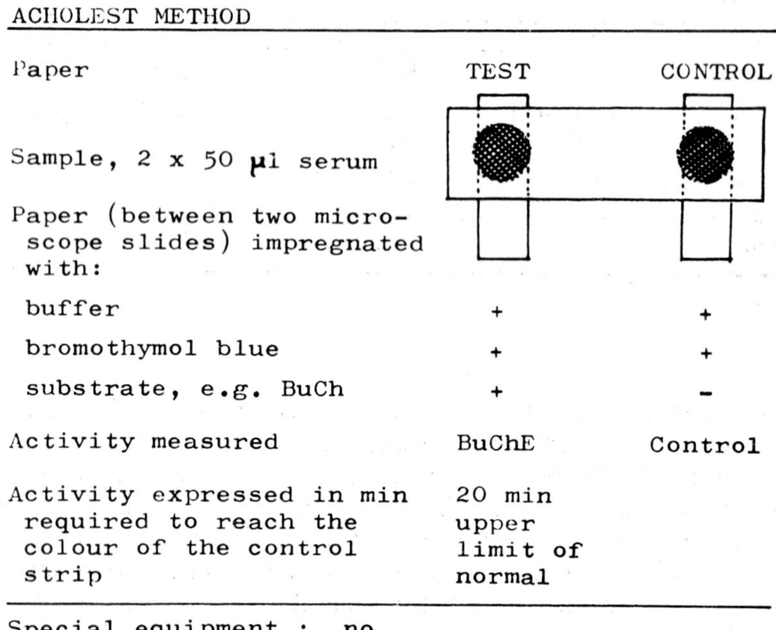

ACHOLEST METHOD		
Paper	TEST	CONTROL
Sample, 2 x 50 µl serum		
Paper (between two microscope slides) impregnated with:		
buffer	+	+
bromothymol blue	+	+
substrate, e.g. BuCh	+	-
Activity measured	BuChE	Control
Activity expressed in min required to reach the colour of the control strip	20 min upper limit of normal	
Special equipment : no		

Fig. 14 – "Acholest" as a screening test of esterase activity based upon the change in colour of an indicator.

"Merchotest" etc. (Fig. 14). The method is based upon the time required for the enzyme activity of a plasma or serum sample to change the colour of the test-paper to match a filter paper similarly impregnated but without substrate. The principal advantages are inexpensive materials with long storage life and very simple equipment requirements. Although the method is not of great precision, it is suitable for routine clinical work. Except for estremely low or high esterase activity values the results obtained correlate fairly well with those obtained by more complicated techniques (Wilhelm *et al.* 1973; Woolfrey 1974).

The "Acholest" method has not been found to be reliable for estimating serum BuChE activity as a preoperative screening test for patients potentially liable to prolonged apnea after use of succinylcholine (Fig. 15). In a study by Dietz *et al.* (1972) 12 of 20 cases at high risk were not detected by this method, as checked by a quantitative assay technique (Ellman method); all these 20 semples had an activity below the lowest normal limit and were from patients with hereditary variants of BuChE strongly predisposed to prolonged apnea after succinylcholine.

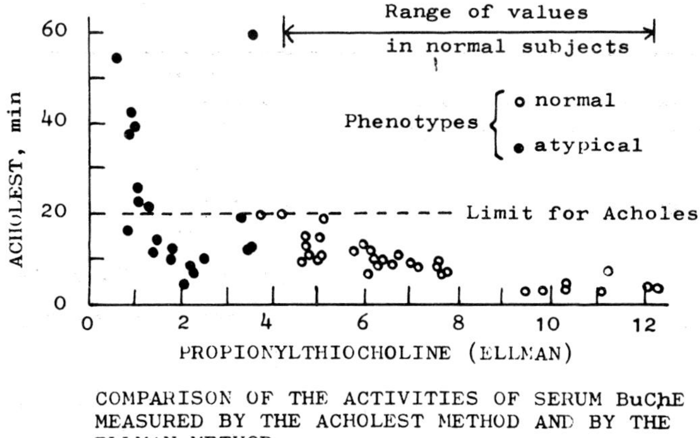

Fig. 15 – Comparison of the activities of serum butyrylcholinesterase measured by the "Acholest" method and by the Ellman quantitative method. The "Acholest" method failed to detect 12 of 20 cases at high risk of prolonged apnea after succinylcholine (after Dietz *et al.* 1972).

Automatic recording titration methods

These procedures are based upon the determination of the acid liberated during the hydrolysis of the ester by titration with standard alkali at constant pH using either an indicator or a potentiometer (Fig. 16). The introduction of an automatic recording titrator has made this [for assaying ChE] one of the most convenient and precise methods, (Keijer 1970). Its great advantage is that the course of acid liberation can be continuously monitored. It is rapid, simple and accurate, and the conditions of the reaction are easily reproduced. The method provides one of the best means to study the kinetics of the initial stages of the reaction of esterases with substrates, inhibitors and activators. At the present time it is probably superseded only by the thiolester method by Ellman.

A promising method, also based upon the production of acid, which makes use of the increase in conductivity of the medium, [was recently described] (Rey & Hanss 1970).

Ellman thiolester method

The use of the spectrophotometric method first described by Ellman *et al.* in 1961 for the determination of ChE activity and recently applied in similar manner for measuring arylesterase (Augustinsson *et al.* 1972) is very attractive because of its simplicity, accuracy and adaptability for micro determina-

Fig. 16 – Titrimetric method of esterase determination employing an automatic recording titrator.

	Reagent	*Volume*
Medium in the reaction vessel with glass electrode	0.1 M NaCl + 0.02 M $CaCl_2$	24.0 ml
Enzyme, suitably diluted	Whole blood, plasma, hemolysate	0.5 ml
pH	adjusted to 7.60 with the titrant	
Gas atmosphere	N_2 blown into the vessel	
Substrate	0.25 M ACh	0.5 ml
Titrant	0.20 M NaOH added automatically to give constant pH 7.60	
Activity measured	Acid production recorded as a straight line for 2-3 min	
Activity in μkat l^{-1}		

Special equipment: pH meter with a titrator unit and titrigraph.

tions (Fig. 17). The method is based on the use of thiocholine esters (or other thiolesters) as substrates and the reaction of the thiocholine formed with 5,5'-dithiobis-(2-nitrobenzoate) (DTNB) to give 5-thio-2-nitrobenzoate, which has an absorption maximum at 405-420 nm. Another disulphide, 2,2'-dithiopyridine (aldrithiol-2), has been suggested to be a better choice than DTNB because the latter was found to activate both AChE and serum BuChE (Brownson & Watts 1973). In a special study we have shown (Augustinsson & Eriksson 1974) that this "activation" could be explained by the fact that the molar absorption coefficient is lowered in the presence of physostigmine, more so with DTNB than with aldrithiol-2 as coupling disulphide. DTNB does not influence the ChE activity, whereas aldrithiol-2 has an inhibitory effect.

A major advantage of the Ellman method is its adaptability for routine analyses suitable for automated procedures (Mersmann & Sanquinetti 1974). These procedures have been applied in a number of studies where esterases or steerase inhibitors are involved. Apart from the advantages already mentioned, this method shows high precision, pH constancy, flexibility, short incubation periods, and continuous increase in colour density as a function of incubation time. The principles of the method represent one of the most straightforward methods developed for ChE experiments.

ELLMAN METHOD	
Sample	50.0 µl enzyme
DTNB-buffer, pH 7.5	3.0 ml
Substrate, thiolester	20.0 µl
Enzyme reaction	$R{-}S{-}CO{-}R_1 \xrightarrow{H_2O} R{-}SH + HOCO{-}R_1$
Yellow reaction product, abs max 405-420 nm	
Activity read manually or recorded	Increase in absorbance at 415 nm
Activity in µkat l^{-1} (ε = 1.36×10^4 $mol^{-1}cm^{-1}$)	$= \frac{\Delta A}{s} \times \frac{10^3}{\varepsilon} \times \frac{\text{total ml}}{\text{sample ml}} \times 10^3$

Special equipment: spectrophotometer with temperature controlled sample compartment for cuvets connected to a log-linear recorder

Fig. 17 – Methods of esterase assay based upon the use of thiolesters according to Ellman (1961).

References

1. Augustinsson K-B: Methods Biochem. Anal., *5*, 1 (1957).
2. Augustinsson K-B: Methods Biochem. Anal., Suppl. 217-273 (1971).
3. Augustinsson K-B, Axenfors B. & Elander M.: Anal. Biochem., *48*, 428-436 (1972).
4. Augustinsson K-B & Eriksson H.: Biochem. J., *139* 123-127 (1974)
5. Augustinsson K-B & Holmstedt B.: Scand. J. Clin. Lab. Invest., *17*, 573-583 (1965).
6. Brownson C. & Watts D.C.: Biochem. J.: *131* 369-374 (1973).
7. Dietz A.A., Rubinstein H. & Lubrano T.: Clin. Chem., *18* 565-566; (1972); ibid, 19 1309-1313 (1973).
8. Ellin R.I. & Vicario P.P.: Arch. Environ. Health, *30*, 263-265 (1975).
9. Keijer J.H.: Anal. Biochem., *37*, 439-446 (1970).
10. Mersmann H.J. & Sanquinetti B.S.: Am. J. Vet. Res., *35*, 579-583 (1974).
11. Rey A. & Hanss M.: Clin. Chim. Acta, *30*, 207-214 (1970).
12. Tanaka K.R.: in "Erythrocyte Structure and Function", Alan R. Liss, Inc., New York 1975, pp. 269-282.
13. Wilhelm K., Vandekar M. & Reiner E.: Bull. Org. mond. Santé, *48*, 41-44 (1973).
14. Woolfrey J.: Can. J. Med. Techn., *36*, 188-203 (1974).

HUMAN PLACENTAL AMINE OXIDASE AND PLASMA DAO LEVELS IN PREGNANCY AND CANCER

I.V. Scott, R.E. Childs, M.J.C. Crabbe, V.R. Tindall and W.G. Bardsley

Summary

Controversy surrounds the identity and clinical relevance of placental and pregnancy plasma diamine oxidase (DAO, Histaminase, 1.4.3.6) some claiming value for serial DAO in high risk pregnancy and others being more critical. Elevated plasma DAO has also been claimed in cancer. We have used a new direct spectrophotometric DAO assay in the analysis of amniotic fluid (150 specimens), term placenta (100), cancer sera (60 patients), pregnancy plasmas (100 patients followed serially), post-partum elimination and further detailed kinetic studies of the purified enzyme using new methods of analysis leading to new insight into the physiological function of DAO.

From over 2,000 assays, we conclude that plasma DAO rises steadily throughut pregnancy largely due to increased diffusion into the plasma, placental concentration remaining steady, and at term 0.2 i.u.hr^{-1} are being synthesised in each placenta giving a concentration of 43.5 ± 26.5 i.u.l^{-1} (total amount of ~6.5 i.u. per placenta). From there it diffuses into the plasma (k_D ~ 0.03 hr^{-1}) giving a concentration there of 2.75 ± 1.4 i.u.l^{-1} and being eliminated with a half life of 23 hr (k_e = 0.032 ± 0.02 hr^{-1}). Plasma and placental DAO levels correlate at term but even serial values have little diagnostic value. Occasional extreme values occur in cancer and pregnancy but this is not statistically significant.

One of the principal research projects in our laboratories is concerned with methods for interpreting complex kinetics. At present, there are over 400 distinct enzymes reported as showing pronounced deviations from Michaelis Menten kinetics and the number would be over 700 if different tissues were counted. Our interest in this subject stems directly from the demonstration that detailed investigations of the initial velocity data for diamine oxidase show such deviations and we have attempted to develop both the theoretical basis for understanding such behaviour and new techniques for the pratical study of these phenomena. More recently, we have worked with a number of well characterised enzymes and having always found such behaviour, have concluded that probably very few enzymes can be represen-

---

Department of Obstetrics and Gynaecology, University of Manchester, St. Mary's Hospital, Whitworth Park, Manchester, M13 OJH, U.K.

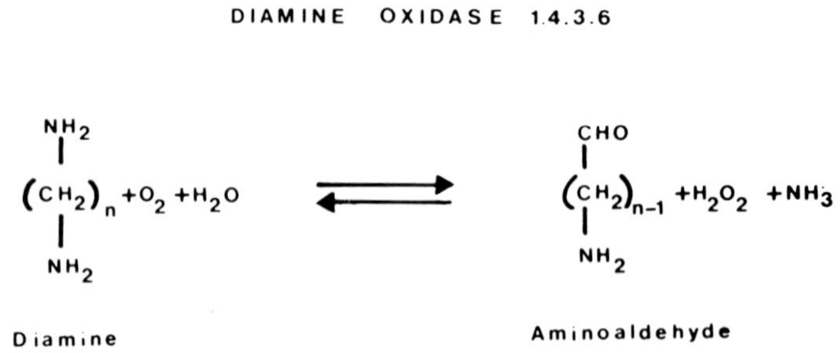

DIAMINE OXIDASE 1.4.3.6

$$\underset{\text{Diamine}}{NH_2-(CH_2)_n-NH_2} + O_2 + H_2O \rightleftharpoons \underset{\text{Aminoaldehyde}}{CHO-(CH_2)_{n-1}-NH_2} + H_2O_2 + NH_3$$

Fig. 1 – *The reaction catalysed by diamine oxidase (DAO).*

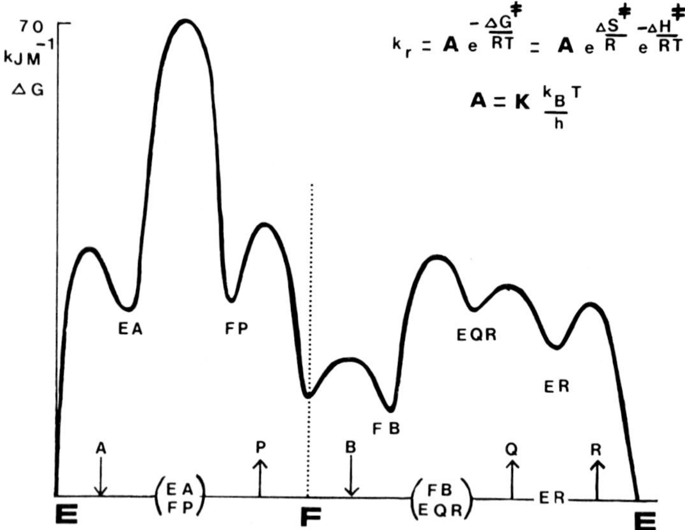

Fig. 2 – *The reaction coordinate for the DAO reaction.* The order of addition of amine (A), oxygen (B) and release of aldehyde (P), hydrogen peroxide (Q) and ammonia (R) was obtained by initial rate and product inhibition studies and partial reactions at the molecular level. The activation energy barriers were suggested by kinetic isotope studies and chemical arguments and the whole scheme is an approximation reasonably valid at the otpimum pH, ionic strength and O_2 concentration (air) for the best substrates such as cadaverine, putrescine and p-dimethylaminomethylbenzylamine.

ted with an acceptable degree of experimental accuracy by a simple rate equation with kinetic constants if sufficient attention is given to experimental detail. However, at this conference, the interest is mainly centred

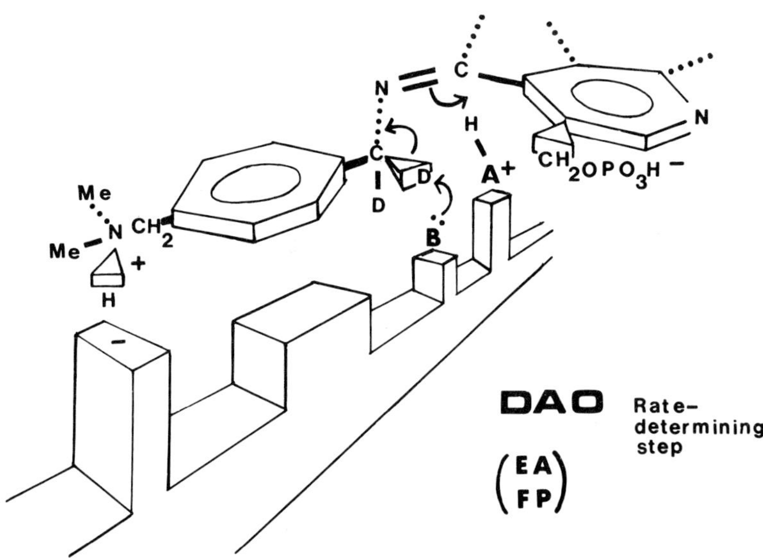

Fig. 3 – *The rate limiting step for the DAO reaction.* This scheme incorporates results obtained from studies conducted with many substrates and inhibitors and illustrates the general acid-base catalysed rate limiting prototropic shift involving the Schiff base between p-dimethylaminomethylbenzylamine and pyridoxal phosphate although the presence of pyridoxal phosphate is by no means certain.

upon clinical enzymology and I am very grateful to be given the opportunity to review the recent developments in diamine oxidase chemistry and to present some of our recent unpublished findings concerning the possible medical relevance of plasma diamine oxidase levels (W.G. Bardsley).

Diamine oxidase (DAO) is an enzyme found in kidney[1,2,3], placenta[4,5,6], intestine[7] and other tissues[8] catalysing the oxidation of diamines by molecular oxygen to aminoaldehydes, ammonia and hydrogen peroxide according to the stoichiometry of figure 1. The reaction mechanism for the hog kidney[9], human kidney[10] and placental enzymes[6,11] predominantly involves binary complexes as shown by kinetic studies with purified enzyme preparations according to the sequence described by the free energy profile illustrated in figure 2.

The enzyme is soluble and cytoplasmic in origin and carries the enzyme commission number 1.4.3.6. It is regarded as identical to histaminase[12] and is usually thought to contain pyridoxal phosphate[13,14] or at least a carbonyl group[15] at the active site and also to require cupric ions for activity[16] in a reaction mechanism as schematically illustrated in figure 3 which is based upon substrate and inhibitor specificities and primary kinetic isotope effects[3,6,9] studied using a a novel synthetic substrate. Normally the plasma levels are low but measurable and widely varying, some individuals, for instance, having extremely high levels[17] but in pregnancy there is a rapid initial rise in concentration[18,19] to an elevated level which is maintained throughout the later states[20] followed by a rapid post-partum decay returning to normal within a few days[21,22]. Monoamine oxidase (MAO) levels in plasma do not fluctuate during this period[23]. However, elevated plasma levels are also associated with medullary carcinoma[24] and rapidly occur following heparin injections[21,22,25].

Now the enzyme has recently been purified from human placenta[6] and pregnancy plasma[26] and since the placenta is regarded as the source of the plasma enzyme[27] it would seem reasonable to expect that plasma levels might reflect placental levels and hence placental function. This has excited the hope that plasma levels might be clinically valuable but opinion is currently divided on this issue. Some groups[28,29] regard plasma DAO as a valuable index of fetoplacental functionality and base fetal salvage programmes on routine assays whilst others are somewhat reserved[22] regarding assays as valuable in predicting gestational age[30] or diagnosing ruptured membranes[31] and some authorities are critical of the value of routine DAO assays[32,33]. In such circumstances, it seems wise to consider alternative assay methods to the ones usually used[34,35] to see if such controversy can thus be resolved and we have recently introduced a new direct spectrophotometric DAO assay that has been developed especially for kinetic studies[9,36] based upon the oxidation of an optically transparent synthetic amine substrate to a UV absorbing aldehyde (λ_{max} = 250 nm). This assay uses p-dimethylaminomethylbenzylamine as substrate as shown in figure 3 as a Schiff-base with pyridoxal phosphate at the active site and it is extremely convenient for working with purified enzymes as it is direct and continuous but it has certain disadvantages with unpurified plasma samples which absorb strongly at 250 nm. However, we have now modified the assay[37] in such a way that it compares favourably with other assays and continued development by others[38] enables measurement in the UV region to be avoided. Using a precision spectrophotometer, the rates of absorbance change are linear over many hours and reproducibility excellent for assays with 0.2 ml of plasma provided that our novel synthetic analogue, 1,4 bis-dimethylaminomethylbenzene, is included in the blank cell to balance out non-specific absorbance changes[37]. Recently, we have used a thermostatically controlled rotating 5-cell turret which considerably facilitates analysis and this has the added advantage that any deviations from linearity due to artefacts can be detected and corrected for in a way that is not possible with discontinuous assay procedures.

At this point it seems appropriate to pose several questions concerning DAO activity in the plasma and placenta.

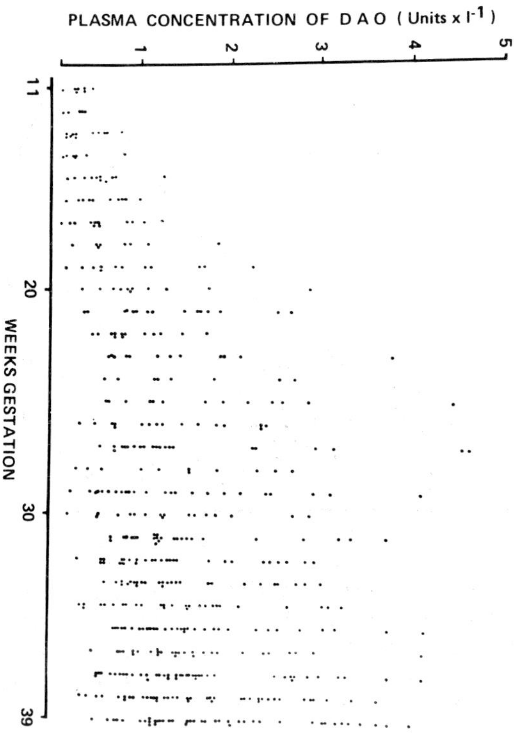

Fig. 4 – *The scatter of plasma DAO levels.* A selection of assays is shown mainly on patients who were clinically normal. The feature of importance is the extremely wide spread of experimental determinations which is believed to genuinely represent biological variation and not experimental error.

1. Does pregnancy plasma DAO come from the placenta and, if so, do plasma levels reflect placental levels?
2. What is the relationship between DAO in amniotic fluid, placenta lymph and plasma and how is the enzyme partitioned between these compartments?
3. Do plasma levels indicate fetoplacental wellbeing and therefore have a clinical value?
4. What is the physiological function of DAO and how does it fit into the overall scheme of placental biochemistry?

The first point we would like to make concerns the spread of normal values, a selection of which are shown in figure 4. Our total sample size was comparable or larger than most studies and we find the actual scatter of values to be wider than usually reported[20, 28, 29]. The rapid initial rise in

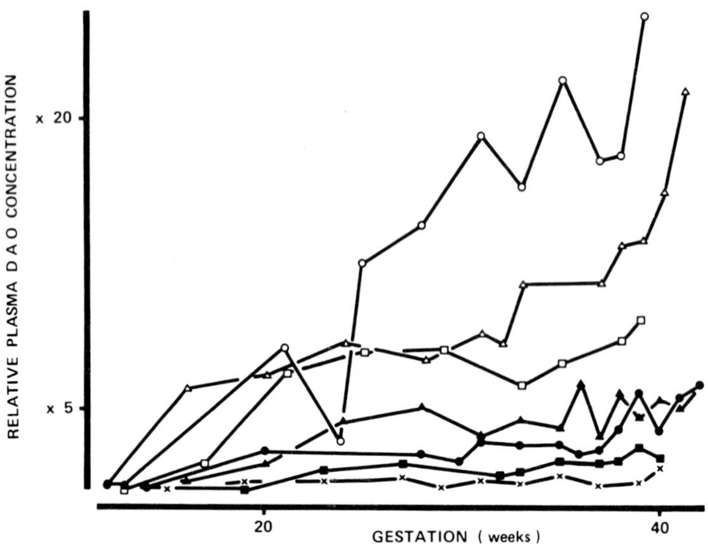

Fig. 5 – *Serial DAO assays.* A random selection of normal serial assays is shown in which all values were standardised to unit activity at the 12th week. Clinically normal pregnancies can be associated with values that increase very little after the 12th week or increase up to 20 times. Our sample had no examples of precipitously falling levels near term which is supposed to indicate placental insufficiency.

plasma concentration was confirmed to be of some limited value as an indicator of gestational age as has been claimed[30] but we could not confirm the suggestions of unusual values associated with abnormalities[28, 29] since these were not found to lie significantly outside the normal range in our sample.

A second point concerns our conclusions based upon over 100 cases followed serially, some of which are shown in figure 5, all normalised to unit activity at 12 weeks. The usual trend is a rapid increase in the period between weeks 6 to 12 followed by a smooth continuous increase to a plateau level in mid trimester which can be anywhere up to 20 times the 12 week level. Some patients showed oscillations about a mean position but with no detectable clinical consequences and diurnal variations were small in all cases examined.

Turning now to the source of the plasma DAO in pregnancy, it seems that the strongest evidence for a placental origin is the high concentration in the maternal decidua compared to myometrium[27], the chemical similarity of purified placental[6] and pregnancy plasma[26] enzymes and the rapid postpartum decay[22]. In the non-pregnant subject, lymphatic fluid has a higher DAO activity than plasma but this seems to be reversed in pregnancy[22], eli-

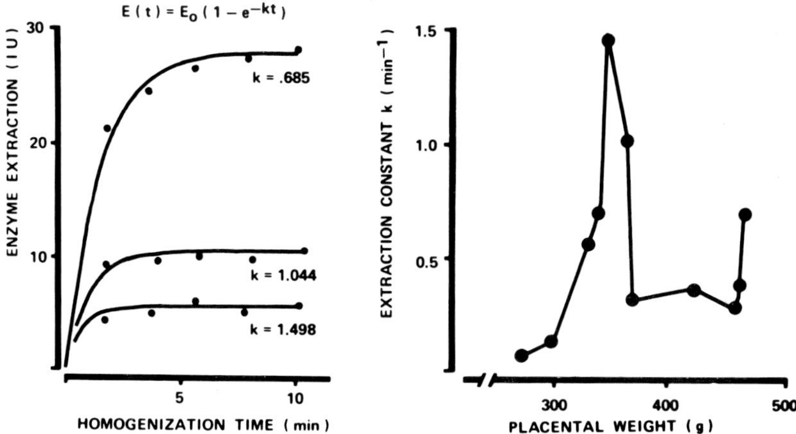

Fig. 6 – *Extraction of DAO from placenta.* A standard homogenisation time of 10 minutes in a Waring blendor was adopted as being sufficient to extract over 95% of the enzyme content into the supernatant even for very small or very large placentae which, for purely physical reasons, were the most difficult to homogenise in our apparatus.

minating this as a possible source. The heparin induced DAO activity returns to normal with a much shorter half life than the post-partum decay[21, 25] even in pregnancy and, since there are also electrophoretic differences, it seems that pregnancy provoked DAO and heparin induced DAO are probably distinct isoenzymes[21].

We have attempted to decide whether the plasma enzyme comes from the placenta by relating the concentration in plasma and placenta at term and figure 6 shows the rate of extraction of DAO when placentae are homogenised in a Waring blendor. 10 minutes is a sufficient extraction time to ensure virtually complete extraction of enzyme into the supernatant from the centrifuged homogenate and the figure also shows that the efficiency of extraction is lower for extremely small or large placentae as seen by the graph of extraction constant plotted against placental weight and as would be expected on purely physical grounds. In all cases, the concentration of enzyme in placental homogenate supernatant was greater than the corresponding parturition plasma level and usually it was some twenty fold greater. This can readily be seen in figure 7 where results from forty eight normal pregnancies are collected. Parturition plasma level is plotted against activity in the individual placentae, the upper line representing equipartition and the lower line the least squares regression line. Again, the extremely wide scatter of points should be noted but it seems reasonable to conclude that, since the concentration of enzyme in placental tissue is considerably higher than

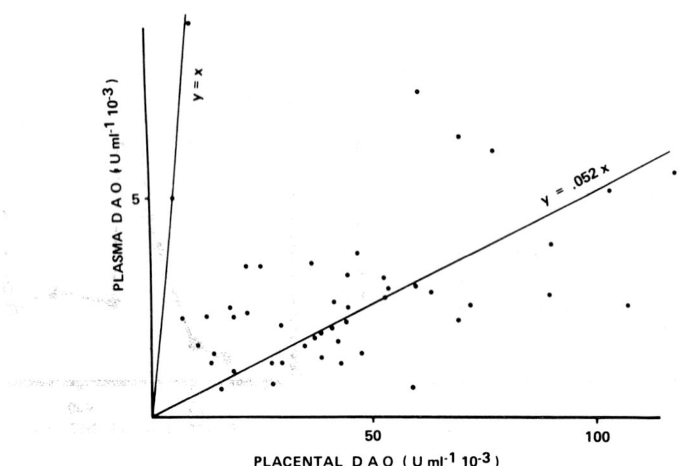

Fig. 7 – *Plasma DAO versus placental DAO.* 48 normal term placental enzyme concentrations are shown together with the corresponding term plasma levels. In all cases, the placental levels were greater than plasma levels as shown by the accumulation of points to the right of the line of equipartion (y = x).

plasma, movement down a concentration gradient is occurring and, in the absence of evidence for active transport, we can assume diffusion and proceed with further analysis. From several analyses on preterm placentae, we conclude that a gradual build-up takes place in the placentae but increased diffusion occurs, presumably as a result of increased surface area, leading to a final steady-state in which synthesis occurs in the placenta, followed by diffusion into the plasma and elimination, all occurring at such a rate as to maintain constant concentrations in the system.

Such a scheme is illustrated in figure 8 and can be described by the diffusion constant k_0 accounting for diffusion between placenta and plasma in both directions and an elimination constant k_1 accounting for elimination from the plasma. At term, the placenta as a source is removed and we would have an interesting situation in which elimination from the plasma is isolated and hence k_1 or rather k_1/V_1 measurable experimentally from the slope of semilogarithmic plots. Typical semilogarithmic elimination graphs are displayed in figure 9 and, although these do eventually become convex indicating a higher order compartmental elimination model, this was not pursued, pseudo first order kinetics was assumed and the elimination constants obtained by least squares fitting of regression lines. Some of the date obtained in this way is tabulated in table 1 and we merely mention that, as we found amniotic fluid level comparable to plasma values as others have done[39], this could easily be fitted into such a scheme in line with contemporary ideas on the origin of the soluble protein in amniotic fluid[40].

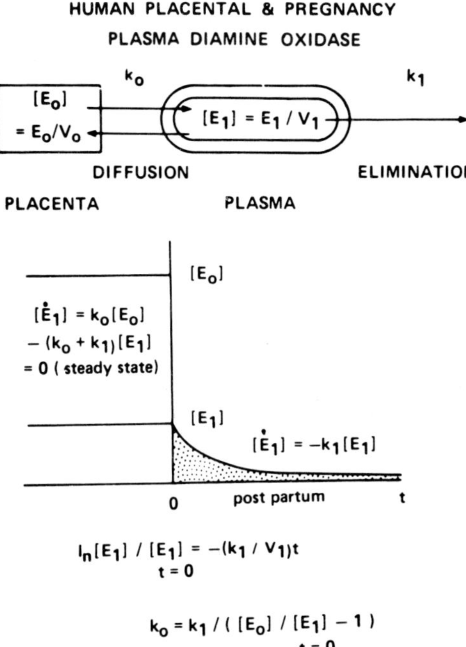

Fig. 8 – *The source of plasma DAO.* E_o is the total enzyme content of a placenta and V_o the volume of the placental homogenate supernatant. E_1 is the total enzyme in a plasma volume of V_1. k_o is a diffusion constant for movement between plasma and placenta and k_1 and elimination constant. E_o, E_1, V_o, V_1 and k_1 are measured experimentally and thus k_o determined as indicated.

Table I. *Post partum elimination kinetics and diffusion of DAO between plasma and placenta for 7 normal patients.* V_o is placental volume and V_1 plasma volume. k_o is the plasma/placenta diffusion constant and k_1 the elimination constant.

Placental DAO		*Term Plasma DAO*		k_o/V_o	k_1/V_1
$1\ U \times 1^{-1}$	$Vol(ml) = V_o$	$1\ U \times 1^{-1}$	$V_1 = Vol(ml)$	hr^{-1}	hr^{-1}
.164	136	6.2	4043	.06	.042
.132	214.2	4.5	5740	.03	.022
.038	138	10.4	4830	.05	.025
.068	86	2.2	4305	.08	.029
.039	139	1.5	5082	.13	.067
.08	127.6	3.2	4032	.02	.01
.137	155.2	5.7	5816	.04	.02

The mean kinetic diffusion and elimination constants were calculated to be as follows:
$k_D = k_o/V_o = 0.06 \pm 0.034\ hr^{-1}$ $k_E = k_1/V_1 = 0.03 \pm 0.077\ hr^{-1}$

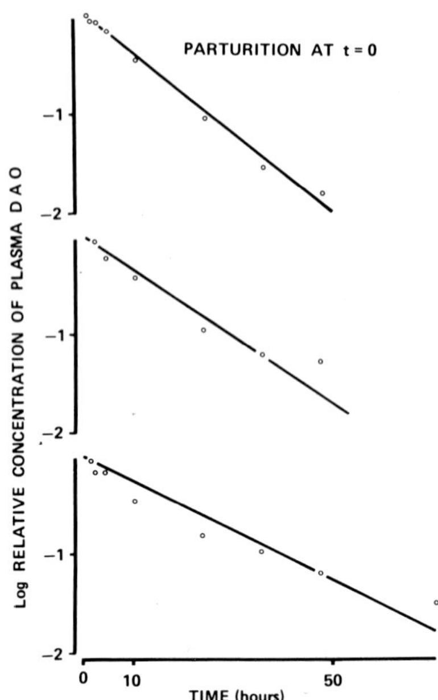

Fig. 9 – *The post partum decay of plasma DAO activity.* Elimination kinetics were approximately first order as indicated giving semilogarithmic plots with slope $-k_1/V_1$. Deviations from pseudo first order elimination kinetics occur at longer time intervals indicating a more complex elimination mechanism but this was not investigated.

From the seven determinations in table 1 k_1/V_i values were obtained by least squares regression lines fitted to the individual post partum decay data. Individual placenta were investigated for volume (V_o) and totale enzyme (E_o) and blood volume (V_1) was calculated from the patient's weight and this information then used to give k_o. The mean kinetic diffusion constant k_D and kinetic elimination constant k_E were found to be as follows:

$$k_D = k_o/V_o = 0.06 \pm 0.034 \text{ hr}^{-1}$$
$$k_E = k_1/V_1 = 0.03 \pm 0.077 \text{ hr}^{-1}$$

However, we turn now to the possible physiological function of placental DAO.

We have recently obtained highly purified preparations of placental DAO with a specific activity of >6.0 i.u.mg^{-1} by affinity chromatography[41]. The enzyme adheres to Con-A Sepharose and is readily eluted by α-methylglucoside suggesting it to be possibly glycoprotein in nature and recent studies with this material have indicated some surprising results. We find the enzyme to consist of an active monomer of molecular weight 70.000 Daltons containing one mole each of manganese and cupric copper by e.p.r. spectroscopy and chemical analysis. This polymerises to form less active dimers especially at low oxygen concentration and also, at extremely high enzyme concentra-

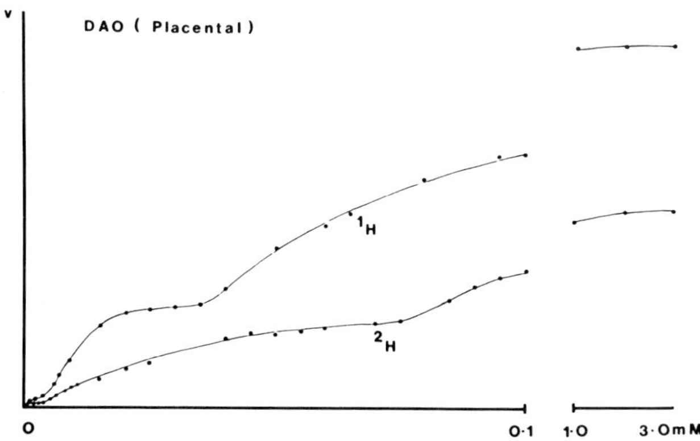

Fig. 10 – *The v versus S plot.* The amine and deuterio analogue give v versus S plots with clearly defined inflexions at low substrate concentrations and at high substrate levels, inhibition is seen.

Fig. 11 – *The v versus both substrates surface.* From over 400 velocity determinations at different concentrations of amine and oxygen, the surface was constructed to show the complex dependence of velocity on substrate concentration. A is ln (1 + amine) and B is ln (1 + oxygen), the vertical axis being initial velocity. Further studies of this data reveal a rate equation at least 4:4 in oxygen and 4:6 in amine.

tion, there is evidence from ultracentrifugation studies of the formation of tetramers. This explains why, at substrate and enzyme concentrations near the physiological levels, the enzyme exhibits properties normally associated with regulatory enzymes, namely, pronounced deviation from Michaelis-Menten kinetics and highly complex velocity versus substrate concentration profiles[11] as shown for velocity versus amine or deuterio amine in figure 10.

We have constructed many such profiles at several fixed oxygen concentrations and analysed the results using new techniques for analysing complex kinetics[42] leading to the conclusion that the initial rate equation is of minimum degree 4:4 in oxygen and 4:6 in amine. The complex nature of the steady-state data is shown in figure 11 which is a model of the surface of velocity as a function of both substrate concentrations, A representing ln (1 + amine) and B, ln (1 + oxygen) for convenience. Further, we find the enzyme to be fairly stable in concentrated ureas solutions[15, 41] and to be inhibited by β-aminoproprionitrile (BPN). Also, it oxidises lysyl peptides, e.g. lysyl vasopressin, lysylarginine and collagen precursors. This behaviour is not found for the classifical DAO of hog kidney, but is much more reminiscent of those enzymes known as lysyl oxidases[43, 44] which oxidise lysyl to alysyl residues and thus are of importance in collagen biosynthesis. It has been suggested that the function of plasma MAO may be in the realm of collagen biosynthesis[45] but this must be considered together with the recent evidence that purified lysyl oxidase has no MAO activity[46]. Finally, we find that, like the hog kidney enzyme[47], the purified placental enzyme does not oxidise benzylamine or p-nitrobenzylamine appreciably and this is difficult to reconcile with recent reports that p-nitrobenzylamine is oxidised by DAO at a comparable rate to diamines[48] and can be thus used as a DAO assay[38].

Now the earliest idea for the biological function of placental DAO was as a general amine scavenging enzyme which prevented local accumulation of histamine and adverse vasomotor effects. In fact, a similar role has been suggested for placental monoamine oxidase which is thought to be low in toxaemia[49]. Amniotic fluid contains 5HT[50] and injection of MAO inhibitors produces abortion[51] suggesting the presence of a protective enzyme. It should, however, be pointed out that MAO inhibitors do not seem to exacerbate pregnancy toxaemia[49] and these substances also are known to inhibit placental DAO to a certain extent[52]. There are some contradictions also in the scheme which represents DAO as protecting against histamine intoxication and to illustrate these we turn finally to an outline of histamine catabolism as shown in figure 12 with some minor pathways omitted for clarity.

Injected histamine is eliminated by the liver[53] and kidneys[54] mainly following methylation, oxidation by MAO and conjugation and capacity to metabolise histamine does not appreciably increase in pregnancy[55, 56] although some have claimed increased resistance to injected histamine in pregnancy[57]. Aminoguanine, a powerful inhibitor of purified DAO[15] inhibits direct oxidation of histamine in vitro[58] and in vivo[59] leading to increased methylation and oxidation by MAO but, although histamine excretion increases in pregnancy[60] and increased histamine levels may be associated with conditions or rapid growth[57], the adequate metabolism of histamine by pathways not involving DAO does seem to argue against the physiological function of placental DAO as a histamine scavenger. Also, the fact that histamine is by no means one of the best substrates of DAO[3, 6, 61] and, in addition, shows

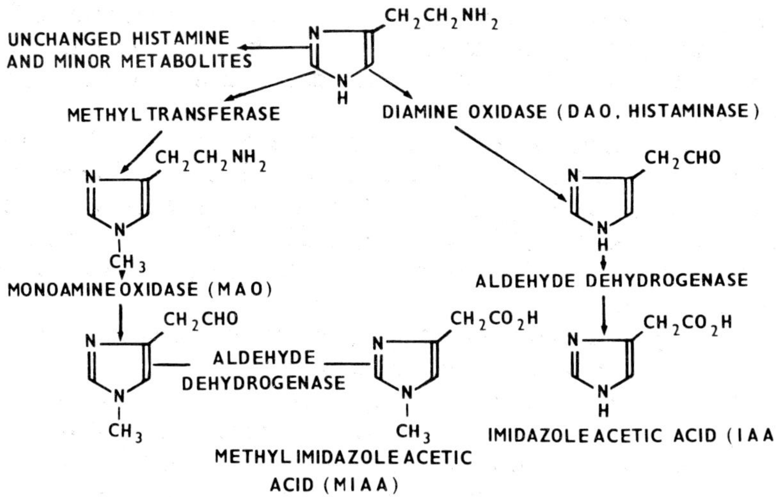

Fig. 12 – *Histamine catabolism.* The main pathways by which exogenous histamine is metabolised are shown and some minor metabolites are omitted for clarity.

extremely pronounced substrate inhibition[47, 61] not given by diamines further reinforces this conclusion.

There is currently much interest in the subject of polyamines[62] which are thought to be of importance in rapidly growing tissue and, although polyamines have been measured in placental tissue[63], there is, as yet, no definitive work available as the placental biochemistry of polyamines. Polyamines such as spermine are oxidised in vivo to cytotoxic aldehydes which are being evaluated as anti-tumour agents[64] and the basic unit from which polyamines are biosynthesised is putrescine originating from ornithine under the action of ornithine decarboxylase. Now the purified placental DAO, like the human semen enzyme[65], oxidises spermine and spermidine albeit rather weakly but putrescine and cadaverine are amongst the best substrates and it might be that the biological function might be sought in this direction, namely, control of polyamine levels and hence tissue growth. However, as yet, the in vivo metabolism of diamines in the placenta is not fully understood and it may be, for instance, that alternative pathways exist such as acetylation and oxidation by MAO as found for brain tissue[66].

To summarise. From recent work on purified human placental diamine oxidase and from over 1,000 DAO assays performed on pregnancy plasma (100 patients followed serially), amniotic fluid (150 specimens), term placentae (100), cancer sera (60 patients mainly ovarian carcinoma) using a new direct, continuous spectrophotometric assay, we reach the following conclusions:

1. The enzyme diamine oxidase is synthesised in the placenta (0.2 i.u.hr^{-1} per placenta at term) and transported, probably by diffusion ($k_D = k_o/V_o$ approximately 0.06 hr^{-1}) into the maternal plasma down a steep concentration gradient.
2. The levels of activity in term plasma (2.75 ± 1.4 i.u.$l.^{-1}$) correlate well with levels in the placenta (43.5 ± 26.5 i.u.$l.^{-1}$), $p < 0.001$) and plasma levels do reflect placental levels. The half life of the enzyme is of the order of 23 hr ($k_E = k_1/V_1$ approximately 0.03 hr^{-1}) and elimination merely balances synthesis in the latter months resulting in a steady-state with about 6.5 i.u. per placenta.
3. Extremely high and extremely low values can be found in normal pregnancy with no detectable clinical consequences.
4. The values found in abnormal pregnancies are not appreciably distinct from normal on account of the extremely wide spread of normal values. In ovarian cancer over 40 % of patients showed elevated levels but this is of no diagnostic and, as yet, no obvious prognostic value.
5. Apart from limited value in predicting gestational age in early pregnancy or diagnosing ruptured membranes, DAO assays even performed serially, are not useful indicators of placental function or of progression or regression of ovarian carcinoma.
6. A new direct spectrophotometric assay for DAO activity has been proved convenient and reproducible and has given results in substantial agreement with those experimental workers who, using other assay methods, have reached similar rather negative conclusions concerning the clinical value of plasma DAO assays.
7. It seems reasonable that an enzyme which is virtually absent from the plasma except during pregnancy should have a specific physiological function and our results indicate that placental diamine oxidase has many unusual properties not found with the classical hog kidney enzyme. If a physiological functions exists, then it is probably in the realm of control of polyamine or diamine levels or, less likely, in collagen biosynthesis and not in histamine metabolism.

Acknowledgement

We gratefully acknowledge the assistance of the Medical Research Council in providing funds for maintenance (REC) and for a Cary 118C spectrophotometer used in this work.

References

1. Mondovì B., Rotilio G., Costa M.T., Finazzi-Agrò A., Chiancone E., Hensen R.E. & Beinert H.: J. Biol. Chem., *242,* 1160-1167 (1967).
2. Yamada H., Kumagai H., Kawasaki H., Matsui H. & Ogata K.: Biochem. Biophys. Res. Commun., *29,* 723-727 (1967).

3. Bardsley W.G., Hill C.M. & Lobley R.W.: Biochem. J., *117,* 169-176 (1970).
4. Smith J.K.: Biochem. J., *103,* 110-119 (1967).
5. Paolucci L., Cronenburger R., Plan R. & Pacheco H.: Biochimie, *53,* 735-749 (1971).
6. Bardsley W.G., Crabbe M.J.C. & Scott I.V.: Biochem. J., *139,* 169-181 (1974).
7. Kusche J., Richter H., Hesterberg R., Schmidt J. & Lorenz W.: Agents and Actions, vol. 3/3, 148-156. Birkhauser Verlag Basel (1973).
8. Zeller E.A.: In The Enzymes, vol. 8, 313-335 Ed. Sumner, J.B. & Myrback, K. New York: Academic Press Inc. (1963).
9. Bardsley W.G., Crabbe M.J.C. & Shindler J.S.: Biochem. J., *131,* 459-469 (1973).
10. Shindler J.S. & Bardsley W.G.: unpublished work.
11. Waight R.D. & Bardsley W.G.: unpublished work.
12. Zeller E.A.: Fed. Proc., *24,* 766-768 (1965).
13. Kumagai H., Nagate T., Yamada H. & Fukamai H.: Biochim. Biophys. Acta, *185,* 242-244 (1969).
14. Mondovi B., Costa M.T., Finazzi-Agro, A. & Rotilio G.: Arch. Biochem. Biophys., *119,* 373-381 (1967).
15. Crabbe, M.J.C., Childs, R.E. & Bardsley W.G.: Eur. J. Biochem., *60,* 325-333 (1975).
16. Bardsley W.G., Childs R.E. & Crabbe M.J.C.: Biochem. J., *137,* 61-66 (1974).
17. Gennser G., Soderberg H. & Tryding N.: Amer. J. Obstet. Gynecol., *119,* 1128-1130 (1974).
18. Marcou I., Athanasin-Vergu E., Chiriceanu D., Cosma G., Gingold N. & Parhou C.C.: Presse Med., *46,* 371 (1938).
19. Kobayashi K.: Nature, *203,* 146-147 (1964).
20. Ahlmark A.: Acta Physiol. Scand. 9 suppl. 28-107 (1944).
21. Hansson R.: Scand. J. Clin. Lab. Invest.: *25,* 33-39 (1970).
22. Hansson R.: Scand. J. Clin. Lab. Invest., *31,* suppl. 129, 7-23 (1973).
23. McEwen, C.M. Jr.: J. Lab. & Clin. Med., *64,* 540-547 (1964).
24. Keiser H.R., Beaven M.A., Doppman J., Wells S. Jr. & Buja M.: Annals of Internal Medicine, *78,* 561-579 (1973).
25. Baylin S.B., Beaven M.A., Krauss R.M. & Keiser H.R.: J. Clin. Invest., *52,* 1985-1993 (1973).
26. Baylin S.B. & Margolis S.: Biochim. Biophys. Acta, *397,* 294-306 (1975).
27. Swanberg H.: Acta Physiol. Scand. *23,* suppl. 79,7- (1950).
28. Weingold A. & Southren L.: Obstet. Gynecol., *32,* 593-606 (1968).
29. Schari G., Achari K. & Rao K.K.: Japan. J. Pharmacol., *21,* 33-40 (1971).
30. Elmfors B. & Tryding N.: Brit. J. Obstet. Gynaecol., *83,* 6-10 (1976).
31. Elmfors B., Tryding N. & Tufvesson G.: J. Obstet. Gynaecol. Br. Commonw., *81,* 361-362 (1974).
32. Resnik R. & Levine R.J.: Amer. J. Obstet. Gynecol., *104,* 1061-1066 (1969).
33. Ward H., Rochman H., Varnavides L.A. & Whyley G.A.: Amer. J. Obstet. Gynecol., *116,* 1105-1113 (1973).
34. Tryding N. & Willert B.: Scand. J. Clin. Lab. Invest., *22,* 29-32 (1968).
35. Southren A.L., Kobayashi Y., Carmody N.C. & Weingold A.B.: Amer. J. Obstet. Gynecol., *95,* 615-620 (1966).
36. Bardsley W.G., Crabbe M.J.C. & Shindler J.S.: Biochem. J., *127,* 875-879 (1972).
37. Bardsley W.G., Crabbe M.J.C. & Scott I.V.: Biochem. Med., *11,* 138-146 (1974).
38. Tourkov M.I., Klimova G.I., Davydova G.A., Yermolaev K.M. & Gorkin V.Z.: Analyt. Biochem., *64,* 177-185 (1975).

39. Ward H., Whyley G.A. & Millard M.D.: J. Obstet. Gynaecol. Br. Commonw., *80,* 525-530 (1973).
40. Sutcliffe R.G.: Biol. Rev.. *50,* 1-33 (1975).
41. Crabbe M.J.C., Waight R.D., Bardsley W.G., Barker R.L., Kelly I.D., Knowles P.F. Biochem. J. In Press (1976).
42. Bardsley W.G.: Biochem. J., *153,* 101-117 (1976).
43. Shieh J.J., Tamaye R. & Yasunobu K.T.: Biochim. Biophys. Acta, *377,* 229-238 (1975).
44. Vidal G.P., Shieh J.J. & Yasunobu K.T.: Biochem. Biophys. Res. Commun., *64,* 989-995 (1975).
45. McEwen, C.M. Jnr.: Adv. Biochem. Psychopharmacol., *5,* 151-165, Raven Press, New York (1972).
46. Narayanan A.S., Siegel R.C. & Martin C.R.: Arch. Biochem. Biophys., *162,* 231-237 (1974).
47. Hill C.M. & Bardsley W.G.: Biochem. Pharmacol., *24,* 253-257 (1975).
48. Zeller E.A.: Adv. Biochem. Psychopharmacol., *5,* 167-180 (1972).
49. Sandler M. & Coveney J.: The Lancet, *1,* 1096-1098 (1962).
50. Koren Z., Pfeifer Y. & Sulman F.G.: Amer. J. Obstet. Gynaecol., *93,* 411-415 (1965).
51. Koren Z., Pfeifer Y. & Sulman F.G.: J. Reprod. Fertil, *12,* 75-79 (1966).
52. Crabbe M.J.C. & Bardsley W.G.: Biochem. Pharmacol., *23,* 2983-2990 (1974).
53. Lindell S.E. & Westling H.: Scand. J. Clin. Lab. Invest.: *18,* 268-272 (1966).
54. Helander C.G., Lindell S.E. & Westling H.: Scand. J. Clin. Lab. Invest., *17,* 524-528 (1965).
55. Nisson K., Lindell S.E., Schayer R.W. & Westling H.: Clin. Science, *18,* 313-319 (1959).
56. Granerus G.: Scand. J. Clin. Lab. Invest., *22,* suppl. 104, 59-68 (1968).
57. Kahlson G. & Rosengren E.: Biogenesis & Physiology of Histamine. E. Arnold (London) (1971).
58. Lindberg S.: Acta Obstet. Gynecol. Scand. XLII, suppl. 1, 26-33 (1962).
59. Lindell S.E., Nilsson K. & Westling H.: Brit. J. Pharmacol. Chemother., *15,* 351-355 (1960).
60. Harrison V.C., Peat G. & Heese, H. de V.: J. Obstet. Gynaecol. Brit. Commonw., *81,* 686-690 (1974).
61. Bardsley W.G., Ashford J.S. & Hill C.M.: Biochem. J., *122,* 557-567 (1971).
62. Kusche J., Lorenze W. & Schmidt J.: Hoppe-Seyler's Z. Physiol. Chem., *356,* 1485-1496 (1975).
63. Southren A.L., Kobayashi Y., Jung W., Carmody N.C. & Weingold A.B.: J. Clin. Endocrinol., *26,* 1005-1009 (1966).
64. Southren A.L., Kobayashi Y., Jung W. & Weingold A.B.: J. Clin. Endocrinol. Metab., *28,* 1724-1729 (1968).
65. Tabor H. & Tabor C.W.: Advances in Enzymology, Ed. A. Meister, *36,* 203-268 (1972).
66. Gunaga C.K., Sheth A.R., Gunaga K.P. & Raeo S.S.: Ind. J. Biochem. Biophys., *9,* 272-274 (1972).
67. Israel M., Zoll E.C., Muhammad N. & Modest E.J.: J. Med. Chem., *16,* 1-5 (1973).
68. Holtta E., Pulkinnen P., Elfving K. & Janne J.: Biochem., J., *145,* 373-378 (1975).
69. Seiler N., Al-Therib M.J.: Biochem., J., *144,* 29-35 (1974).
70. Weingold A.B., Southren A.L. & Lee B.O.: Int. J. Fertil, *16,* 24-35 (1971).
71. Ward H., Whyley G.A. & Millar M.D.: Acta Obstet. Gynecol. Scand., *55,* 63-68 (1976).

LECITHIN: CHOLESTEROL ACYLTRANSFERASE (LCAT). BIOCHEMISTRY AND CLINICAL SIGNIFICANCE

E. Kaiser and R. Paula

Summary

The plasma lecithin: cholesterol acyltransferase (LCAT)* mainly transfers fatty acids from lecithin to cholesterol of plasma lipoproteins. Properties of LCAT and reliability of assay procedures are discussed. Observations on the diagnostic significance in liver diseases are presented.

The clinical biochemist is constantly met with the demand to develop new tests which might be useful in clinical and experimental hepatology. Recently two new tests in hepatology were extensively studied: the determination of lipoprotein X (LP – X) and the assay of Lecithin : Cholesterol Actyl-transferase, the esterifying enzyme of cholesterol in blood. Whereas LP–X proved to be useful in the diagnosis of cholestasis, the reliability of cholesterol esterifying activity in the diagnosis of different liver diseases is still under discussion.

The purpose of this paper is to review the information available on cholesterol esterification as it relates to liver damage in man and experimental animals.

It is well known that in various diseases of the liver abnormal patterns of plasma lipids may be found. In 1926 Thannhauser and Schaber[54] demonstrated that in cases of liver damage the serum concentration of cholesterol esters is low and is almost lacking in severe cases. This observation called "Estersturz" was extensively studied by several investigators and Epstein and Greenspan[9,10] claimed that the alterations in the amount of esterified

Dept. of Medical Chemistry, School of Medicine, University of Vienna, Austria.

*Abbreviations: VLDL: Very Low Density Lipoproteins; LDL: Low Density Lipoproteins; HDL: High Density Lipoproteins; ACD-Plasma: Acidum Citricum Dextrose Plasma: LCAT: Lecithin Cholesterol Acyltransferase; LP–X: Lipoprotein X; SGOT: Serum Glutamic Oxaloacetic Transaminase; SGPT: Serum Glutamic Pyruvatic Transaminase; AP: Alkaline Phosphatase; FC: Free Cholesterol; EC : Esterified Cholesterol.

cholesterol could have diagnostic and even prognostic significance in diseases of the liver.

The increase of free cholesterol in plasma in obstructive jaundice was explained by the assumption that bile, which is rich in cholesterol, entered into the blood after the rupture of distended bile ductules[9]. Apart from this explanation it was assumed that the reduced cholesterol ester fraction was due to a failure of the diseased liver to synthesize cholesterol-esters.

In 1934 Sperry[47] demonstrated that in normal serum and plasma cholesterol is esterified on incubation in vitro. He suggested that the reaction was caused by a plasma cholesterol esterase, the fatty acids involved coming from the hydrolytic break-down of lecithin released by a plasma lipase. In 1962 it was shown by Glomset and co-workers[22, 27] that free cholesterol is esterified by an enzyme called LCAT that is secreted into plasma by the liver[32, 44]. LCAT is a transacylase that catalyses the transfer of one fatty acid from position two of lecithin to unesterified cholesterol of plasma lipoproteins, producing lysolecithin and cholesterylester. This reaction provides the main source of cholesterylesters in plasma. The lysolecithin is transported in large part on plasma albumin[49]. The cholesterylesters are transported in part in the HDL and are transferred to the low and very low lipoprotein classes[26].

Though LCAT is most probably produced in the liver, it is a specifically plasma enzyme and its activity in various tissues is considerably lower[30, 31]. It circulates as an HDL complex acting upon smaller HDL's; HDL 3 serving as principal substrate[12, 24]. Whether apolipoprotein A 1[11, 13] or apolipoprotein A 3[32] act as a co-factor is still discussed. As a result of this reaction the concentration of free cholesterol in plasma is reduced.

The rate of reaction mainly depends on the concentration and composition of the plasma lipoprotein substrate. Factors like age, sex and diets apparently do not influence the enzyme activity significantly.

In addition to its role as an esterifying enzyme for plasma unesterified cholesterol, LCAT reduces the concentration of free cholesterol of erythrocyte membranes in vitro. The enzyme is present as well in peripheral lymph and hence probably in interstitial fluid[23]. This suggests that LCAT may even act on plasma lipoproteins in interstitial fluid and serves as a possible mechanism to remove unesterified cholesterol from plasma and from peripheral cells.

LCAT is present in plasma of primates, dogs, rats, rabbits, calves, sheep and chickens[49, 58] but only minimal enzyme activity was found in the plasma of some marine animals[37].

Information on the esterification of cholesterol has been highly stimulated by the discovery of familial LCAT deficiency[15, 19, 33, 34] Familial LCAT deficiency is an inborn error of lipoprotein metabolism with complete lack of plasma LCAT and with a variety of secondary abnormalities in lipoprotein concentration, composition and distribution. Patients suffering from this

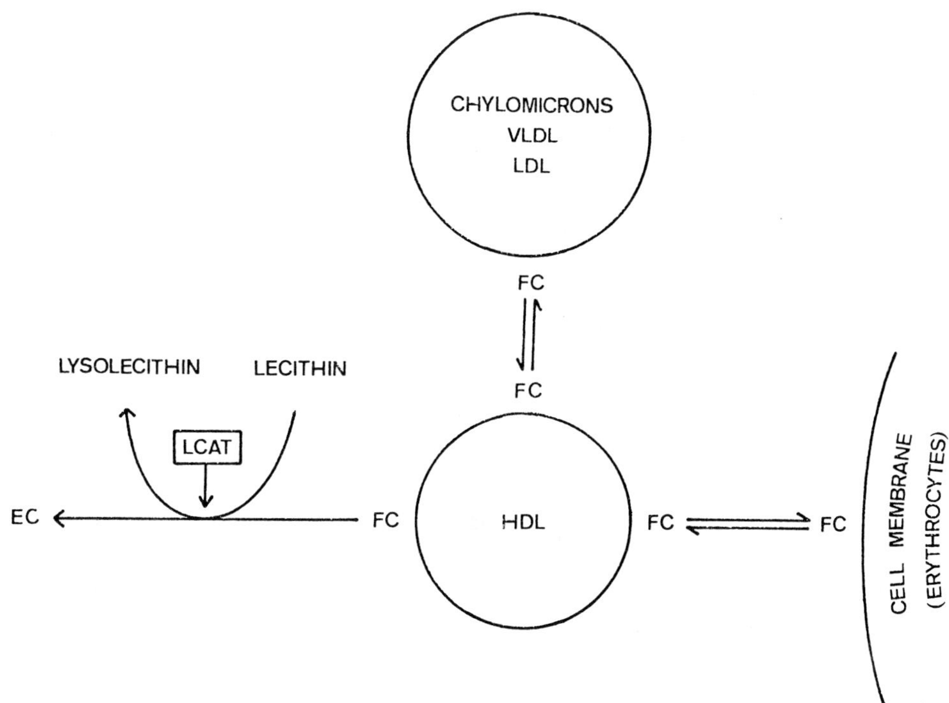

Fig. 1 – Postulated equilibrium mechanism etween free cholesterol of plasma lipoproteins and plasma membranes. The reaction between LCAT and HDL cholesteryl esters.

disease show clinical signs such as turbid plasma, corneal opacities, anemia, proteinuria with late renal insufficiency and have "sea blue" histocytes in bone marrow and spleen. Their plasma is abundant in free cholesterol and lecithin and plasma concentrations of esterified cholesterol and lysolecithin are abnormally low. Plasma of LCAT deficient patients contain virtually all lipoprotein families and apoprotein peptides of normal subjects but in reduced concentrations[1]. The presence of LP–X, the abnormal lipoprotein of early biliary obstruction, has been demonstrated in these patients.

Unesterified cholesterol and lecithin are present in increasing amounts in VLDL, LDL and HDL. The few cholestrylesters present in the patients, plasma are probably derived from chylomicrons produced in the intestinal mucosa. Each of the major lipoprotein classes is abnormal by several criteria.

In VLDL of LCAT deficient plasma 3 subfractions are found which contain increasing amounts of free cholesterol and lecithin as the size of the VLDL particle decreases. All three subfractions are also abnormal by electrophoretic mobility. They migrate rather as β-lipoproteins than as pre β-lipoproteins.

Table I. *Assay of cholesterol esterifying activity in blood*

Substrate	*Incubation time*	*Author*
1. *Net cholesterol esterification*		
Patient's serum	24 hrs.	Turber et al. (1953)
Patient's serum	5 hrs.	Jones et al. (1971)
2. *LCAT-activity*		
Inactivated normal ACD-plasma + 7-^{3}H-Cholesterol	4 hrs.	Glomset & Wright (1964)
Normal Serum + 4-^{14}C-Cholesterol	3 hrs.	Simon & Scheig (1970)
Patient's plasma + 7-^{3}H-Cholesterol	4 hrs. (preincubation) 1 hr.	Stokke & Norum (1961)

Analogus abnormalities as regards size and composition have been reported for LDL. A large molecular weight particle fraction appearing as flat structures upon electron microscopy[15] is rich in free cholesterol, lecithin and triglycerides. Particles of the subfraction of intermediate size contain less lipids but more protein and form stacks on electron microscope pictures. These particles have a lipid composition identical to LP–X. Particles of the smallest subfraction of LDL look similar to normal LDL but contain more triglycerides.

HDL are abnormal in size and composition as well. A large particle population containing mainly unesterified cholesterol and lecithin appears as stacked discs upon electron microscopy, whereas the smaller HDL are rather round particles, considerably smaller than normal HDL.

Determination of cholesterol esterifying activity in plasma (cf. table 1)

Cholesterol esterification reaction in plasma can easily be performed by incubating fresh plasma at 37° and measuring the increase in cholesterol esters[29, 57]. The results of this procedure, giving an information on net cholesterol esterification in plasma, do not only depend on the activity of the esterifying enzyme but also on hydrolytic enzymes and on the distribution of plasma lipoproteins. The rate of the reaction is much faster if high density lipoproteins, the preferred substrate for LCAT, are present[24].

Currently most investigators make use of radioactive cholesterol as a substrate. Several assay procedures are now available. Glomset and Wright's[28] assay uses normal ACD plasma as substrate. Labeled cholesterol, dispersed in albumin, is added to heat in activated plasma and shaken in a water bath at 36° for 4 hours. Patients' plasma is incubated for 4 hours with this substrate. Free cholesterol and cholesterol esters are determined after separation by thin layer chromatography. According to Stokke and Norum[52] labelled cholesterol is added to patients' plasma. During a preincubation of 4 hours LCAT is inhibited by a disulphide. Following reactivation by excess of mercaptoethanol, samples are incubated for an additional hour.

Simon and Scheig[46] incubated labeled cholesterol with serum lipoproteins using a modification of Avigan's method[2], where the labeled substrate is incubated with the patients' serum for three hours.

Blomhoff and coworkers[4] compared the method of Stokke and Norum[52] with Glomset's procedure[28]: in most cases of liver disease identical results were obtained. Simon and co-workers[45] compared net cholesterol esterification, determined according to Jones[29], with simultaneously determined LCAT-activity according to their own method[52], in 25 patients with paren-disease, and found a close correlation between both results.

Plasma cholesterolesterifying activity in human liver disease: (cf. table 2)

Most investigators agree that cholesterol esterification is reduced in human parenchymal liver disease. Contradictory results on cholesterol esterification

Table II. *Cholesterol esterification in liver disease*

Liver Disease	*Cholesterol esterification in plasma*	*Author*
Parenchymal liver disease	low	Gjone, Norum (1970)
		Simon, Scheig (1970)
		Gjone et al. (1971)
		Jones et al. (1971)
		Calandra et al. (1971)
		Wengeler et al. (1972)
		Cooper et al. (1972)
		Simon et al. (1974)
Cholestasis	low	Gjone, Norum (1970)
	normal/high	Simon, Scheig (1970)
	normal/low	Calandra et al. (1971)
	normal/high	Kepkay et al. (1973)
	low/normal/high (not correlated to LP-X)	Ritland et al. (1973)
Extrahepatic (LP-X negative)	normal	Wengeler et al. (1972)
Intrahepatic (LP-X positive)	low	Wengeler et al. (1972)

in cholestasisare show in the literature: low normaland high cholesterol esterification has been reported. These discrepancies can partly be explained by differences in the assay procedure, partly by differences in the progress and severity of the patients' diesease.

Some yerars ago Seidel and co-workers[41, 42] found LP–X to be a characteristic lipoprotein for cholestasis. LP–X has an uniquely high concentration of free cholesterol and lecithin and is made largely responsible for the lipid abnormalities in obstructive jaundice. Since this time the determination of LP–X has proved to be useful in the diagnosis of liver disease. Concerning a possible correlation between LCAT-activity and LP–X, special attention should be paid to the observation of Seidel and co-workers[59]. These authors reported that in hepatitis the activity of LCAT is low in cases with cholestasis, proved by positive LP–X, and normal in LP–X negative sera from patients without any signs of cholestasis. In extrahepatic biliary obstruction, however, the activity of LCAT is normal though LP–X is positive. The authors came to the conclusion that combined determination of LCAT and LP–X might be useful in differentiating between intra and extrahepatic cholestasis. Ritland and co-workers[38] carefully reinvestigated this question. In acute hepatitis they found reduced LCAT activity in both LP–X positive and LP–X negative patients. In primary biliary cirrhosis and large bile duct obstruction, LP–X being positive, low normal or increased LCAT activity was detected, and they could not demonstrate any specific LCAT – LP–X pattern in liver disease.

Plasma cholesterol esterifying activity in experimental liver damage

Experiments from Sugano and co-workers[53] demonstrated that plasma LCAT activity was reduced following the administration of hepatotoxic agents like ethionine or carbon tetrachloride in rats. Katterman and Wolfrum[30] performed a comparative study on cholesterol metabolism and LCAT activity in experimental hepatitis (galactosamine) and cholestasis (α-naphtylisothiocyanate).

In rats with galactosaminhepatitis, a rapid fall in LCAT activity to about 10% of normal activity was observed. In accordance with the decrease in LCAT activity in plasma, cholesterol esters were reduced to about 15% of the normal level. Following α-naphtylisothiocyanate, which is known to produce cholestasis in experimental animals, a relative decrease in plasma cholesterol esters from 67 % to 43% was observed; however, LCAT activity showed an increase rather than a decrease.

Pathogenesis of reduced cholesterol esterification in liver damage

1. *Presence of inhibitors in plasma*

In 1937 Sperry and Stoyanoff[48] demonstrated that bile salts were able to reduce esterification of plasma-free cholesterol in vitro. This observation has been confirmed by other investigators[8, 16, 28, 40]. However Kalandra and co-workers[5] showed that bile salt levels in patients' blood were below those necessary to inhibit esterification in vitro. The absence of a correlation between elevated bile salt levels in blood and reduced LCAT activity is supported by the results of experimental bile duct obstruction, leading to an increase in cholesterol esterification[3, 6].

2. *Absence or decrease of activators in plasma*

The inability of normal plasma to increase the activity of LCAT in plasma from patients with liver disease suggests that a decrease in activators of esterification cannot be responsible for the decreased reaction[5].

3. *Presence of abnormal substrates in plasma*

Whether or not the presence of abnormal substrates is responsible for the diminished esterification reaction is not clear. As has been pointed out by Simon[43], the presence of abnormal substrate lipoprotein is not a sufficient explanation for the reduced cholesterol esterification in liver disease.

It should be added that Simon and Scheig[46] put forward the hypothesis that LP–X might be a good substrate for the esterification reaction. Wengeler and Seidel[60] examined this question and concluded that LP–X does not act as a substrate for the esterification reaction.

4. *Increased cholesterol esterase in plasma*

The decrease in cholesterol esterification might be the result of an increased cholesterol esterase activity in blood. In normal human serum this enzyme cannot be detected. However, in liver lysosomes an acid hydrolase has been found which could be released into the blood in cases of liver damage[50, 51]. This possibility has found some support from Jones and co-workers[29] who performed combined determinations of cholesterol esterifying and cholesterol ester hydrolyzing activities in blood from patients with liver disease. These authors concluded from their experiments that net cholesterol esterification in blood is the result of both reactions. However, according to Simon and co-workers[45], who measured net cholesterol esterification and LCAT activity simultaneously in patients with liver disease, net cholesterol esterification correlated fairly well to LCAT activity, but neither direct nor indirect evidence of circulating cholesterol ester hydrolase was found.

5. *Decreased synthesis or release of esterifying enzymes*

According to the negative results of the above-mentioned speculations it is more likely that low cholesterol esterification in liver diseases in due to decreased synthesis or to an impaired release of esterifying enzyme by the damaged liver cell. However, this explanation is only acceptable under the assumption that LCAT activity in blood originates from the liver[23].

This has been only indirectly proved by the demonstration that isolated perfused rat liver released LCAT activity into the perfusate.

In a study with 53 patients with various liver diseases Blomhoff and co-workers[4] found that the activity of LCAT closely correlated with the plasma protein concentration. In acute hepatitis LCAT activity followed the proteins with short half life (vitamin K dependent clotting factors II, VII and X; prealbumin; a-lipoprotein). In chronic liver diseases LCAT activity was also correlated with plasma albumin. This observation again is an indirect proof of a reduced synthesis of LCAT in the damaged liver.

6. *Liver test parameters and LCAT activity*

Gjone and co-workers[20] examined the relationship between liver test parameters and LCAT activity. They found that LCAT activity was reduced to about 20% of normal when liver parameters such as SGOT, SGPT, AP, Bilirubin and Thymol Turbidity Test are elevated. In addition, they found a close correlation between erythrocyte cholesterol and serum unesterified cholesterol in patients with chronic hepatitis and liver cirrhosis. In general LCAT activity is reduced in liver damage but no evidence is yet available that the determination of LCAT might be a valuable test in diagnosis of any specific liver disease. Since there are several simple and reliable methods for detecting reduced protein synthesis in liver disease the assay of cholesterol esterification in blood is unwarranted in the routine diagnosis of liver disease. It must be pointed out, however, that the study of LCAT has highly improved the current knowledge on cholesterol metabolism.

References:

1. Alaupovic P., Mc Conathy W.J., Curry M.D., Magnani H.N., Torsvik H., Berg K.: Apolipoproteins and Lipoprotein Families in Familial Lecithin: Cholesterol Acyltransferase Deficiency. Scand. J. Clin. Lab. Invest., *33*, Suppl. 137 (83) (1974).
2. Avignan J.: A method for incorporating cholesterol and other lipids into serum lipoproteins in vitro. J. Biol. Chem., *234*, 787 (1959).
3. Bergan A., and K.T. Stokke: Scand. J. Gastroenterol., *8*, Suppl. 20, 48 (1973). (from Ritland and Gjone (36)).
4. Blomhoff J.P., S. Skrede and S. Ritland: Lecithin: cholesterol acyltransferase and plasma proteins in liver disease. Clin. Chim. Acta, *53*, 197 (1974).
5. Calandra S.M. J. Martin and N. McIntyre: Plasma lecithin: cholesterol acyltransferase activity in liver disease. Europ. J. clin. Invest., *1*, 352 (1971).
6. Calandra S., M.J. Martin, M.J.O'Shea and N. McIntyre: Effect of experimental biliary obstruction on the structure and lipid content of rate erythrocytes. Biochim. Biophys. Acta, *260*, 424 (1972).

7. Cooper R.A., M. Diloy-Puray, P. Lando and M.S. Greenberg: An analysis of lipoproteins, bile acids and red cell membranes associated with target cells and spur cells in patients with liver disease. J. clin. Invest., *51*, 3182 (1972).
8. Cooper R.A. and J.H. Jandl: Bile salts and cholesterol in the pathogenesis of target cells in obstructive jaundice. J. clin. Invest., *47*, 809 (1968).
9. Epstein, E.Z.: Cholesterol of the blood plasma in hepatic and biliary diseases. Arch. Int. Med., *50*, 203 (1932).
10. Epstein E.Z. and E.B. Greenspan: Clinical significance of the cholesterol partition of the blood plasma in hepatic and biliary diseases. Arch. Int. Med., *58*, 860 (1936).
11. Fielding C.J.: Phospholipid substrate specificity of purified human plasma lecithin: cholesterol acyltransferase. Scand. J. clin. Lab. Invest., *33*, Suppl. 137, 15 (1974).
12. Fielding C.J. and P.E. Fielding: Purification and substrate specificity of lecithin: cholesterol acyltransferase from human plasma. FEBS Lett., *15*, 355 (1971).
13. Fielding C.J., V.G. Shore and P.E. Fielding: A protein cofactor of lecithin:cholesterol acyltransferase. Biochem. Biophys. Res. Comm., *46*, 1493 (1972).
14. Forte T., K.R. Norum, J.A. Glomset and A.V. Nichols: Plasma lipoprotein in LCAT deficiency. Structure of low and high density lipoproteins as revealed by electron microscopy. J. Clin. Invest., *50*, 1141 (1971).
15. Gjone E.: Familial lecithin: cholesterol acyltransferase deficiency. A clinical survey. Scand. J. clin. Lab. Invest., *33*, Suppl. 137, 73 (1974).
16. Gjone E., and J.P. Blomhoff: Plasma lecithin-cholesterol acyltransferase in obstructive jaundice. Scand. J. Gastroent., *5*, 305 (1970)
17. Gjone E., J.P. Blomhoff and A.J. Skarbovik: Possible association between an abnormal low density lipoprotein and nephropathy in lecithin: cholesterol acyltransferase deficiency. Clin. Chim. Acta, *54*, 11 (1974).
18. Gjone J.P., Blomhoff and I. Wiencke: Plasma lecithin: cholesterol acyltransferase activity in acute hepatitis. Scand. J. Gastroent., *6*, 161 (1971).
19. Gjone E. and K.R. Norum: Familial serum colesterol ester deficiency. Clinical study of a patient with a new syndrome. Acta Med. Scand., *183*, 107 (1968).
20. Gjone E. and K.R. Norum: Plasma lecithin-cholesterol acyltransferase and erythrocyte lipids in liver disease. Acta Med. Scand., *187*, 153 (1970).
21. Glomset J.A.: The mechanism of the plasma cholesterol esterification reaction: plasma fatty acid transferase. Biochim. Biophys. Acta, *65*, 128 (1962).
22. Glomset J.A.: The plasma lecithin: cholesterol acyltransferase reaction. J. Lipid Res., *9*, 155 (1968).
23. Glomset J.A., E. Jansen, R. Kennedy and J. Dobbins: Role of plasma lecithin: cholesterol acyltransferase in the metabolism of high density liproproteins. J. Lipid Res., *7*, 638 (1966).
24. Glomset J.A., A.V. Nichols, K.R. Norum, W. King and T. Forte: Plasma lipoproteins in familial LCAT deficiency. Further studies of very low and low density lipoprotein abnormalities. J. Clin. Invest., *52*, 1078 (1973).
25. Glomset J.A., K.R. Norum and W. King: Plasma lipoproteins in familial LCAT deficiency. Lipid composition and reactivity in vitro. J. Clin. Invest., *49*, 1827 (1970).
26. Glomset J.A., F. Parker, M. Tjaden and R.H. Williams: The esterification in vitro of free cholesterol in human and rat plasma. Biochim. Biophys. Acta, *58*, 398 (1962).
27. Glomset J.A. and J.L. Wright: Some properties of cholesterol esterifying enzyme in human plasma. Biochim. Biophys. Acta, *89*, 266 (1964).
28. Jones D.P., F.R. Sosa, J. Sharsis, P.T. Shah, E. Skoomak and W.T. Behr: Serum cho-

lesterol esterifying and cholesterol ester hydrolyzing activities in liver disease: relationships to cholesterol, biilirubin and bile salt concentrations. J. Clin. Invest., *50*, 259 (1971).
29. Kattermann R. and D.I. Wolfrum: Cholesterinstoffwechsel und Lecithin-Cholesterin-Acyltransferase in Plasma bei experimenteller Hepatitis und Cholestase and der Ratte. Z. Klin. Chem. u. klin. Biochem., *8*, 413 (1970).
30. Kepkay D.L., R. Poon and J.B. Simon: Lecithin-cholesterol acyltransferase and serum cholesterol esterification in obstructive jaundice. J. Lab. clin. Med., *81*, 172 (1973).
31. Kostner G.: Studies on the cofactor requirements for lecithin: cholesterol acyltransferase. Scand. J. Clin. Lab. Invest., *33*, Suppl. 137, 19 (1974).
32. Norum K.R. and E. Gjone: Familial plasma lecithin: cholesterol acyltransferase deficiency. Biochemical study of a new inborn error of metabolism. Scand. J. Clin. Lab. Invest, *20*, 231 (1967).
33. Norum K.R. and E. Gjone: Familial serum cholesterol esterification failure. A new inborn error of metabolism. Biochim. Biophys. Acta, *144*, 698 (1967).
34. Norum K.R., J.A. Glomset, A.V. Nichols and T. Forte: Plasma lipoproteins in familial LCAT deficiency: physical and chemical studies of low and high density lipoproteins, J. Clin. Invest., *50*, 1131 (1971).
35. Osuga T. and O.W. Portman: Origin and disappearence of plasma lecithin:cholesterol acyltransferase. Amer. J. Physiol., *220*, 735 (1971).
36. Puppione D.L.: A physical and chemical Characterization of the Serum Lipoproteins of Marine Mammals. Dissertation, University of California at Berkely (1969).
37. Ritland S., J.P. Blomhoff and E. Gjone: Lecithin-cholesterol acyl-transferase and lipoprotein-X in liver disease. Clin. Chim. Acta, *49*, 251 (1973).
38. Ritland S. and E. Gjone: Quantitative studies of lipoprotein-X in familial LCAT deficiency and during cholesterol esterification. Clin. Chim. Acta, *59*, 109 (1975).
39. Rowen R. and J. Martin: Enhancement of cholesterol esterification in serum by an extract of group-A streptococcus. Biochim. Biophys. Acta, *70*, 396 (1963).
40. Seidel D., P. Alaupovic and R.H. Furman: A lipoprotein characterizing obstructive jaundice. I. Method for quantitative separation and identification of lipoproteins in jaundiced subjects. J. Clin. Invest., *48*, 1211 (1969).
41. Seidel D., P. Alaupovic, R.H. Furman and W.I. McConathy: A lipoprotein characterizing obstructive jaundice. II. Isolation and partial characterization of the protein moieties of loq density lipoproteins. J. Clin. Invest., *49*, 2396 (1970).
42. Simon J.B.: Lecithin: cholesterol acyltransferase in human liver disease. Scand. J. Clin. Lab. Invest, *33*, Suppl. 137, 107 (1974).
43. Simon J.B. and J.L. Boyer: Production of lecithin: cholesterol acyltransferase by the isolated perfused rat liver. Biochim. Biophys. Acta, *218*, 549 (1970).
44. Simon J.B., D.L. Kepkay and R. Poon: Serum cholesterol esterification in human liver disease: role of lecithin-cholesterol acyltransferase and cholesterol ester hydrolase. Gastroenterol., *66*, 539 (1974).
45. Simon J.B. and R. Scheig: Serum cholesterol esterification in liver disease. Importance of lecithin--cholesterol acyltransferase. N. Engl. J. Med., *283*, 841 (1970).
46. Sperry W.M.: Cholesterol esterase in blood. J. Biol. Chem., *111*, 467 (1935).
47. Sperry W.M. and V.A. Stoyanoff: The influence of bile salts on the enzymatic synthesis and hydrolysis of cholesterol esters in blood serum. J. Biol. Chem., *121*, 101 (1937).

48. Stefanovich V.: Cholesterol Esterification in Rabbit Plasma. Biochem. J., *115,* 555, (1969).
49. Stokke K.T.: The existence of an acid cholesterol esterase in human liver. Biochim. Biophys. Acta, *270,* 156 (1972).
50. Stokke K.T.: Subcellular distribution and kinetics of the acid cholesterol esterase in liver. Biochim. Biophys. Acta, *280,* 329 (1972).
51. Stokke K.T. and K.R. Norum: Determination of lecithin:cholesterol acyltransferase in human blood plasma. Scand. J. Clin. Lab. Invest., *27,* 21 (1971).
52. Sugano M.K. Hori and M. Wada: Hepatotoxicity and plasma cholesterol esterification by rats. Arch. Biochem. Biophys., *129,* 588 (1969).
53. Switzer S., Eder H.A.: Transport of Lysolecithin by Albumin in Human and Rat Plasma. J. Lipid Res., *6,* 506 (1965).
54. Thannhauser S.J. and H. Schaber: Uber die Beziehungen des Gleichgewichtes Cholesterin und Cholesterinester im Blut und Serum zur Leberfunktion. Klin. Wschr., *5,* 252 (1926).
55. Torsvik H., K. Berg, H.N. Magnani, W.J. McConathy, P. Alaupovic and E. Gjone: Identification of the abnormal cholestatic lipoprotein (LP-X) in familial lecithin: cholesterol acyltransferase deficiency. FEBS Lett., *24,* 165 (1972).
56. Turner K.B., G.H. McCormack and A. Richards: The cholesterol-esterifying enzyme of human serum. I. In liver disease. J. Clin. Invest., *32,* 801 (1953).
57. Wengeler H., H. Greten and D. Seidel: Serum cholesterol esterification in liver disease. Combined determination of lecithin:cholesterol acyltransferase and lipoprotein-X. Europ. J. Clin. Invest., *2,* 372 (1970).
58. Wengeler H., and D. Seidel: Does lipoprotein-X (LP-X) act as a substrate for lecithin: cholesterol acyltransferase (LCAT)? Clin. Chim. Acta, *45,* 429 (1973).
59. Wells I.C., Rongone E.L.: Dietary Cholesterol and Serum Cholesterol Esterifying Activity in Rabbits. Proc. Soc. Exp. Biol. & Med., *130,* 661 (1970).

LECITHIN-CHOLESTEROL ACYL TRANSFERASE (LCAT) ACTIVITY AS A DETERMINING FACTOR IN LIPOPROTEIN CATABOLISM: A STUDY ON HYPERLIPOPROTEINEMIC SUBJECTS

L. Ferri, F. Ursini, M. Valente and C. Gregolin

Summary

Lecithin-Cholesterol Acyl Transferase (LCAT) activity was studied in hyperlipemic patients in correlation with the serum lipid concentration, the lipoprotein phenotype according to Fredrickson and the fatty acid profile of serum lecithins and cholesterol esters. Moreover the distribution of labelled cholesterol and the cholesterol esterification rate in high-density lipoproteins (HDL) and low density plus very low density lipoproteins (LDL + VLDL) were determined. On the basis of the activity the minimal turnover and the clearance of cholesterol esters in serum were calculated. The enzyme activity was normal or slightly increased in correlation with the total cholesterol concentration in type IIa hypercholesterolemia and was markedly increased in dependence of triglyceride concentration in type IIb and IV hyperlipemia. As a consequence, in type IIa hypercholesterolemia the minimal turnover of cholesterol esters was in the normal renge or slightly above the normal and the clearance was decreased, while in type IIb and IV hyperlipemia the minimal turnover was clearly increased and the clearance was normal or increased. The enzyme activity was proportional to the esterification rate in HDL, which suggests that HDL play a central role in the turnover and the clearance of cholesterol esters and thereby of the catabolic flow of VLDL to LDL following the removal of triglycerides by lipoprotein lipase (LpL). The fatty acid profile of lecithins and cholesterol esters in serum is correlated with the LCAT activity in that the higher enzyme activity is found when the ratio 18 : 2/18 : 1 in lecithins is higher.

The results support the idea that LCAT activity – together with LpL – is a determinant factor in lipoprotein catabolism and that the atherogenic potential of the different types of hyperlipemia is correlated with a relatively impaired enzyme activity, which leads to an increased free cholesterol/esterified cholesterol ratio in serum in the presence of an increased level of lysolecithin.

Introduction

In a previous paper[1] we discussed the cholesterol esterification reaction catalyzed in human plasma by Lecithin Cholesterol Acyl Transferase (LCAT), and reported our results on the determination of enzyme activity in various diseases with serum lipid disorders.

Istituto di Chimica Biologica dell'Università di Padova e Centro per lo Studio della Fisiologia dei Mitocondri del CNR.

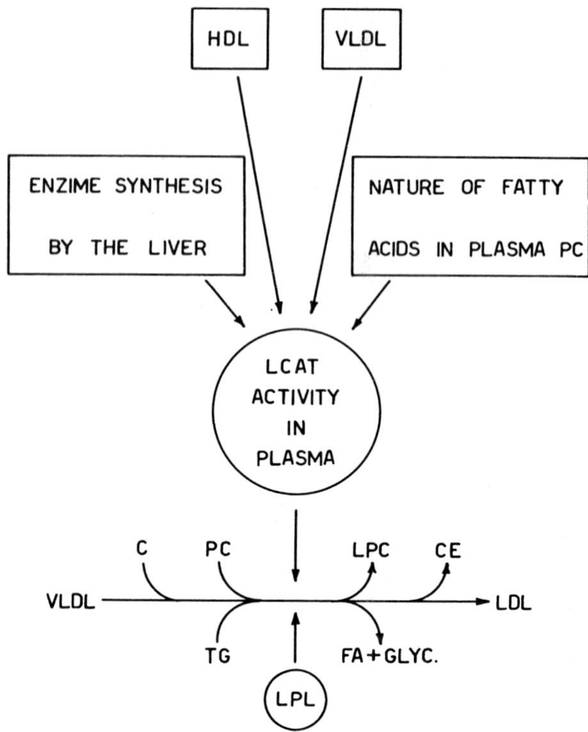

Fig. 1 – Some determinants and metabolic effect of LCAT activity in plasma. Abbreviations: C = cholesterol, CE = cholesterol esters, PC = phosphatidyl choline, LPC = lysophosphatidyl choline, TG = triglycerides, FA = fatty acids, GLYC = glycerol.

We reported that in liver disease enzyme activity decreases proportionally to the total cholesterol concentration, so that the ratio of esterified to unesterified cholesterol is not significantly modified except in more serious cases. On the contrary, in hyperlipemic states and in the nephrotic syndrome, enzyme activity increases, but it does not keep up with the increment of serum cholesterol. Therefore, in these cases the percentage of esterified cholesterol is usually reduced, compared to normal, and *viceversa* the percentage of unesterified cholesterol is usually increased.

The purpose of the present study is to correlate LCAT activity in hyperlipoproteinemic states with the influence of different factors, which have been found to affect the enzyme in vitro and can be expected to modulate the physiological and pathological effects of substrates and products of the reaction.

On the basis of current knowledge, LCAT activity in human plasma is probably influenced by at least the factors summarized in Fig. 1:

1. concentration of the *enzyme,* which is produced by continuous *synthesis* taking place *in the liver*[2,3] and is probably regulated by the level of circulating plasma cholesterol[1].
2. *the concentration* and possibly the *fine structure of high density lipoproteins* (HDL), since the substrates proper of the enzyme are cholesterol and lecithin carried by these particles, as shown by Akanuma and Glomset[4].
3. *the concentration of triglycerides* transported by very low density lipoproteins (VLDL) (and chylomicrons when present). A stimulatory effect of these agents on LCAT activity was demonstrated *in vitro* by Marcel and Vezina[5]. A similar conclusion was deduced from our studies on patients with chronic liver parenchymatous disease or hyperlipemia[1]. A transfer of triglycerides from VLDL to HDL is most likely not only an indirect secondary effect, but a basic requirement for full expression of enzyme activity. However, it is also possible that VLDL act by modifying the steric conformation of HDL and by favoring the interaction between enzyme and HDL. Moreover, it is known that VLDL contain an apopeptide (Apo C-I) which stimulates LCAT activity[6].
4. *the nature of the fatty acid in position two of lecithin,* and, consequently, of fatty acid present in cholesterol esters. Indeed, human plasma LCAT preferentially transfers fatty acids from the position two of lecithin[7]. More than 50 per cent of fatty acids present in cholesterol esters are in the form of linoleic acid, and it appears probable that a modification in the fatty acid profile in position two of lecithin, in particular on HDL, reflects itself in a different ability of the enzyme to transfer the fatty acid to cholesterol.

To determine the influence of these factors on the LCAT reaction in sera of hyperlipoproteinemic subjects, enzyme activity was measured by using the serum of each subject both as a source of enzyme and as a substrate and the results were then correlated with different factors, as described in the following sections.

Samples were obtained from 15 persons of both sexes, presenting hyperlipoproteinemias classified as type II a, II b and IV, according to Fredrickson[8] by agarose-gel electrophoresis of serum.

Total serum cholesterol was measured by the Lieberman-Burchard reaction. Unesterified cholesterol was determined following precipitation as the digitonin complex. Other methods are described below. No significant deviation from normal parameters of liver function was observed in the patients studied. Eight normal subjects (age 18 to 45) served as controls.

Assay of LCAT activity

A flow-sheet of the assay of enzyme activity is presented in Fig. 2. The serum from each subject was equilibrated with an albumin suspension of ra-

Fig. 2 – Flow-sheet of the assay of LCAT activity in serum. See text for comments.

0.5 ml test serum
0.05 ml 7-^{3}H-Cholesterol-albumin suspension (7-^{3}H-Cholesterol from the Radiochemical Centre, Amersham; specific activity: 500 mCi/nmole)
– hr at 37°C, in Dubnoff shaker, 60 cycles/min
– Lipid extraction with alcohol-acetone (1:1) at 60°C
– Thin layer chromatographic separation of cholesterol and cholesteryl esters (solvent: hexane-ethyl ether-methanol-acetic acid, 90:20:2:3, v:v) at room temperature.
– Counting by liquid scintillation spectrometry of cholesterol and cholesteryl ester spots (scintillation fluid: PPO, 4 g; POPOP, 100 mg; toluene, 1 liter) and determination of percent esterification of cholesterol.
– Calculation of LCAT activity as μg of cholesterol esterified/ml test serum/hr.

dioactive cholesterol, and incubated for 60 min. at 37°C. The reaction was stopped by adding lipid extraction mixture. Lipids were extracted and chromatographed on silica gel plates to separate cholesterol and cholesterol esters. These were eluted from the gel, and radioactivity present in cholesterol and cholesterol esters was measured. The percentage of esterification was converted into enzyme activity by multiplying by the amount of unesterified cholesterol present in the serum volume at the start of the assay.

This assay takes advantage of a rapid equilibration of the serum lipoproteins with an albumin suspension of radioactive cholesterol, which is prepared by mixing together a small volume of a solution of radioactive cholesterol in acetone, as suggested by Norum[9], and a 5 % solution of human albumin. Acetone is then evaporated under a nitrogen stream. When the cholesterol suspension in albumin was added to the test serum, radioactive cholesterol almost instantaneously equilibrated with cholesterol present in lipoproteins. This was demonstrated with normal and hyperlipoproteinemic sera by analyses performed following preparative electrophoretic separation of serum proteins with 0,9% agarose gel in 25 mM veronal buffer, pH 8.2. The use of this cholesterol suspension simplifies the equilibration step, which is very critical and renders cumbersome various other methods employing radioactive cholesterol described in the literature[10, 11, 12].

The validity of the method is confirmed by the cholesterol esterification rates which were measured in the sera of normal subjects.

The normal value of LCAT activity was determined in 8 healthy subjects and was 48.85 ± 12 μg cholesterol esterified/ml/hr of serum, which compares quite favorably with the cholesterol esterification rate in man as determined in vivo by Nestel and Mongel (46.40 μg/ml/hr)[13]. Therefore this assay procedure measures esterification rates which are closer to physiological rates than those we obtained with the assay system prevously employed[1].

However, this type of assay offers yet another advantage. Measurement

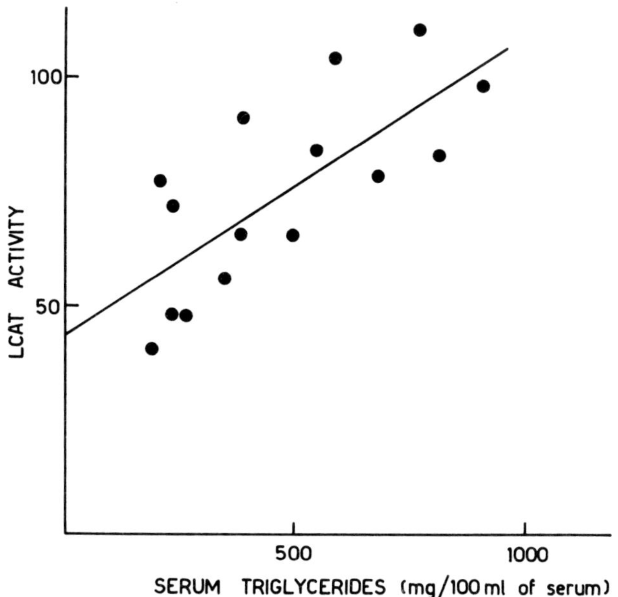

Fig. 3 – Correlation between serum triglyceride concentration and LCAT activity in hyperlipemic patients ($r = 0.73$, $n = 15$, $p < 0.01$).

of enzyme activity using the serum of each subject both as enzyme source and substrate means that the influence of the different individual determinants of *enzyme activity* is amplified. On the other hand, the assay which requires a small volume of test serum and a large volume of standard substrate plasma[1] gives a closer approximation of the absolute *enzyme concentration,* isolated from other enzyme activity determinants, which can be of major importance in individual cases.

Effect of serum triglyceride concentration on LCAT activity

In the work reported at the previous Symposium[1] we had shown that triglyceride concentration in serum stimulates LCAT activity. However, this correlation, established by assay with a standard substrate plasma in which individual factors are minimized, was less definite than that obtained with the present assay. When the results of LCAT activity determinations on sera of hyperlipoproteinemic subjects were plotted as a function of triglyceride concentration as determined enzymatically on the basis of glycerol content, the diagram reported in Fig. 3 was obtained. It can be concluded that, while the enzyme concentration is probably dependent in each case on the total cholesterol concentration in plasma[1], in hyperlipemic subjects the activity of the enzyme is stimulated by the increased triglyceride concentration pos-

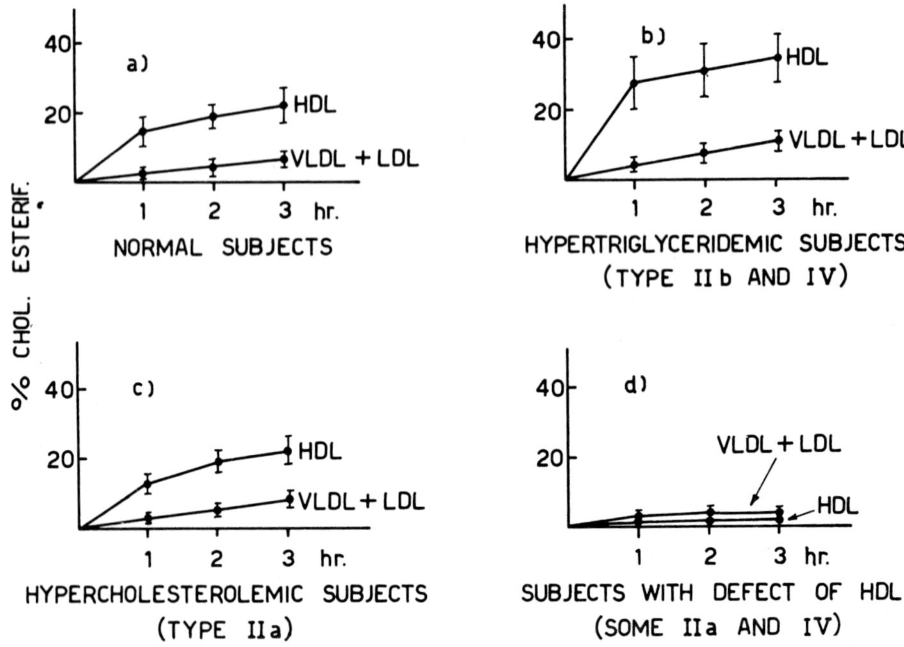

Fig. 4 – Cholesterol esterification activity in HDL and VLDL + LDL. The indicated values are mean values obtained on the following number of subjects: a) 8; b) 8, c) 4, d) 3.

sibly as a part of the enzymatic mechanism devoted to removal of VLDL from plasma, which involves both LCAT and lipoproteins lipase. When triglycerides are hydrolized by lipoprotein lipase, cholesterol leaves VLDL, is taken up by HDL where it is esterified and is given back to VLDL during their transformation to LDL. In this way, increased LCAT activity favors a more rapid turnover of both triglycerides and cholesterol in circulating plasma of hyperlipemic subjects.

Influence of HDL on enzyme activity

The effect of HDL on LCAT activity is shown in the experiments reported in Fig. 4. Sera of normal and hyperlipoproteinemic subjects were incubated at 37°C for different times up to three hours following equilibration with radioactive cholesterol. Sera were then fractionated by preparative 0.9%agarose-gel electrophoresis. The HDL and VLDL + LDL bands were extracted with dimetyl sulphoxide and petrol ether (1:1, v:v). Lipids present in both bands were fractionated by thin layer chromatography (system as in Fig. 2) and radioactivity was measured in the cholesterol and cholesterol ester frac-

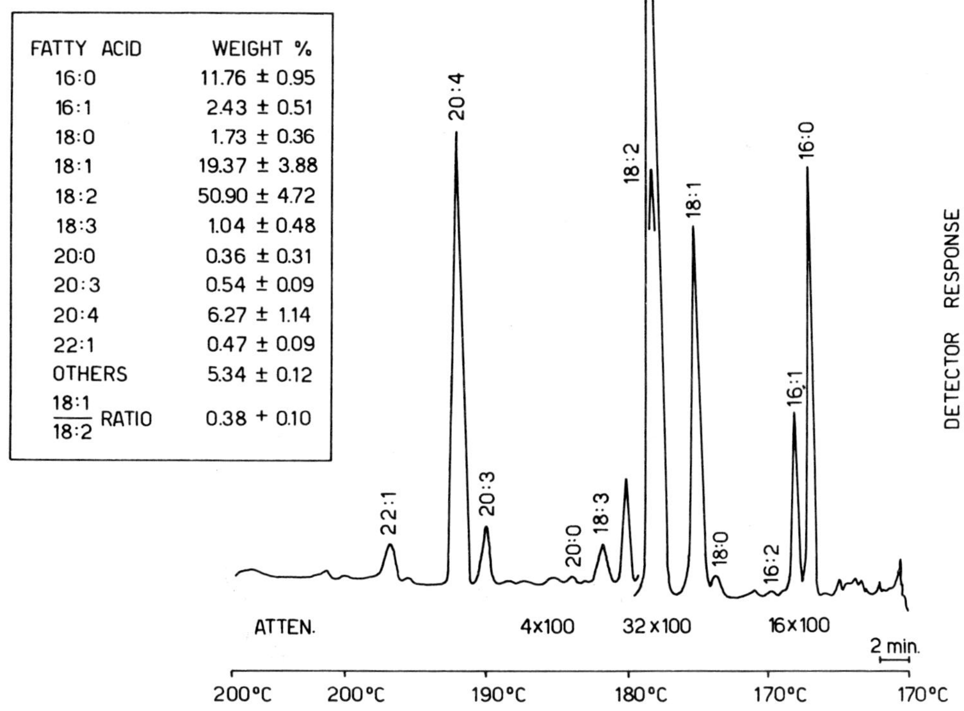

FATTY ACID	WEIGHT %
16:0	11.76 ± 0.95
16:1	2.43 ± 0.51
18:0	1.73 ± 0.36
18:1	19.37 ± 3.88
18:2	50.90 ± 4.72
18:3	1.04 ± 0.48
20:0	0.36 ± 0.31
20:3	0.54 ± 0.09
20:4	6.27 ± 1.14
22:1	0.47 ± 0.09
OTHERS	5.34 ± 0.12
$\frac{18:1}{18:2}$ RATIO	0.38 + 0.10

Fig. 5 – Fatty acid composition, and ratio of oleic to linoleic acid in serum lecithin of normal subjects.

tions. The percentage of esterification activity in normal subjects is high in HDL and low in LDL + VLDL, as expected since HDL are the substrate for the enzyme. In HDL, equilibrium is reached rather early, while in LDL+ VLDL the accumulation of cholesterol esters proceeds linearly within the limit of the experiment time.

In the majority of type IIb and IV hyperlipoproteinemic subjects, the percent esterification in the HDL fraction is twice as high as in normal subjects, and equilibrium is reached more quickly.

In type IIa hyperlipoproteinemia, esterification in HDL follows normal pattern and rates. This, however, leads to an esterification activity that does not keep pace with the increased cholesterol concentration present in these subjects, and therefore, a decrease in cholesterol eschange between VLDL-HDL-LDL can be assumed. In some cases of type IIa and type IV hyperlipoproteinemia, which are accompanied by increased cholesterol : cholesterol ester ratio, esterification activity is very low on HDL, which suggests a profound alteration in the synthesis and structure of this lipoprotein class.

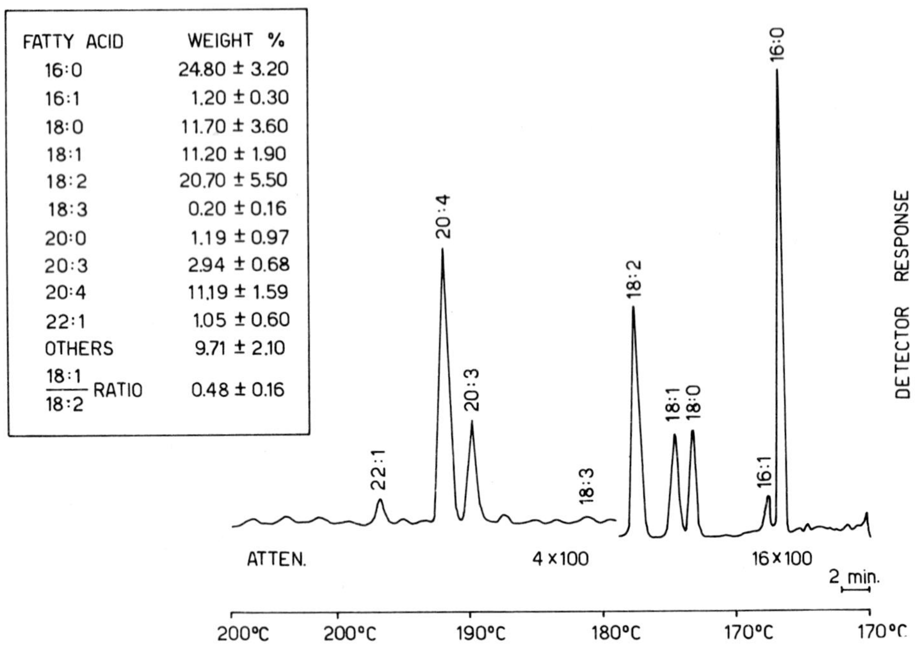

FATTY ACID	WEIGHT %
16:0	24.80 ± 3.20
16:1	1.20 ± 0.30
18:0	11.70 ± 3.60
18:1	11.20 ± 1.90
18:2	20.70 ± 5.50
18:3	0.20 ± 0.16
20:0	1.19 ± 0.97
20:3	2.94 ± 0.68
20:4	11.19 ± 1.59
22:1	1.05 ± 0.60
OTHERS	9.71 ± 2.10
18:1/18:2 RATIO	0.48 ± 0.16

Fig. 6 – Fatty acid composition and ratio of oleic to linoleic acid in serum cholesterol esters of normal subjects.

Fatty acid in serum and LCAT activity

The influence of the fatty acid profile in serum on LCAT activity has been studied as follows. Lipids were extracted from sera of normal and hyperlipoproteinemic patients, and lecithin and cholesterol esters were then isolated through thin layer chromatography. Following saponification in alcoholic potassium hydroxide, fatty acids were methylated with boron trifluoride and methyl esthers analyzed by gas liquid chromatography (10% diethylene glycol succinate on Chromosorb W). The normal profile of fatty acids in lecithin is shown in Fig. 5. On the basis of the results obtained with hyperlipoproteinemic subjects (see below) particular attention was devoted to the ratio between oleic and linoleic acids, which in normal cases was equal to 0.48 ± 0.16. Fig. 6 presents data obtained in normal subjects for the fatty acid composition of cholesterol esters. This is not exactly identical with that of lecithin, because the method employed in this study determined both the fatty acids in position 1 and 2 of lecithin. Moreover, on the basis of different experiments performed by Glomset[14], it may be expected that LCAT acts only on some molecular classes of lecithins, and not on other classes. In addition, intestinal synthesis also contributes to some extent (10-

15%) to the pool of cholesterol esters circulating in plasma. The normal ratio between oleic and linoleic acid in cholesterol esters was equal to 0.38 ± 0.10. In hyperlipoproteinemic subjects the percent composition of fatty acids was not substantially altered, neither in lecithins nor in cholesterol esters, with the only exception of the oleic/linoleic ratio in lecithin. Fig. 7 shows a significant negative correlation between this ratio and LCAT activity. We could not find a clear dependence of the ratio alteration on either form of hyperlipoproteinemia studied.

"Minimal turnover of cholesterol esters" and "cholesterol ester clearance"

In the presence of a constant concentration of cholesterol and cholesterol esters in plasma, some very important parameters of cholesterol metabolism may be calculated on the basis of LCAT activity, as indicated in Table I. The first parameter is the "minimal turnover of plasma cholesterol esters", namely the amount of cholesterol which is first esterified through the LCAT reaction and then removed from plasma. In a normal subject this amount corresponds to 117.24 ± 30.43 mg/hr, assuming a mean plasma volume of 2400 ml. The removal of this amount of cholesterol should be proportional to the catabolism of lipoproteins which carry most of the cholesterol esters, namely VLDL and LDL. A second parameter which derives from the first, is the "clearance of the cholesterol esters", namely the volume of plasma from which cholesterol esters, and therefore LDL, are removed per hour. The normal values is 74.5 ± 19.3 ml/hr. A third parameter, which is

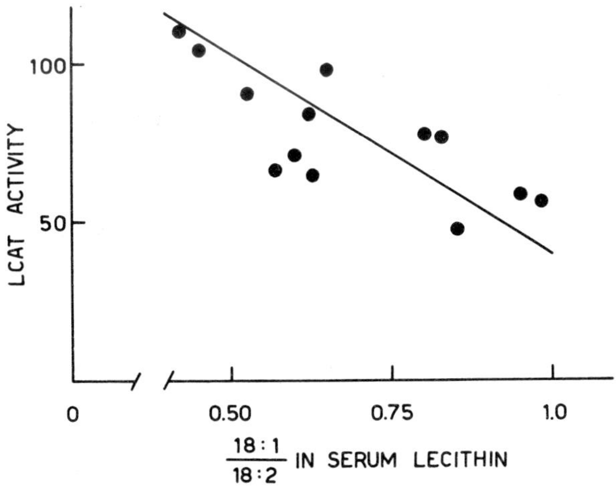

Fig. 7 – Correlation of LCAT activity and ratio of oleic to linoleic acid in serum lecithin ($r = 0.76$, $n = 13$, $p < 0.01$).

Table I. *Calculation of the minimal turnover of cholesterol esters, cholesterol esters clearance and the amount of lysolecithin formed in plasma*

Minimal turnover of plasma cholesterol esters (mg/hr)	=	LCAT Activity x plasma volume (mg chol. esterif./ml/hr x 2400 ml)
Normal value	=	117.24 ± 30.43
Cholesterol esters clearance (ml/hr)	=	$\frac{\text{Minimal turnover}}{\text{mg cholesterol in ester form/100 ml}} \times 100$
Normal value	=	74.5 ± 19.3
Lysolecithin formed in plasma (µmoles/hr)	=	$\frac{\mu\text{g cholesterol esterif/ml/hr x 2400 ml}}{\text{M.W. of cholesterol}}$
Normal value	=	303.2 ± 78.7 µmoles/hr

Table II. *Effect of LCAT activity on parameters of cholesterol metabolism*

	Normal *(8)*	*II a* *(4)*	*II b* *(3)*	*IV* *(8)*
Total cholesterol (mg/100 ml of serum)	210,7± 31,5	406,6 ±31,0	360,0± 78,3	298,8± 56,3
Unesterified cholesterol mg/100 ml of serum	47,04± 5,73	132,4 ±37,6	98,5± 18,6	95,4± 32 3
LCAT activity µg chol. est./ml/hr	48,58± 12,01	55,36 ±13,65	84,71± 11,63	76,45± 8,48
Minimal turnover of plasma cholesterol esters – mg/hr	117,24± 30,43	132,84±19,02	203,27± 27,93	190,61± 20,59
Clearance of cholesterol esters ml/hr	74,5 ± 19,3	49,16 ±11,38	79,5± 11,57	94,85± 12,82
Lysolecithin formed umoles/ hr/2400 ml plasma	302,2 ± 78,7	343,56±80,94	525,77± 70,68	451,3± 47,96

Figures in parentheses refer to the number of subjects.

simply the counterpart of the cholesterol esters formed through LCAT activity, is the "amount of lysolecithin formed in plasma "in a given time". In a normal subject, this amount is 303.2 ± 78.7 µmoles/hr. per 2400 ml of plasma. It should be emphasized that these parameters are different from the *cholesterol metabolic clearance rate* and the cholesterol *daily clearance fraction* which have been proposed and determined by Nestel et al. from an analysis of the kinetics and distribution of cholesterol in the two exchangeable body pool model[15, 16]. Indeed Nestel's values are correlated with the rate of decay of specific activity of radioactive cholesterol injected endovenously, while the parameters calculated in the present work are correlated with the rate of the cholesterol esterification reaction, which can act more than once on the same cholesterol molecules, provided they are de-esterified in the liver and secreted again intact into the blood stream.

If we now analyze how these different parameters vary in the different hyperlipoproteinemias, we may achieve a better insight into the alteration that takes place in these states (see Table II, in which the data have been calculated by assuming an ideal plasma volume of 2400 ml for all patients). In type IIa hyperlipoproteinemia, where LCAT activity is slightly increased, possibly in dependence of the stimulus due to the high cholesterol level, the minimal turnover of plasma cholesterol esters is slightly elevated. However, due to the high concentration of circulating cholesterol, the clearance of cholesterol esters is reduced. The amount of lysolecithin is slightly increased. More striking, however, are the phenomena that accompany type IIb and type IV hyperlipoproteinemia, where the stimulatory effect of triglyceride concentration is marked. Here, the minimal turnover of plasma cholesterol esters is clearly increased, and, since cholesterol concentration is not as high as in type IIa, the clearance of cholesterol esters is normal or higher than normal. Obviously, lysolecithin formation is also elevated in these cases.

From these data we think it is reasonable to conclude that the biological mechanisms which come into play in these different hyperlipoproteinemic states are different. In type IIa, enzyme level and activity do not keep pace with the elevation of plasma cholesterol, and therefore the proportion of unesterified to esterified cholesterol increases. Since unesterified cholesterol can easily exchange with the intimal cell membranes, as with any other tissue membrane, in type IIa hyperlipoproteinemia this might represent the first cause of intimal alteration, which leads to atheroma. In type IIb and IV hyperlipoproteinemia, increased LCAT activity assures a more correct ratio of unesterified to esterified cholesterol, and therefore intimal cell membranes and other membranes are more protected from cholesterol exchange, while they are more exposed to insult from an elevated concentration of lysolecithin. In these cases the *primum movens* in atherogenetic events could be the lesion brought about by lysolecithin, even though this may be associated with cholesterol infiltration if LCAT activity does not follow exactly the rate of cholesterol increase (see also Ref. 17).

Conclusion

On the basis of information currently available, the role of LCAT in physiological and pathological states can be interpreted as summarized in Fig. 8. In the schema, LCAT is seen as a determining factor in the catabolism of VLDL to LDL, together with lipoprotein lipase and HDL cooperation. It is worth emphasising that LDL metabolism is severely altered in the inheri-

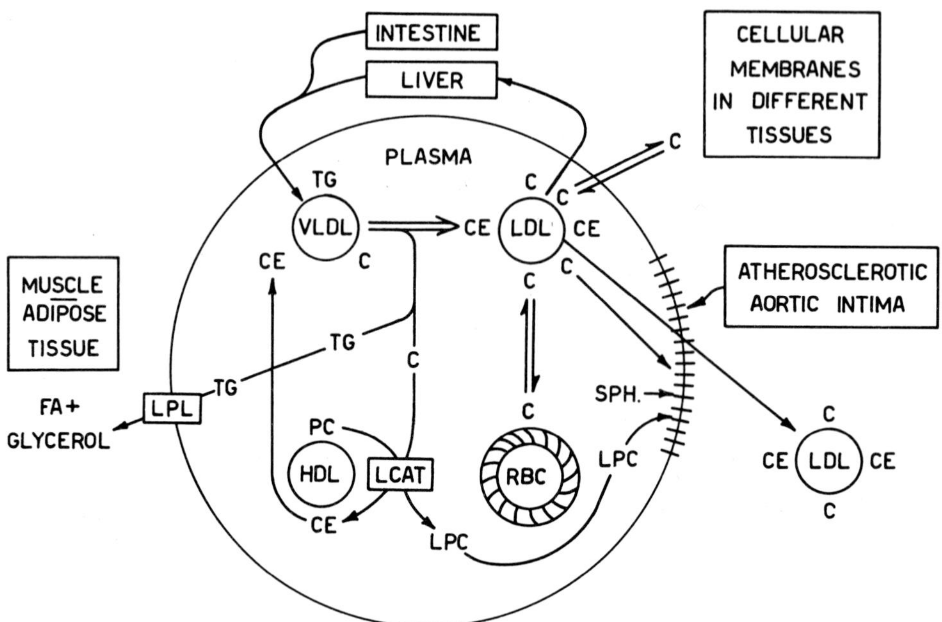

Fig. 8 – Schematic diagram of LCAT effects on lipoprotein catabolism and cholesterol exchange between plasma lipoproteins and cellular membranes. Abbreviations: TG = triglycerides, C = cholesterol esters, PC = phosphatidyl choline, LPC = lysophosphatidyl choline, SPM = sphyngomyelin, FA = fatty acids, RBC = red blood cells, LPL = lipoprotein lipase. Indicated on the right side of the figure are the phenomena which are postulated to accompany the atherosclerotic degeneration of the aortic wall. The lesion brought about by an excess of cholesterol, lysolecithin and sphyngomyelin permits infiltration of entire LDL molecules. (see also Ref. 17).

ted deficiency of α-lipoproteins or Tangier disease[18], and in acquired forms of α-lipoprotein reduction[19], as well as in familial LCAT deficiency or Norum's disease[20, 21, 22]. Alterations in VLDL structure were observed in galactosamine-induced rat liver injury, where LCAT activity is decreased and the α-lipoprotein fraction is absent[23]. Moreover, by assuring a normal cholesterol/cholesterol ester ratio, LCAT confers proper configuration to LDL, and thus renders them suitable to be taken up for degradation by the liver and other tissues[24, 25, 26, 27]. Finally, an important aspect of enzyme function is the maintenance of normal rates of exchange of unesterified cholesterol between LDL and plasma membranes of different cells, with particular regard to erythrocytes and intimal cells of the arterial wall.

References

1. Ursini F., Ferri L., Esposito C., Valente M., Gregolin C., Cortesi S. and Rigoni G.: in Proc. 6th Int. Symp. on Clin. Enz., 1974 edit. by A. Burlina, Editrice Kurtis s.r.l., Milano, p. 439-459 (1976).
2. Simon J.B. and Boyer J.L.: Biochim. Biophys. Acta, *218,* 549 (1971).
3. Osuga T. and Portman O.W.: Amer. J. Physiol., *220,* 735 (1971).
4. Akanuma Y. and Glomset J.A.: J. Lip. Res., *9,* 620, 1968.
5. Marcel Y.L. and Vezina C.: J. Biol. Chem., *248,* 8254, 1973.
6. Garner C.W. Jr., Smith L.C., Jackson R.L. and Gotto A.M. Jr., Circulation 45/46 (Suppl. II), 958 (abstr.), 1972.
7. Glomset J.A.: in Blood Lipids and Lipoproteins: Quantitation, Composition and Metabolism, Edit. by G.J. Nelson, Wiley-Interscience, New York, 1972, p. 825-880.
8. Fredrickson D.S., Levy R.I. and Lees S.S.: N. Engl. J. Med., *276,* 34, 94, 148, 215 and 273, 1967.
9. Norum K.R.: Scand. J. Clin. Lab. Invest., *33,* suppl. 137, 7, 1974.
10. Simon J.B. and Scheig R.: New England J. Med., *238,* 841, 1970.
11. Stokke K.T. and Norum K.R.: Scand J. Clin. Lab. Invest., *27,* 21, 1971.
12. Lacko A.G., Rutenberg H.L. and Soloff L.A.: Biochemical Medicine, *7,* 178, 1973.
13. Nestel P.J. and Monger E.A.: J. Clin. Invest., *46,* 967, 1967.
14. Glomset J.A.: J. Lip. Res., *9,* 155, 1968.
15. Nestel P.J., Whyte H.N. and Goodman De W.S.: J. Clin. Invest., *48,* 982, 1969.
16. Nestel J.P.: in Advances in Lipid Research v. 8, Edit. by R. Paoletti and D. Kritchevsky D., Academic Press, New York, 1970, p. 1-39.
17. Portman P.W. in Advances in Lipid Research, vol. 8, Edit. by R. Paoletti and D. Kritchevsky, Academic Press, New York, 1970, p. 41-114.
18. Fredrickson D.S., Levy R.I. and Lindgren F.T.: J. Clin. Invest., *47,* 2446, 1968.
19. Levy R.I., Lees R.S. and Fredrickson D.S.: J. Clin. Invest., *45,* 63, 1966.
20. Glomset J.A., Norum K.R. and King W.: J. Clin. Invest., *49,* 1827, 1970.
21. Norum K.R., Glomset J.A., Nichols A.V. and Forte G.M.: J. Clin. Invest., *50,* 1131 (1971).
22. Forte G.M., Norum K.R., Glomset J.A. and Nichols A.V.: J. Clin. Invest., *50,* 1141, (1971).
23. Sabesin S.M., Kuiken L.B. and Ragland J.B.: Science, *190,* 1302 (1975).
24. Sniderman A.D., Carew T.E., Chandler J.G. and Steinberg D.: Science, *183,* 526 (1974).
25. Bierman E.L., Stein O. and Stein Y.: Circ. Res., *35,* 136, 1974.
26. Goldstein J.L. and Brown M.S.: J. Biol. Chem., *249,* 5153, 1974.
27. Weinstein D.B., Carew T.E. and Steinberg D.: Biochim. Biophis. Acta, *424,* 404 (1976).

AN ASPECT OF THE APPLICATION OF COMPUTER AIDED QUANTITATIVE CYTOPHOTOMETRY IN HEMATOLOGY

H. Aus, U. Gunzer and V. ter Meulen

Summary

An application of computer aided image analysis in hematology, especially early detection of blood diseases, is presented. A clinical example is included to demonstrate the complexity of the problems involved and the possible solution.

Introduction

Since leucocytes play a dominant role in the diagnoses of inflammatory and malignant blood diseases, visual observation of human blood smears in the light microscope is an important research and diagnostic tool in hematology. For these observations histochemical methods have been developed which enable the hematologist to differentiate visually among the various types of white blood cells. The most commonly used parameters for visual differentiations include size and shape of cell and nucleus, N/C ratio, amount and intensity of granules, and color variations in the various subcellular particles. Subject to operator training, subjectivity and fatigue, it has always been of interest to automate and standardize these visual diagnostic tests. Only recently, however, have the cost and technical developments of laboratory computers enabled the practical application of space-age computer aided picture analysis in the hematology laboratory. Recent literature and commercially available systems already demonstrate that computerized screening can be applied relatively easily and successfully to the analysis of Basophils, Eosinophils, Lymphocytes, Monocytes and Neutrophils from a normal blood smear. For example, Neurath has demonstrated that these 5 types can be correctly classified with probably better than 90% accuracy if the immature, atypical and abnormal leucocytes are *not* included in the measured population[1]. However, it is precisely the pathological leucocytes which need to be measured and analysed if such image analysis systems are efficiently and accurately to detect such blood diseases as: acute and chronic myeloid and lymphatic leucemia.

Computer Aided Cytophotometry Group, University Medical Clinic and Institute for Virology, Würzburg, G.F.R.

The early detection of these types of diseases depends to a great extent upon the localization of minute and subtle morphological and color differences in a relatively small number of cells in a peripheral blood smear. Since color is one of the most important fingerprints in hematological diagnosis, the standardization of the diagnostic procedure with the help of computer aided color and shape discrimination is a new and important addition to the hematology laboratory. An example of the need for such methods is described below.

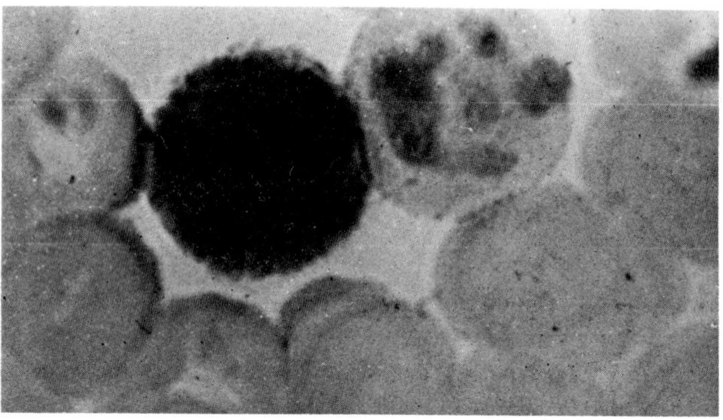

Fig. 1 – 2 neutrophil granulocytes stained by the method of Kaplow to reveal alkaline phosphatase.

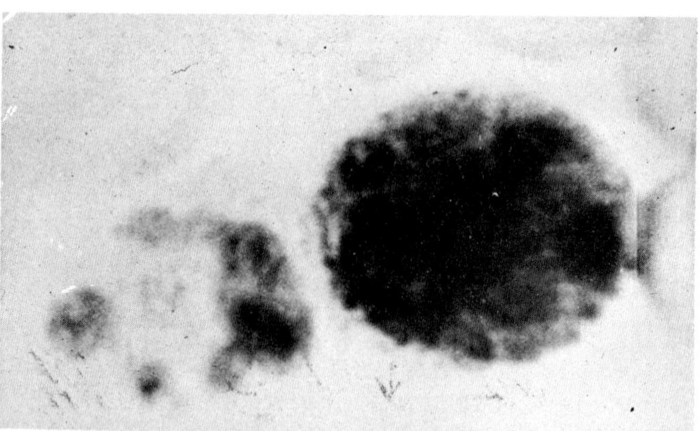

Fig. 2 – Digitized image of the two granulocytes in Fig. 1 as scanned with a green Kodak Wratten filter.

Alkaline phosphatase in Neutrophil Granulocytes

Surrounded by erythrocytes in Fig. 1 are two neutrophil granulocytes stained by the method of Kaplow to reveal alkaline phosphatase as brown stained granules[2]. Taken from a patient with symptoms suggestive of a leucemic process, this staining procedure is used to differentiate between a reactive leucocytosis and chronic myeloid leucemia. The intensity of the alkaline phosphatase as well as the morphology of the nucleus determines whether or not these cells are malignant neutrophil granulocytes. Since the distribution of the brownish granules is inhomogeneous throughout the cytoplasm including the cytoplasm covering the nucleus, substantial parts of the nucleus of the left hand granulocytes in Fig. 1 are not visible. Nevertheless, an image analysis system must correctly differentiate such cells from, for example, uniformly shaped blast-cells.

Computer Aided Color Separation in granulocytes

Using a slide of Fig. 1, a test was conducted using the Spatial Data Computer Eye System to demonstrate that such complex problems can be correctly analyzed by multicolor image processing[3]. The Computer Eye System, complete with a specially manufactured Vidicon TV Camera, TV monitor, film recorder and PDP 11, stored the digitized images on a digital disk for on-line processing.

A black and white photomicrograph of the digitized image in Fig. 1 scanned in green light is shown in Fig. 2. As seen by visual inspection, the single scan image does not contain optical density contrasts which correspond to the original color differences. As calculated by the computer, Fig. 3 is a photomicrograph of the pixel by pixel difference between the optical density of the image scanned in red light and in green light. As an intermediate step, both red and green images had to be stored on the disk. In Fig. 3 a composite image of a cell is displayed revealing a light cytoplasm and darker polymorph-shape nuclei. The picture processing has transposed the brown stain to white and the blue stain to dark and has thereby enhanced the contours of the nuclei. In contrast to Fig 2, the hematologist is now able to recognize the left hand cell as a polymorph-shape neutrophil, which indicated that the requirements for a correct automated differentiation are being achieved.

Comments

Taking advantage of the difference in the blue and brown stains as a function of wavelength, the interfering stains have been used to enhance the optical density contrast of the nucleus and cytoplasm. As seen in Fig. 3, one could perhaps conjecture that various multi-color digital methods could be used to measure and extract extremely subtle morphological differences as

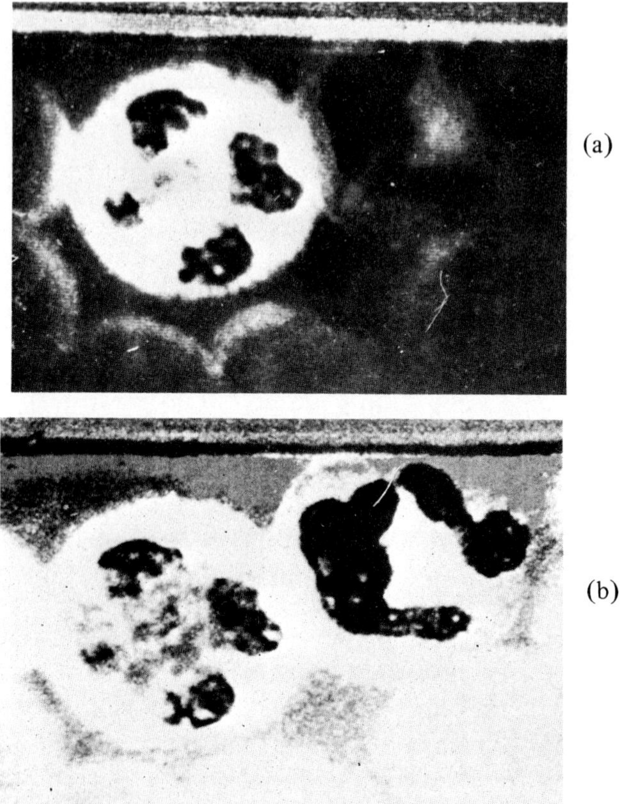

Fig. 3 – Computer calculated red minus green scan difference image with the thresholds set slightly different in (a) and (b).

well as to separate touching cells. Only with the development and application of such methods is it possible to achieve, on a computer cytophotometric basis, the same cell differentiations as the hematologist can. Continuing investigations will hopefully increase the understanding of how image analysis technology can be effectively used in early detection of similar and even more complex blood diseases.

Acknowledgements

This work is supported in part by the Deutsche Forschungsgemeinschaft, Sonderforschungsbereich 105 Würzburg, and the Bundesministerium für Forschung und Technologie. We wish to thank Spatial Data Group of Groleta, California for Figs 2 and 3.

References

1. Brenner J.F., Gelsema E.S., Necheles T.F., Neurath P.W., Selles W.D. and Vastola E.: Automated Classification of Normal and Abnormal Leucocytes. J. Histo. Chem. Cytochem., *22,* 697-706 (1974).
2. Kaplow L.S., Blood *10,* 1028 (1955).
3. Aus H.M., Gunzer U. and ter Meulen V.: A Note on the Usefulness of Multi-Color Scanning and Image Processing in Cell Biology. The Microscope *24,* 39-44 (1976).

ASSAY OF URINARY ENZYMES FOR THE ASSESSMENT OF ACUTELY IMPAIRED TUBULAR FUNCTION

M. Werner and D. Gabrielson

Summary

The determination of urinary enzymes for diagnostic purposes has been hampered by three main difficulties: (1) interferences with the assays by low molecular weight inhibitors, activators and non-enzymatic (heat-stable) activities; (2) the wide variability of normal values; and (3) our limited understanding of the pathophysiology of urinary enzyme excretion.

To eliminate interferences with the assays, dialysis of urine against water is frequently used. However, this approach does not assure reliable results. Therefore, we have developed a simple and rapid gel filtration procedure to separate reliably interferences from leucine aminopeptidase, arylsulphatase, β-glucuronidase, lactate dehydrogenase, alkaline and acid phosphatase activities.

The normal biological patterns of urinary enzyme excretion become recognizable after reducing the methodological error by gel filtration. When enzyme excretion is estimated in a number of consecutive 12-hour periods, subjects can be characterized by the central level as well as the dispersion of results. Results expressed either as enzyme activity/urinary volume, or excreted urinary enzyme activity / 12 hours, or urinary enzyme activity / mg of creatinine all conformed to logarithmically normal distributions. This allows analysis of variance to estimate the significant contributions of intra- and inter- individual differences to total variability.

Understanding of the pathophysiology of urinary enzyme excretion can be improved through the study of appropriate biological models. Along these lines we have assayed urinary enzymes before, during and after administration of a number of clinically used drugs to assess their possible effects on the renal tubule. Since enzymes are not distributed uniformly within the tubules, the pattern of excreted enzymes can indicate which part of the nephron a drug effect migh involve. Conversely, once the pattern of enzyme excretion can be linked to such localizable tubular effects, diagnosis of tubular damage caused by other mechanisms is facilitated.

Division of Laboratory Medicine, The George Washington University Medical Center 901 Twenty-third Street, N.W., Washington, DC 20037.

Introduction

There are well tested and universally accepted assays for the assessment of renal glomerular function, but the laboratory evaluation of acutely impaired tubular function is difficult. Analysis of urinary sediment and special clearance studies are the most generally available such assays. From a diagnostic viewpoint both have shortcomings, and alternative approaches would be desirable.

Many enzymes regularly excreted in urine are thought to originate from "wear and tear" breakdown of renal cells. The reasons for this assumption are: First, molecular size restricts glomerular filtration of many enzymes[1]. (Amylase is a notable exception to this). Second, no consistent dependence between urinary and serum enzyme activity, or between enzymuria and proteinuria exists[2]. Third, even the urinary and serum isozymes often differ. Together this evidence, although indirect, points to a tubular origin of certain urinary enzymes. These enzymes would be of the greatest clinical interest, should their excretion pattern reflect acute tubular irritation.

Apart from the fact that the lower urogenital tract may contribute to the enzymes excreted in urine, three main difficulties have hampered their diagnostic use: First, low molecular weight inhibitors, activators and non-enzymatic, heat-stable, catalytic activities interfere with the assays[3, 8]. Second, normal or reference values vary widely[9, 10]. Third, there is limited understanding of the pathophysiology of urinary enzyme excretion. This paper focuses on experimental work addressing these three problems. On the other hand, urinary enzymes of clinical interest but unrelated to the diagnosis of tubular damage are not considered. (These include digestive enzymes such as amylase and uropepsinogen, enzymes involved in blood clotting such as urokinase and plasminogen, and enzymes lacking in certain inborn errors of metabolism such as α-glucosidase and α-galactosidase).

Methodology

The details of selected urinary enzyme assays used by us are listed in Table 1. Others have reported on additional methods[2, 11-13]. All require previous treatment of urine to remove interferences. Since the sources and amounts of interfering substances in "native" urine are varied among different enzymes and unpredictable, changing from day to day even in the same individual, it is thus important that they be eliminated entirely[14]. The apparent activity measurable in native urine may represent between 25 and 1700 % of the true value in the case of acid phosphatase, between 50 and 250 % in the cases of alkaline phosphatase and leucine aminopeptidase, and between 15 and 100 % in the cases of β-glucuronidase, N-acetyl-β-glucosaminidase, and β-galactosidase.

Table I. *Enzyme assay methods. Partly modified micro versions of standard techniques to which references are given*

Enzyme	*Sample µl*	*Incubation Mixture*	*Assay*	*Reference*
Lactate dehydrogenase (E.C. 1.1.1.27)	50	1.0 ml Taps buffer (pH 8.55) with NAD (6.1 mM) and L−+− latic acid (47.9 mM)	37° kinetic 340 nm	Gay, R.J., McComb, R.B., and Bowers, G.N., Clin. Chem., **14**:740-753, 1968.
Alkaline phosphatase (E.C.3.1.3.1)	50	1.0 ml AMP buffer (805 mM, pH 10.1) with p-nitrophenol phosphate (15.8 mM)	37° kinetic 405 nm	Bowers, G.N., and Mc Comb, R.B., Clin. Chem. **18**: 97-104, 1972.
Acid phosphatase (E.C. 3.1.3.2)	50	1.0 ml Citrate buffer (30 mM citric acid, 50 mM sodium citrate, pH 5.0) with α-naphthyl phosphate (3 mM) and Fast Red TR (1 mM)	37° kinetic 450 nm	Amador, E., Price, J.W., and Marshall, G., Am. J. Clin. Path. **51**: 202-206, 1969.
Arylsulfatase (E.C. 3.1.6.1)	200	01. ml Acetate buffer (1.0 M containing 0.5 MM $Na_4P_2O_7$ and 10% NaCl) 0.1 ml 2-hydroxy-5-nitrophenyl sulfate (0.02 M)	37° 60 min inc. 515 nm	Baum, H., Dogson, K.S., and Spencer, B., Clin. Chim. Acta **4**: 453-455, 1959.
β-Galactosidase (E.C. 3.2.1.23)	100	0.35 ml Citrate buffer (0.09 M, pH 4.8) 0.05 ml 4-methylumbelliferyl-β-D-galactoside (0.4 mM)	37° 30 min incub. fluorometric 360 nm – 450 nm	Woolen, J.W., and Walker, P.G., Clin. Chim. Acta *12*: 647-658, 1965.
N-Acetyl-β-glucosaminidase (E.C. 3.2.1.30)	100	0.35 ml Citrate buffer (0.09 M, pH 4.8) 0.05 ml 4-methyl-umbelliferyl-N-acetyl-β-D-glucosaminide (1 mM)	37° 30 min incub. fluorimetric 360 nm – 450 nm	Woolen, J.W., and Walker, P.G., Clin. Chim., Acta *12*:647-658, 1965.
β-Glucuronidase (E.C. 3.2.1.31)	100	0.35 ml Acetate buffer (0.1 M, pH 4.6) 0.05 ml 4-methyl-umbelliferyl-β-D-glucuronide (1 mM)	37° 30 min incub. fluorometric 360 nm – 450 nm	Mead, J.A. R., Smith, J.N., and Williams, R.T. Biochem. J. *61*: 569-574, 1955.

Table I. cont.

Enzyme	*Sample µl*	*Incubation Mixture*	*Assay*	*Reference*
N-Acetyl-β-galactosaminidase (E.C. 3.2.1.53)	100	0.35 ml Citrate buffer (0.09 M, pH 4.8) 0.05 ml 4-methylumbelliferyl-N-acetyl-β-D-galactosaminide (1 mM)	37° 30 min incub. fluorometric 360 nm – 450 nm	Woolen, J.W. and Walker, P.G., Clin. Chim. Acta **12**: 647-658, 1965. Woolen, J.W., Heyworth, R., and Walker, P.G., Biochem. J. 78: 111-116, 1961.
Leucine aminopeptidase (E.C. 3.4.11.2)	50	1.0 ml Phosphate buffer (100 mM, pH 7.2) with L-leucine-p-nitroanilide (0.8 mM)	37° kinetic 405 nm	Nagel, W., Willig, F., and Schmidt, F.H., Klin. Wschr. **42**: 447-449, 1964.

Dialysis is the classical method for removing low molecular weight substances interfering with enzyme assays in urine[15, 19]. Even though this procedure does not reliably remove all interfering substances, most published investigations continue to be based on it. Figure 1 illustrates the effect of dialysis on the activity of six urinary enzymes[3]. With varying length of time, apparent activities increase, then diminish again. For different enzymes, transient peak activities are found at different times, and so the same dialysis procedure cannot be used to prepare a single specimen for the assay of multiple enzymes.

Gel filtration on Sephadex G-50 fine (Pharmacia, Uppsala, Sweden) has been developed to overcome the shortcomings of dialysis[3]. Figure 2 shows the two well separated major fractions obtained upon elution of urine from a small column: A first fraction contains all assayed enzymatic activities which are destroyed by heating to 95°C., and a second fraction contains substances absorbing at 280 nm, inorganic phosphate and heat-stable, spurious lactate dehydrogenase, alkaline and acid phosphatase activities.

For routine use, sample applications is facilitated by placing a rubber sponge of known volume on top of the gel bed. This also prevents the column from running dry between applications of eluant. Elution can be reduced to only three steps, with the volumes of applied eluant so adjusted that the first step elutes the so-called "external" volume of the the gel bed, the second step elutes all enzymatic activity, and the third step rinses all low molecular weight substances from the column. The first and third portions of

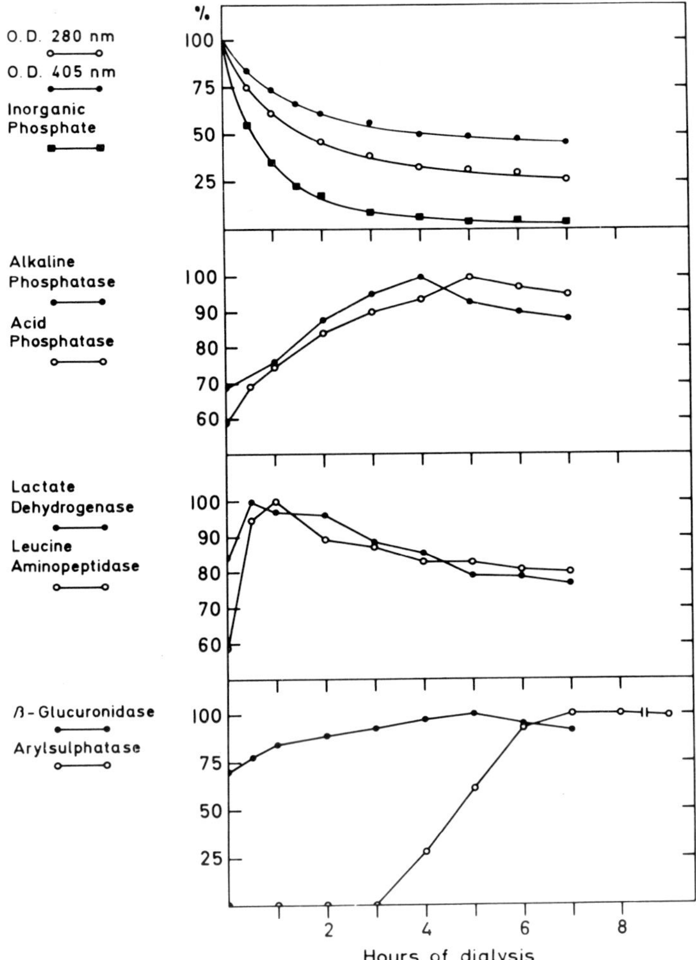

Fig. 1 – Effect of varying the length of dialysis time for urine. There was an almost exponential removal of all inorganic phosphate but substances absorbing light at 280 nm and at 405 nm (urochromes) were only partly removed. The apparent activity of the six assayed enzymes initially increased. Lactate dehydrogenase and leucine aminopeptidase reached peak activity during the first hour, alkaline phosphatase, acid phosphatase and β-glucuronidase from the fourth to the fifth hour, and arylsulfatase at the seventh hour. Subsequently the activities of all enzymes, except arylsulfatase, decreased. Even at the activity peak, there is no assurance that all interferences have been removed. (Taken from M. Werner, D. Maruhn, and M. Atoba, J. Chromatog. **40**: 254-263, 1969).

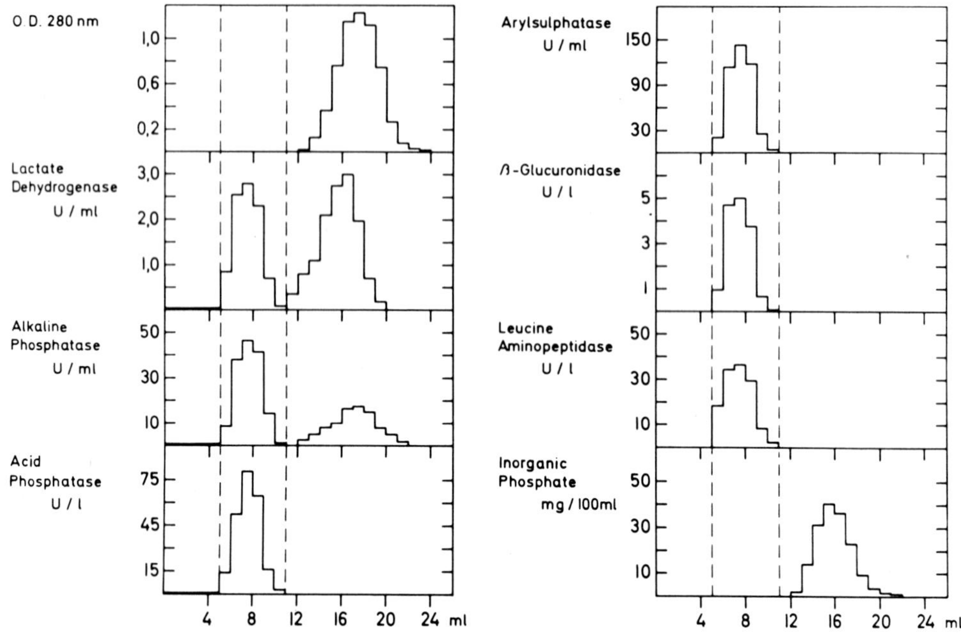

Fig. 2 – Gel Filtration of Urine. A Sephadex G-50 fine gel bed (Pharmacia, Uppsala, Sweden) 1 cm in diameter and 14 cm in height was used. Optical density at 280 nm (eluates diluted 1:20), apparent activities of six enzymes and concentration of inorganic phosphate in the eluates are shown. The first 5 ml of the eluate are free of enzyme activity (volume "external" to the gel bed). The activity of all assayed enzymes reaches a peak in the next 6 ml with a maximum in the eighth ml. This fraction is indicated by dashed lines and is used for assay. Lactate dehydrogenase and alkaline phosphatase show a second peak of apparent, heat-stable, activity with a maximum at the 16th and 17th ml, where substances absorbing light at 280 nm and inorganic phosphorus also emerge. (Taken from M. Werner, D. Maruhn, and M. Atoba, J. Chromatog. **40**: 254-263, 1969).

eluate are discarded, and only the second fraction is used for assay. Despite its relative simplicity such bulk elution is both accurate and precise.

Gel filtration can be considered a reference method for the purification of urine from substances interfering with urinary enzyme assays. Compared to dialysis, the activities of acid phosphatase, β-glucuronidase and leucine aminopeptidase are usually *higher* after gel filtration. On the other hand, the activities of lactate dehydrogenase and alkaline phosphatase are usually *lower* after gel filtration, since dialysis does not remove all heat-stable, spurious activity. Such positive errors can range to over 300% of true activity in the case of lactate dehydrogenase and to over 100% in the case of alkaline phosphatase. For arylsulphatase the difference between the two methods tends to be smallest and is sometimes positive and sometimes negative. These findings clearly indicate that dialysis should be abandoned.

Ultrafiltration. Gel filtration, even when bulk elution is used, is somewhat cumbersome, and a simpler procedure for busy service laboratories appeared desirable. Ultrafiltration provides such ease while retaining accuracy[20]. Figure 3 shows the polycarbonate ultrafiltration cell we use (catalogue number XX42 013 10, Millipore Corp., Bedford, Massachusetts 01730). At the bottom of the cell is a fritte support for the ultrafilter. Pellicon membranes (Millipore, catalogue number PTGC 013 10) of 13 mm diameter and with 10,000 dalton nominal separation cutoff are used. The filter consists of a thin polymeric film of proprietary composition bonded onto a highly porous support of cellulose acetate and is used with the film facing the fluid to be filtered. Urine is placed in the sample compartment which has a 3 ml capacity. Filtering pressure is applied through a Luer-lock connection on the cell cap. Low molecular weight substance and water pass the filter for collection from the exit tubing.

We use a 1 ml sample and a pressure of 65 lbs applied from a nitrogen tank. Typically, ultrafiltration is completed within five to ten minutes, but certain samples with a high content of protein or other solid matter require more time. Following filtration, the apparatus is uncapped, 1 ml of saline and a magnetic stirring bar are introduced into the sample compartment, and

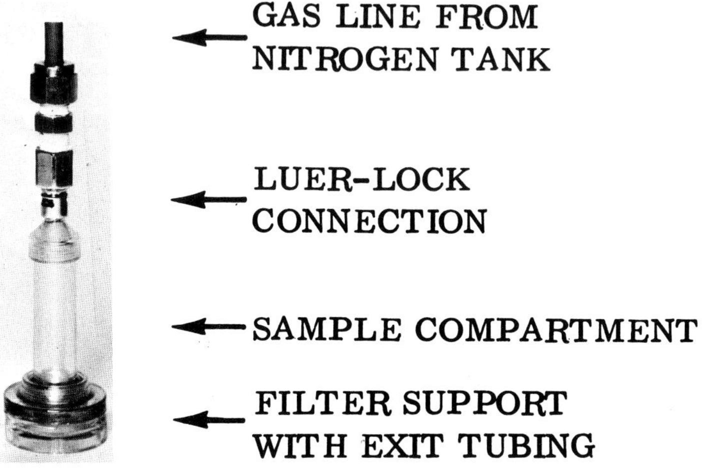

Fig. 3 – Ultrafiltration cell (Millipore Corp., Bedford, Mass. 01730, catalogue number XX42 013 10). The sample compartment has a capacity of 3 ml. Filtering pressure is applied from a nitrogen tank through a Luer-lock connection at the top. The ultrafiltrate can be collected from the exit tubing at the bottom.

the substances retained on the filter are solubilized by stirring for five minutes on an electrical mixer. Thus, ultrafiltration produces two fractions: an ultrafilter eluate and an ultrafiltrate. Enrichment of low enzyme activities can be achieved by eluting the filter into a lesser solvent volume than that of the original sample.

Figure 4 compares apparent urinary activities of lactate dehydrogenase, alkaline and acid phosphatase in native urine, following ultrafiltration and following gel filtration. In native urine all three enzymes usually contain both heat-labile and heat-stable activities. Following gel filtration, heat-stable activities are recovered quantitatively in the low molecular weight fraction, where no heat-labile activities are found. On the other hand, the heat-labile activities found in the high molecular weight fraction typically are larger than in native urine since inhibitors have been removed. Ultrafiltration produces a separation into two fractions free from cross contamination analogous to that of gel filtration. All true enzyme activity present in urine is recovered in the filter eluate, and the inhibitors, activators and non-protein, heat-stable, spurious catalytic activities are removed in the ultrafiltrate.

Normal values

Biochemical Individuality. While homeostatic mechanisms closely guard such biochemical systems serving vital functions, as plasma osmolality, urinary enzyme excretion falls among those parameters which are not rigidly regulated since no apparent physiological function is attached to their excretion. Thus, wide physiological variability occurs between subjects and with time for the same subject (Figure 5) and has hampered the discrimination between normal and abnormal findings[9]. A circadian fluctuation has been inferred from few and isolated observation to explain part of the biological variability of urinary enzyme excretion[21, 22]. While some individuals regularly excrete more of a given enzyme during the day or during the night, a pertinent study found no consistent circadian variations for any enzyme[9].

Obviously, the definition of reference values for urinary enzymes only could be attacked after reliable methods had been developed. In an attempt to define the key traits of biochemical individuality the assay results were transformed in various numerical ways. The best fit to normal (gaussian) distributions was obtained when the logarithms of the assay results were used (Figure 6). Results from each individual were plotted as separate, continuous distribution curves against a probability scale, which transforms gaussian distributions into straight lines. Findings from different individuals indeed approximated gaussian distributions, with some showing little variability of excretion as indicated by a steeper slope of the cumulative distribution curve, others showing more variability as indicated by a flatter slope,

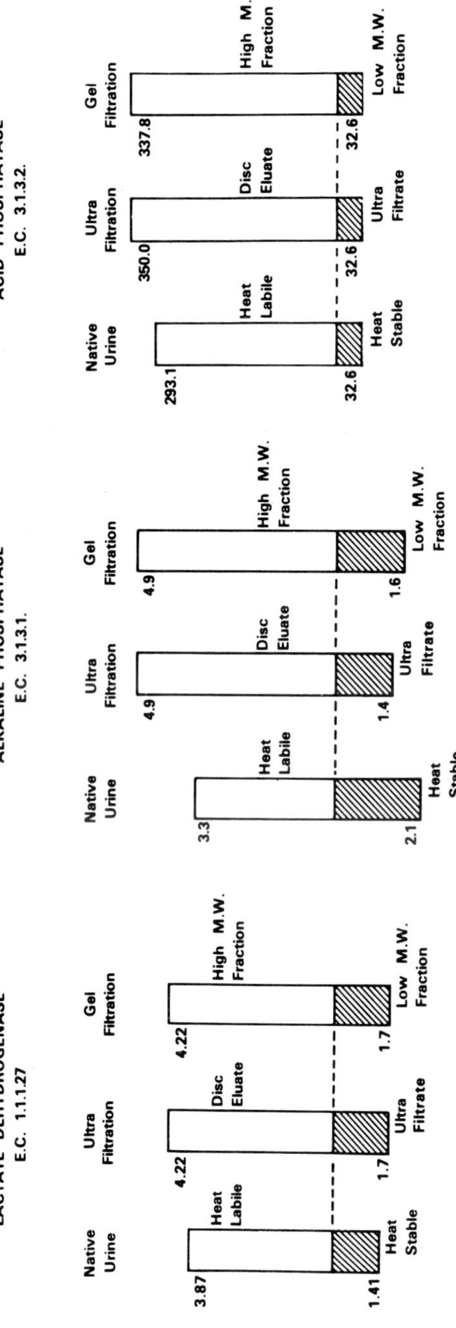

Fig. 4 – Ultrafiltration of urine. Comparison of lactate dehydrogenase, alkaline phosphatase, and acid phosphatase activities assayed in native urine (heat labile and heat stable fractions), in urine subjected to ultrafiltration (disc eluate and ultrafiltrate), and urine subjected to gel filtration (high and low molecular weight fractions).

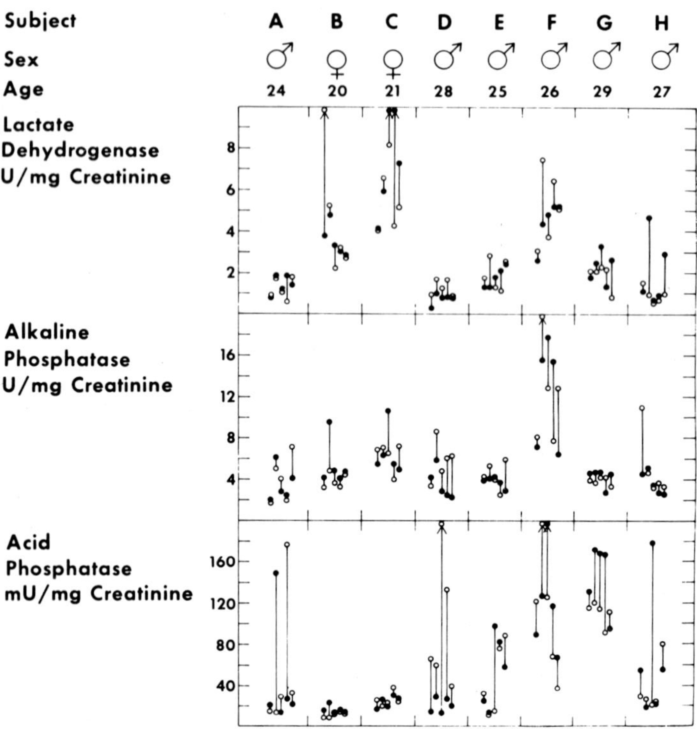

Fig. 5 – Urinary enzyme excretion. Urine was collected in eight subjects for five days in 12 hour periods. Open circles: day. Closed circles: night. (Taken from M. Werner, D.C. Heilbron, D. Maruhn, and M. Atoba, Clin. Chim. Acta 29: 437-449, 1970).

and typical cases clustering between the extremes. Further, a relationship appeared to exist between the scatter of values and the typical level of excretion (for which the 50 percentile value can be considered indicative). Thus, a low excretion level was accompanied by little scatter, a high excretion level by more scatter. Since both the value of central tendency and variability of enzyme excretion differ between subjects, a reasonable hypothesis concerning its distribution was that it is log-normally distributed with means and standard deviations depending on the individual. To substantiate this hypothesis the logarithmically transformed values, x, were corrected for each subject's own mean, $\bar{x}$, and standard deviation, S.D., to produce "normalized deviates", z:

$$z = \frac{x - \bar{x}}{S.D.}$$

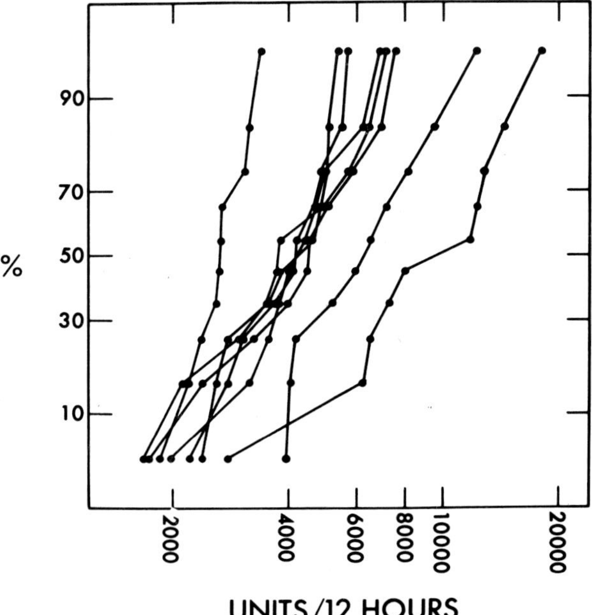

Fig. 6 – Cumulative distribution curves of alkaline phosphatase excretion for different subjects plotted on a probability scale (y axis). Results are expressed in units/12 hr. and are plotted on a logarithmic scale (x axis). Note individual differences both in mean level and scatter of results (slope of cumulative curve). (Taken from M. Werner, D.C. Heilbron, D. Maruhn, and M. Atoba, Clin. Chim. Acta **29**: 437-449, 1970).

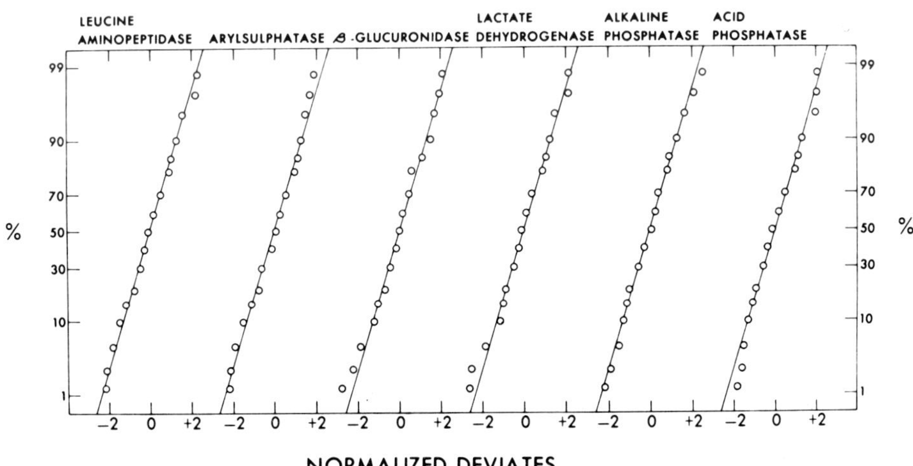

Fig. 7 – Cumulative distributions of logarithms of urinary enzyme activity per milligram creatinine (x-axis) plotted on a probability scale (y axis) on which gaussian distributions are shown as straight lines. (Taken from M. Werner, D.C. Heilbron, D. Maruhn, and M. Atoba, Clin. Chim. Acta **29**: 437-449, 1970).

Figure 7 shows cumulative frequency plots of the collected normalized deviates from *all* subjects pooled together. Once the effect of individual means and deviations were eliminated, a surprisingly excellent fit to the log-normal distribution model was seen for all enzymes. Thus, the individual values for central tendency and for variability apperar to fully define "biochemical individuality".

Normalization procedures. Urinary enzyme excretion values can simply be reported as enzyme activity/urinary volume. Alternatively, urinary enzyme activity can be related (1) to urinary volume (multiplying the enzyme concentration by the urine volume), and (2) to urinary creatinine concentration (dividing urinary enzyme activity by urinary creatinine concentration). The intent of both these normalization procedures is to reduce variability. Although urinary volume is readily determined, incomplete collection or loss of enzyme activity during long collection periods may cause errors. Relating results to creatinine excretion does not require timed urine collection, but introduces the error of creatinine determination. In effect, *both* methods relate enzyme excretion to *time*, since creatinine excretion is assumed to be relatively constant with time.

In a previous study the effect of reporting urinary enzyme excretion data in the three ways described above was studied by analysis of variance[9]. In addition to method error, differences between subjects (inter-individual variability) and differences with time for the same subject (intra-individual variability) were considered. Compared to method error, all physiological variables were highly significant. Intra-individual variability was similarly reduced whether enzyme excretion was related to urinary volume or to creatinine excretion, but inter-individual variability was less with the latter correction, as inter-individual differences in creatinine excretion paralleled those in enzyme excretion to some degree. It is known that a significant correlation between urinary creatinine excretion and lean body mass exists[23], and that variation of urinary creatinine is less within a single subject than within a group.

Pathophysiological models

The pattern of excreted enzymes may give some indication as to which part of the nephron kidney damage affects. For instance, lactate dehydrogenase concentration is highest in the proximal convolutions though this enzyme is found throughout the entire nephron. Further, the pattern of excreted enzymes may also allow inferences about the nature of cellular damage. For instance, leucine aminopeptidase and δ-glutamyl transpeptidase[2] are located in the brush border, β-glucuronidase is found throughout the cellular cytoplasm[24] and in lysosomes, while arylsulphatase and acid phosphatase are typically lysosomal enzymes. Thus, the fullest diagnostic use of urinary enzymes probably will come from their use in "batteries" or panels of several different activities the results of which can be integrated into a composite interpretation.

At present the literature does not contain information which would make reliable correlations between urinary enzyme panels and clinical entities possible. On the one hand, the usual approach has been to investigate individual enzymes one at a time. On the other hand, urinary enzymes have been used mainly to diagnose such catastrophic and undifferentiated tissue damage as occurs in the rejection of renal transplants[25-28], renal infarction[29] or kidney surgery[30].

As an alternative to these approaches, we feel that the assay of urinary enzymes panels during drug therapy could be rewarding in two important ways:

1. Results could be utilized to monitor for possible nephrotoxic effects of such drugs as gentamycin, or erythromycin[31].
2. The administration of commonly used drugs could provide pathophysiological models for the study of tubular irritation in man.

In an unreported study, one of the authors[36] has measured six enzymes in the urine of healthy subjects before, during and after giving acetylosalicyl acid or phenacetin and found dissimilar effects of the two drugs. The *mean* excretion of β-glucuronidase and lactate dehydrogenase increased under acetylosalycic acid; the *mean* excretion of lactate dehydrogenase decreased under phenacetin. Drug effects on means of all other measured enzymes were not unequivocal, and their statistical significance differed according to the "normalization" of the excretion data used.

The enzyme excretion pattern under acetylosalicylic acid is not readily explained by tubular cell damage. For instance, leucine aminopeptidase a sensitive indicator of tubular cell damage, remains unaltered. Rather, it was felt that enzyme excretion under aspirin could be explained by induction of increased enzyme synthesis: according to this hypothesis the required glucuronation of drug metabolites during acetylosalicyc acid medication simply induced increased β-glucuronide synthesis. Similarly, the rise of urinary lactate dehydrogenase could be linked to metabolic effects and represent enzyme induction to overcome the inhibition by salicylate involving competition with NAD, $NADH_2$, and NADP.

References

1. Rigas D.A. and Heller C.G.: J. clin. Invest., *30,* 853 (1951).
2. Thiele K.G.: Klin. Wschr., *51,* 339 (1973).
3. Werner M., Maruhn D., and Atoba M.: J. Chromatog., *40,* 254 (1969).
4. Abu-Fadl, M.A.M.: Biochem. J., *65,* 16P, (1957).
5. Hilliard S.D., O'Donnell J.F. and Schenker S.: Clin. Chem., *11,* 570, (1965).
6. Marsh C.A.: Biochem. J., *86,* 77, (1957).
7. Schoenenberger G.A. and Wacker W.E.C.: Biochemistry *5,* 1375, (1966).
8. Wacker W.E.C. and Schoenenberger G.A.: Biochem. Biophys. Res. Commun., *22,* 291, (1966).
9. Werner M., Heilbron D.C., Maruhn D. and Atoba M.: Clin. Chim. Acta, *29,* 437 (1970).

10. Maruhn D., Fuchs I., Mues G. and Bock K.D.: Clin. Chem., *22,* 1567 (1976).
11. Peters J.E., Schneider I., and Haschen J.: Clin. Chim. Acta, *36,* 289 (1972).
12. Richterich R., Cantz B. and Dauwalder H.: Clin. Chim. Acta, *29,* 295 (1970).
13. Szasz V.G.: Klin. Biochem., *8,* 1 (1970).
14. Paul W., Shapiro A., and Gonick H.: Enzymol. Biol. Clin., *8,* 47 (1967).
15. Amador E., Zimmerman T.S. and Wacker W.E.C.: J. Am. Med. Assoc., *185,* 953 (1963).
16. Butterworth P.J., Moss D.W., Pitkanen E., and Pringle A.: Clin. Chim. Acta, *11,* 212 (1965).
17. Dorfman L.F., Amador E. and Wacker W.E.C.: J. Am. Med. Assoc., *184,* 1 (1963).
18. Schmidt J.D.: Invest. Urol., *3,* 405 (1966).
19. Schmidt J.D.: Invest. Urol., *3,* 405 (1966).
20. Werner M. and Gabrielson D.: Clin. Chem. in press.
21. Dubach U.C. and Paclina G.: Klin. Wschr., *44,* 180 (1960).
22. Dubach U.C. and Rediger R.: Urol. Intern., *17,* 65 (1964).
23. Doolan C.P., Alpen E.L., and Thiel G.B.: Amer. J. Med., *32,* 65 (1962).
24. Fishman W.H., Goldman S.S., and DeLellis R.: Nature, *213,* 457 (1967).
25. Hansen N.E. and Weeke E.: Acta Med. Scand., *188,* 317 (1970).
26. Shehadeh I.H., Carpenter C.B., Monterio C.H. and Merrill J.P.: Arch. Intern. Med., *125,* 850, (1970).
27. Wellwood J.M., Ellis B.G., Hall J.H., Robinson D.R., and Thompson A.E.: Br. Med. J., *2,* 261, (1973).
28. Sandman R., Margules R.M. and Kountz S.L.: Clin. Chim. Acta, *45,* 349 (1973).
29. London I.L., Hoffsten P., Perkoff G.T. and Pennington T.G.: Arch. Intern. Med., *121,* 87 (1968).
30. Price R.G., Dance N., Richards B. and Cattell W.R.: Clin. Chim. Acta, *27,* 65 (1970).
31. Wright P.J. and Plummer D.T.: Bioch. Pharm., *23,* 65 (1974).
32. Prescott L.F.: Lancet *ii,* 91, (1965).
33. Scott J.T.: Amer. Heart J., *71,* 715 (1966).
34. Scott J.T., Dewman M.A., and Dorling J.: Lancet *i,* 344, (1963).
35. Shelley J.H.: Clin. Pharm. and Therap., *8,* 427 (1967).
36. Werner M., Heilbron D.C., and Maruhn, D.: unpublished.

EFFECT OF BROMELIN ON THE JEJUNAL MUCOSA STRUCTURE AND ENZYMES IN NORMAL ANIMALS AND IN HETEROIMMUNE AND CYTOSTATIC ENTEROPATHY

F. Barbarino, E. Neumann, P. Szabo, N. Parau and S. Toader

Summary

The authors studied experimentally the influence of Bromelin (EC = 3.4.4.24), a proteolytic enzyme obtained from Ananas Comosus, on the structure and function of the jejunal mucosa.

The activity of the enzymes participating in the absorption processes (LAP, ATP-ases, alk. phosphatase, G_6P-ase) was enhanced after intragastric instillation of 0.50 mg/kg Bromelin for 3 days in 25 healthy guinea pigs. ^{131}I-triolein, ^{14}C-stearin and ^{3}H-leucine showed the decrease of radioactivity remaining in the content of the isolated and cannulated duodenojejunal segment.

In "hetero-immune" enteropathy (induced in 62 guinea pigs with antiserum from rabbits treated with a homogenate from guinea pig jejunal mucosa and Freund's adjuvant), Bromelin association protected them for 4 weeks against enterocyte dystrophy and enzyme depression, without diminishing the mesenchymal cell proliferation in the lamina propria.

In acute "cytostatic" enteropathy (intragastric instillation of 40 mg/kg Methotrexate in 80 Wistar rats) Bromelin reduced the great severity of the morphologic and histoenzymatic lesions.

A hypothetic explanation of these results is that metabolic activation in the enterocytes could influence their reactivity and consequently enhance their resistance against the noxious action of hetero-antiserum and of the cytostatic drug, Methotrexate.

Introduction

After its empiric application in American popular medicine, the proteolytic activity of the fresh juice from pine-apple fruits was discovered by V. Marcano in 1891; the active substance was concentrated and studied by R.H. Chittenden in 1894 and the "stem bromelin" was isolated from Ananas Comosus, Cayenne variety, by R.M. Heinicke and W.A. Görtner in 1957 (Berndt[6]).

Bromelin (EC = 3.4.4.24) was integrated in several drugs, e.g. 50 mg = 1000 U in Nutrizym (Fa E. Merck 1965) together with 400 mg Pankdreatin

Institute of Hygiene and Public Health, Ist Surgical Clinic and Nuclear Medicina Institute, Cluj-Napoca, Romania.

(= 1500 U Protease, 8000 U. Lipase and 1600 U. Amylase) and 30 mg dried ox bile (Hennrich[16]).

The proteolytic activity of Bromelin exerted between pH 2 and 8 establishes a bridge over the physiologic range of the gastric enzymes (at pH 3-4) and that of the intestinal ones (at pH 6) and may substitute their digestive function (Berndt[6], Hennrich[16, 17], Lang[18]). The clinical and experimental data regarding its action on lipid digestion and absorption are controversial (Berndt[7]). The local anti-inflammatory effect is applied in surgery. There are justified doubts concerning the documentation of the "systemic" anti-edematous, analgesic and thrombolytic action through fibrinolysis and proteolysis (Lang[19, 20]).

In these conditions, radioisotope studies were performed to differentiate the digestion and absorption of proteins and lipids. Histochemical reactions were applied for the detection of the consequences of digestive stimulation by Bromelin on the structure and enzymes of enterocytes, both in healthy animals, and in 2 experimental models (moderate and severe enteropathy).

"Hetero-immune" enteropathy was induced according to the model used by Dumitrascu[10] and Fodor[13]. The degree of the enzymatic depression overcomes the moderate enterocyte dystrophy, and there is an evident mesenchymal hyperplasia in the lamina propria. The cytostatic-immuno-depressor treatment (Goia[14]) and even more the aspartate (Barbarino[5]) restricted the enterocytic and mesenchymal lesions.

"Cytostatic" enteropathy was produced with Methotrexate, as folic acid antagonist. The pathogenesis consists in a primary mitosis arrest in the crypts (after 2-4 hours), and consequently the lack of enterocyte regeneration and migration of maturing cells towards the villi tips. The lining epithelium degenerates vacuolarly and then only some necrobiotic enterocytes persist on the denuded intestinal surface. A rich inflammatory cell infiltrate develops in the lamina propria mucosae. As a consequence of the severe morphologic lesions, malabsorption for carbohydrates, amino acids and lipids occurs, with some delay. After 2 days, when necrobiosis is maximal, amino acid resorption is still possible. The mitosis resumption assures the regeneration, with satisfactory structural restoration after 6 days, but the absorption capacity is regained only after 11-15 days. (Achord[1], Antonioli[2], Bernier[8], Eder[11, 12], Hartwhich[15], Millington[22], Pearse[23], Riecken[24], Robinson[25, 26], Schiraldi[27], Trier[28], Vitale[29]).

Material and methods

The effect of Bromelin in normal animals (experiment I) and in hetero-immune enteropathy (experiment II) was studied in the more sensitive guinea pigs, but for severe cytostatic enteropathy (experiment III) the more resistant rats were preferred.

Experiment I (normal animals):

- group 1: 25 control healthy guinea pigs (body weight of 200-500 g);
- group 2: 25 guinea pigs received for 3 days, and 4 hours before their killing, 50 mg/kg body weight Bromelin (E. Merck, Art. 501898 with 20000 U/g), instilled by gastric tube.

Experiment II (hetero-immune enteropathy):

- group 3: In 32 guinea pigs, rabbit antiserum directed against guinea pig jejunal mucosa was inoculated in 1 ml. doses, 6 times, at 4 days intervals, in the course of 3 weeks (3 times s.c. and 3 times i.p.). The antigen was prepared as follows: the jejunal mucosa taken from 3 guinea pigs was washed with saline solution, homogenized in 20 % saline solution, kept for 24 hours at + 4°C with 5000 U.penicillin and 5 micro g/ml streptomycin, then mixed with equal parts of complete Freund's adjuvant and preserved at +4°C. The antigenic solution was injected in rabbits intrapatullarly, 1 ml. in the right anterior and posterior controlateral pad. After 10 days a new inoculation with 1 ml. was made s.c. in the two flancs. Ten days later the rabbits were bled by cardiac puncture. The serum was tested for its antibody titer and specificity by Outcherlony immuno-diffusion, against the guinea pig jejunal antigen. The immune serum was preserved in vials with merthiolat.
- group 4: 30 rats treated with antiserum as in group 3: Bromelin (daily 100 mg/kg) was mixed with flour and, 3 days before killing, as well as 4 hours before it, 50 mg/kg was instilled into the stomach.

Experiment III (cytostatic enteropathy):

- group 5: 10 healthy control rats;
- group 6: 40 rats receiving 72 hours before killing a single sublethal dose of 40 mg/kg Methotrexate (4-amino-N^{10}-methyl pteroylglutamic acid, product No 4561, 4568 and 4567, Fa Lederle Lab. Div. London), administered by gastric tube.
- group 7: 40 rats treated with Methotrexate as in group 6 and with Bromelin as in group 2 (50 mg/kg intragastrically, without mixing it with the cytostatic).

All animals were killed by occipital luxation. The analogous situated jejunal segment was frozen in liquid nitrogen and sectioned in a Slee type cryostate, at –20°C. The juxtaposition on each cryostat holder of jejunal samples taken from an animal with enteropathy, from one treated with Bromelin and from a control, ensures identical conditions during all technical steps and certifies the objectivity of differences found.

The histochemical reactions were carried out on not fixed 8-10 micron sections, postfixed in neutral formalin, and mounted in glycerin gel. The methods were performed in the variants proposed by Arnold[3], Arvy[4], Chayen[9], and Lojda[21].

- histomorphology with haematoxylin-eosin staining (Roberts 1966);
- muccopolysaccharides by PAS reaction (McManus 1946, Hotchkiss 1948);
- nucleic acids with methyl green-pyronine (Brachet 1940, Kirnick 1949);
- leucine-aminopeptidase (EC = 3.4.1.1) (Nachlas 1957, McCabe-Chayen 1965);
- ATP-ase (EC = 3.6.1.4) (Wachstein-Meisel 1960);
- alkaline phosphatase (EC = 3.4.3.1) (Gömöri 1952, Pearse 1961);
- glucose-6-phosphatase (EC = 3.1.3.9) (Wachstein-Meisel 1956)
- alpha-naphthylacetate esterase (Nachlas-Seligman 1949, Gömöri 1952);
- oxidoreductases; SDH (EC = 1.3.99.1), GlDH (EC = 1.4.1.2), beta HBDH (EC = 1.1.1.30), IcDH (EC = 1.1.1.42), LDH (EC = 1.1.1.27), G_6PDH (EC = 1.1.1.49), NADH-tetrazolium reductase (EC = 1.6.4.3) (Sasse 1968, Chayen 1969).

Results

I. *Effect of Bromelin in healthy guinea pigs*

Intragastric Bromelin instillation for 3 days (group 2) amplifies the histo-enzymatic reactions beyond their normal intensity. This finding is valid for LAP in the external protein lamella and ATP-ase in the internal one of the enterocytic microvilli, alkaline phosphatase, which is active in the brush border too, G_6P-ase and nonspecific esterase in the endoplasmic reticulum, oxidoreductases in the cytosol and mitochondria. The individual variations do not reduce the biological significance of the results reproduced 4 times.

The enhancement of enterocyte enzymatic activity corresponds to the difference observed in the radioactivity remaining in the contents of the isolated and cannulated small intestine. The advantage is assured for the absorption of amino acids (^{3}H-leucine = −54,34 %), of fatty acids (^{14}C - stearat = −57,43 %) and less significant for the digestion of lipids (^{131}I - triolein = −23,4%).

II. *Effect of Bromelin in hetero-immune enteropathy*

After the administration of *rabbit serum directed against guinea pig jejunal mucosa* (group 3) histopathologic changes occur: a moderate enterocyte dystrophy (enlarged, settled cells with modified tinctoriality) and inflammatory cell infiltrate (with plasmocytes, lymphocytes and macrophages) in the lamina propria mucosae. The mesenchymal reaction is also present in the regional lymph nodes and in the spleen (hyperplasia of the Malpighi follicles and of the Billroth's cords).

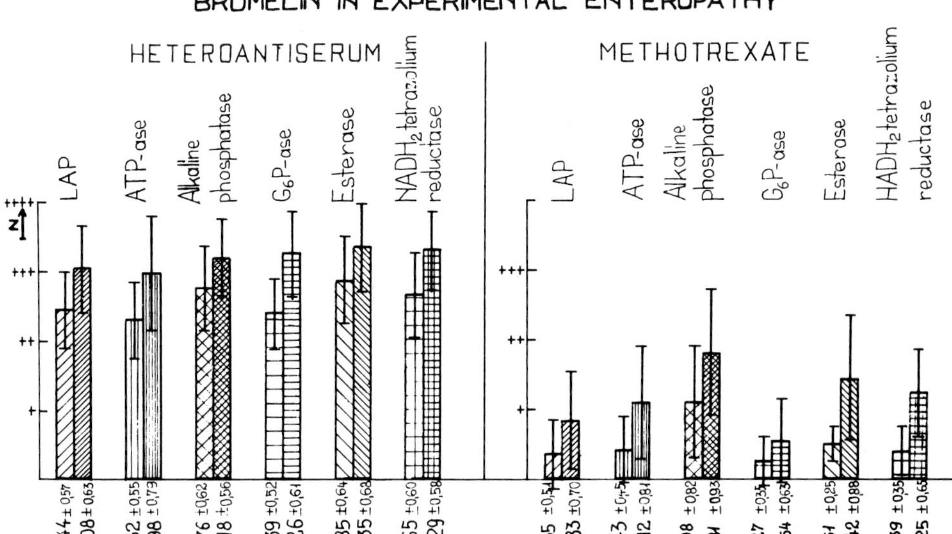

The PAS reaction revealed a marked mucosubstance border covering the villi, numerous large goblet cells and high tinctoriality in Liberkühn's crypts.

The MGP reaction indicated a decrease in the cytoplasmic RNA of the mature enterocytes but a stronger pyroninophilia in the regeneration and maturation zones.

The enterocytic enzymes had an evidently depressed activity as compared to the mild degree of dystrophy:

- LAP generally showed a pale reddish-violet reaction, much diminished in the apical third of the villi, presenting the normal blue colour only in small basal segments.
- ATP-ase and alkaline phosphatase offers a narrow villi circumscription, which was gradually blurred, often to its interruption towards the distal third or half of villi. The smooth muscle cells and the capillaries in the lamina propria had an unusual strong brown-black reaction.
- G_6P-ase presented an unhomogenously depressed activity realising a mosaiclike aspect, in which pale gold villi segments were intricated with others nearly normal brown.
- Nonspecific esterase offered similar results, with pale-pink cells and others marked brick-red.
- Oxidoreductases and NADH-tetrazolium reductase illustrated a reverse of the physiologic advantage of the mature enterocytes: the reaction was diminished and violaceous along the villi, but stronger blue than the normal in the Lieberkühn's crypts.

In the guinea pigs treated with *Bromelin for the duration of the immunization* (group 4) the enterocytic dystrophy was not detected by the light microscope, but the mesenchymal hyperplasia was evident.

The PAS (muccopolysaccharides) and MGP (nucleic acids) reaction becomes normal in the mature cells, but in the enterocyte regeneration zone the pyroninophilia was stronger, as after antiserum.

The enterocyte enzymes showed less depressed activities:

- LAP gave moderately diminished and violaceous reaction only towards the tip with nearly normal bluish colour at the basal half or third of the villi.
- ATP-ase and alkaline phosphatase circumscribed the villi continually, even if in some animals the contour was narrower and paler around the apex.
- G_6P-ase and esterases gave more uniform and strong reactions with a minimal apical decrease.
- Oxidoreductases and NADH-tetrazolium reductase presented blue reactions like the controls in the villi, but more marked in Lieberkühn's glands.

III. *Bromelin in cytostatic enteropathy*

Three days after the *single Methotrexate dose* (group 6) the rats had severe signs of the disease: anorexia, profuse diarrhoea, lethargy, loss of body weight. These symptoms and the alteration of the general state, were less marked in the case of Bromelin association (group 7).

The survival ratio was 18/32 after Methotrexate and 26/30 in the Bromelin treated animals (56,25% and 90,0%; assurance limits for 5% for 37,66-73,64 and 73,17-97,89%).

In histomorphology, profound villi distorsion and atrophy occurred, with denudation of the epithelial lining. Some necrobiotic or dystrophic enterocytes, with vacuolar cytoplasm and nuclear pycnosis may persist, in association with rare fragments of cystically degenerated crypts. The massive inflammatory cell infiltration in the lamina propria was a polymorphous one.

The defficit of histoenzymatic results corresponds to the great severity of the structural lesions.

- LAP reaction was positive, with pale pink colour, only in some cells dispersed in rare villi, but this not in all the rats.
- ATP-ase, alkaline phosphatase and G_6P-ase showed negative reactions, or punctiform remmants, or even small fragments of apical villi outline persisted.
- The esterase reaction was positive only in some reddish cells scattered through the distal half of the villi or whole negative.
- Oxidoreductase and NADH-tatrazolium reductase gave pale violaceous reactions in isolated cells, or a complete loss of enzymatic activity.

Associating Bromelin to cytostatic drug (group 7) the enterocyte necrosis and villi denudation were less extended. More dystrophic cells survived in the villi tips and some Lieberkühn's crypts were partially preserved.

– LAP showed reddish-violet reaction in more numerous, discontinuously distributed, enterocytes.
– ATP-ase, alkaline phosphatase and G_6P-ase revealed segments without enzymatic activity, but in most of the villi linear remnants or a narrow circumscription were present along the distal one or two third.
– Esterase offered pale brick-red reaction with unequal intensity in isolated cells scattered through the villi distal half.
– Oxidoreductases and NADH-tetrazolium reductase gave similar results, violaceous cells being intricated with other bluish ones.

Discussion

In the *control guinea pigs,* intensification of the enzymatic activities was observed after Bromelin instillation into the stomach. This original finding may be interpreted as an expression of the amplification of enterocytic metabolism, corresponding to the stimulation of digestion and resorption, this latter being stated in the clinic and experiment (Berndt[6, 7], Henrich[16, 17], Lang[18]).

It is particularly difficult to interpret the results obtained in enteropathy, namely the structural and enzymatic protection assured by Bromelin treatment.

In *hetero-immune enteropathy* Bromelin prevented enterocyte dystrophy and restricted the enzyme depression. Therefore, it conferred not only a metabolic-functional advantage, but also a resistance against structural lesions, in this mild experimental model.

Even in *cytostatic enteropathy,* Bromelin limited necrobiosis and denudation of the epithelial lining, preserving a more effective enzymatic activity in the remaining enterocytes.

The local anti-inflammatory action of Bromelin is an attractive interpretation. In the presence of severe mucosal lesions, there could also be a noxious action by proteolysis. But in the cytostatic enteropathy a greater number of enterocytes with a partially preserved structure and enzymatic activity persist. So, there is not only an intensification of the enzymatic in a few surviving dystrophic cells, as a consequence of the promotion of digestion and absorption by Bromelin.

As a possible interpretation, it could be expected that the morpho-enzymatic protection derives from a durable stimulation of jejunal mucosal function. The metabolic hyperactivity for the period of the noxious influence of heteroantiserum or of Methotrexate would modify the reactivity of the enterocytes, assuring an increased resistance.

This interpretation is very hypothetical, and the question concerning the mechanism of the protection against the hetero-antiserum and the cytostatic drug, conferred by Bromelin, still remains open. The results may, therefore, constitute a starting point for new researches and justify the therapeutical application of this protelytic ensyme in the maldigestion and malabsorption states.

Acknowledgements

The authors wish express their thanks to Fa E. Merck for his generous gift of histochemical reagents and to Miss Cornelia Brilinschi for her skilful technical assistance.

References

1. Achord J.L.: The effect of Methotrexate on rabbit intestinal mucosal enzymes and flora. Amer. J. Dig. Dis., *14,* 315 (1969).
2. Antonioli J.A., Robinson J.W.L., Fasel J., Vannotti A.: Effet du méthotrexate in vivo sur l'intestin grêle du rat. Gastroenterologia (Basel), *106*, 4, 217-224 (1966).
3. Arnold M.: Histochemie. Springer Verlag, Berlin-Heidelberg-New York, (1968).
4. Arvy L.: Les activités enzymatiques intestinales. In Van Fleet D.S., Montagna W., Ellis R., Wachstein M., Arvy L., Eichner D., Enzyme. Band VII, Teil II., Gustav Fischer Verlag, Stuttgart, pp. 247-260 (1962).
5. Barbarino F., Dumitrascu D., Goia A.: Jejunal and splenic enzymatic changes after hetero-antiserum, cyclophosphamide and asparagine. 9-th Int. Congr. of Chemotherapy, London, 15-18 VII. 1975, in press in
6. Berndt W., Hoffmann U., Müller-Wieland K.: Über die Eigenschaften des Bromelins, einer pflanzlichen Protease aus Ananas comosus. Z. Gastroenterologie, *6,* 3, 185-195 (1968).
7. Berndt W., Pries K., Müller-Wieland K.: Die Wirkung von Bromelin auf die Stuhlfettausscheidung. Z. Gastroenterologie, *9*, 9, 651-658 (1971).
8. Bernier J.J., Bognel J.C., Bognel C., Rainbaud J.C., Etude expérimentale sur la toxicité digestive du méthotrexate. Rev. franç. Et. clin. biol., *12,* 576-580 (1967).
9. Chayen J., Bitensky L., Butcher R., Poulter L., A guide to practical histochemistry. Ed. Oliver & Boyd, Edinburgh, 1969.
10. Dumitrascu D., Fodor O., Caluser I., Iencica R., La participation du facteur immun dans la pathogénèse des entéropathies chroniques. Modern Gastroenterology. Proceedings of VIII-th Int. Congr. of Gastroenterology. Praha, 7-13 VII. 1968, Schattauer Verlag, Stuttgart, 1969, pp. 1073-1074.
11. Eder M., Experimentelle Untersuchungen über Schädigungen der Darmschleimhaut. Verh. d. Dtsch. Ges. f. Path., *49,* 330-333 (1965).
12. Eder M., Rostock H., Vogel G.: Wirkung von Folsäure-antagonisten (Methotrexate) auf die Regeneration der Darmschleimhaut. Virch. Arch. path. Anat., *341,* 164-176 (1966).
13. Fodor O., Dumitrascu D., Caluser I., Iencica R., Parau N.: Recherches expérimentales sur le phénomène auto-immunitaire dans les entéropathies. L'entéropathie par

immuno-sérum hétérologue. Acta gastroent. belg., *30,* 615-630 (1967).
14. Goia A.: Studiul tratamentului imunodepresor in boli digestive, hematologice side colagen. Teza de doctorat, Cluj, (1973).
15. Hartwich G.: Side effects of a cytostatic treatment on the gastro-intestinal tract. Acta Hepato-Gastroenterol., *21,* 89-92 (1974).
16. Hennrich N., Hoffmann A., Lang H.: Eignung der Pflanzenprotease Bromelin für die Substitutionstherapie von Verdauungsstörungen. Arzneim. Forsch. (Drug Res.), *15,* 434-437 (1965).
17. Hennrich N., Klockow M., Lang H., Berndt W.,: Isolation and properties of Bromelin protease. FEBS Letters, *2,* 5, 278-280 (1969).
18. Lang H., Hennrich N.: Komninierte Wirkung von Bromelin und Pankreasenzymen in der Substitutionstherapie. Gastroenterologia (Basel), *107,* 203-208 (1967).
19. Lang H., Breddin K., Rick W., Zur Resorption oral verabreichten Bromelins. Klin. Wschr., *47,* 2, 106-107 (1969).
20. Lang H., Helger R., Lucker P., Breddin K., Hausamen T.U., Rick W.: Zur systemischen Wirkung oral verabreichten Bromelins. Arzneim.-Forsch. (Drug Res.), *19,* 939-944 (1969).
21. Lojda Z., Fric P., Jodl J., Chmelic V.: Cytochemistry of the human jejunal mucosa in the norm and in malabsorption syndrome. Current Topics in Pathology, Springer Verlag, Berlin, 1970, vol. 52, pp. 1-63.
22. Millington P.F., Finean J.B., Forbes O.C., Frazer A.C.: Studies of the effect of aminopterin on the small intestine of rats. I: The morphological changes following a single dose of aminopterin. Expl. Cell. Res., *28,* 162-178 (1962).
23. Pearse A.G.E., Riecken E.O.: Histology and cytochemistry of the cells of the small intestine, in relation to absorption. Brit. med. Bull., *23,* 217-222 (1967).
24. Riecken E.O., Martini G.A.: Die Klassifizierung pathologischer Dünndarmschleimhautbilder. Dtsch. med. Wschr., *98,*, 998-1100 (1973).
25. Robinson J.W.L.: Experimental intestinal malabsorption states and their relation to clinical syndromes. Klin. Wschr. *50,* 173-185 (1972).
26. Robinson J.W.L., Vannotti A.: The effect of oral methotrexate on the rat intestine. Biochem. Pharmacol., *15,* 1479-1489 (1966).
27. Schiraldi O., Marano R.: Enteropatia acuta in ratti trattati con antimetaboliti (aminopterina). Rass. Fisiopat. Clin. Ter., *33,* 920-934 (961).
28. Trier J.S.: Morphologic alterations induced by methotrexate in the mucosa of human proximal intestine. I: Serial observations by light microscopy. Gastroenterology (Baltimore), *42,* 3, 295-305 (1962).
29. Vitale J.J., Zamcheck N., Digiorgio I., Hagsted D.M.: Effect of aminopterin on the respiration and morphology of the gastrointestinal mucosa of rats. J. Lab. clin. Med., *43,* 583-601 (1954).

7th International Symposium on Clinical Enzymology, Venezia 1976

PART 2

CLINICAL AND METHODOLOGICAL ASPECTS

GLYCERALDEHYDE-3-PHOSPHATE DEHYDROGENASE (TOTAL AND ISOENZYME ACTIVITY) IN THE EARLY DIAGNOSIS OF MYOCARDIAL INFARCTION

J. Griffiths and S. Shaw

Summary

With advances in cardiac surgery, new demands have been placed on clinical laboratories to determine precisely both the earliest occurrence and extent of a myocardial infarction.

Presently, enzyme panels utilizing creatine kinase and lactate dehydrogenase are useful indicators of infarction. This study has examined a further enzyme, Glyceraldehyde-3-Phosphate dehydrogenase (GAPDH) by comparison with Creatine Kinase (CK) in the early diagnosis of such infarctions.

Results indicate that total GAPDH appears in the serum before total CK; however, the lack of cardiospecificity relating to GAPDH Isoenzyme Fraction 2 in comparison to the CK BM band is a major disadvantage, as is a relatively poor in vitro stability. It is felt currently that the enzyme does not add significantly to an earlier diagnosis of myocardial infarction.

The capability of cardiac surgery to move with relative impunity in the heart and hence initiate surgical treatment in myocardial infarction (MI) has directed two questions to the clinical pathologist:

1. What is the earliest time after infarction that positive evidence of cellular damage is evident?
2. What data can be obtained from the laboratory to indicate the size of the infarction?

Some evidence has been provided by the use of a limited number of enzymes, with emphasis now placed on creatine kinase (CK) for early diagnosis and lactate dehydrogenase (LDH) for later confirmation. Total activity of enzymes does not give such precise information as specific isoenzyme estimation; however, the use of cardiospecific CK-BM isoenzyme as well as reversal of LDH_1: LDH_2 ratio have added considerable diagnostic accuracy.

In the past year, several other enzymes such as arginase[1], glycogen phosphorylase[2], guanase[3], and elastase[4] have been examined with the intent

Laboratory Service-Veterans Administration Hospital - San Diego; Department of Pathology, University of California - San Diego.

of offering more specific answers to the clinicians' questions. They do not appear to offer significant advantages over the established cardiodiagnostic enzymes.

In the past decade, two studies[5,6] have suggested that Glyceraldehyde-3-Phosphate Dehydrogenase (E.C. 1.2.1.12:GAPDH), total activity may be an early indicator of ischaemic myocytolysis. These authors did not, however, attempt isoenzyme fractionation and hence the present study was designed to:

a. Compare GAPDH with CK as an early indicator of MI.
b. Develop a method for GADPH isoenzymes in human serum, establish normal ranges and further seek a cardiospecific isoenzyme.

Materials & Methods

A. *Establishing Normal Activity:*

The method of Duggleby and Dennis[7] for tissues was modified to estimate serum activity, specifically adjusting the conditions of assay for automation.

1. *Apparatus:*

Beckman TR Automated Enzyme Analyzer (Beckman Instruments Inc., Fullerton, California), which reads changes in absorbance at 340 nm in the "sp" mode.

2. *Reagents:*

Buffer – 0.1 mol/1 Sodium Pyrophosphate adjusted to pH 8.6 with 1 mol/l NaOH; NAD-β-nicotinamide adenine dinucleotide (Sigma-N-7004) 1 x 10^{-2} mol/l (Sigma Chemical Co., P.O. Box 14508, St. Louis, Missouri, 63178); ADP-adenosine-5'-diphosphate (Sigma-A-2754) 5 x 10^{-4} mol/l; EDTA – ethylene dinitrilo tetra-acetic acid (disodium salt) – 2 x 10^{-4} mol/l; Dithiothreitol – (Sigma D-0632) 1 x 10^{-3} mol/l.

PGK-3-phosphoglyceric phosphokinase (Sigma P-7634) Concentration 10^{-2} m U/μl, the original Sigma preparation contain 10 m U/ml, hence, the solution used for actual assay was diluted with 0.1 ml/l phosphate buffer at pH 8.0 accordingly; Magnesium Chloride – 2 x 10^{-4} mol/l; Cysteine – DL – cysteine HCL (Sigma 8256) 3.3 x 10^{-3} mol/l.

GAP – DL – Glyceraldehyde-3-phosphoric acid (diethylacetal, barium salt), (Sigma G-5376). This insoluble solid racemic compound is prepared essentially as described in the information which accompanies the reagent, with the following alterations:

a. 8 ml distilled water (instead of 6 ml) is placed above the 1.5 g of Dowex-50W Hydrogen Form Resin for the first wash. 2.0 ml is used in the second wash. The supernatants of these two washes are combined to yield 10.0 ml solution;

b. 200 mg of the GAP diethylacetal barium salt are added to the first 8 ml water instead of 100 mg.

The combined 10 ml of supernatant solution concentration is 3.75 x 10^{-3} mol/l in the enzymatically active D-isomer. For each single assay, 105 μl was used, yielding a reaction mixture of 7.50 x 10^{-4} mol/l.

Principle of Assay: NAD acts as the hydrogen-acceptor, the concentration of which is monitored in the course of the assay. End product inhibition is relieved by the inclusion of ADP, $MgCl_2$ and PGK.

Dithiothreitol is included to ensure retention of Sulfhydryl groups. Cysteine acts as an enzyme activator, although some authors do not consider this addition as necessary with GAP as the substrate[8]. EDTA is a chelating agent to remove traces of heavy metals, which reportedly inhibit the reaction[8]. The reagents listed were added in the stated order with final addition of GAP.

3. *Procedure:*

Each assay was completed in duplicate. The TR takes 35 μl of sample and 390 μl of reagent. A primer volume was necessary to begin the TR and the following table was useful in calculating the final assay solution:

BUFFER:	390 μl	x	# of assays	+ 3.51 ml
NAD:	3.45 x 10^{-3} g	x	of assays	+ 31 x 10^{-2} g
ADP:	9.42 x 10^{-5} g	x	of assays	+ 8.5 x10^{-4} g
EDTA:	3.76 x 10^{-5} g	x	of assays	+ 3.9 x 10^{-4} g
DITHIO:	8.1 x 10^{-5} g	x	of assays	+ 7.3 x 10^{-4} g
PGK:	30 μl	x	of assays	+ 270 μl
$MgCl_2$:	2.135 x 10^{-5} g	x	of assays	+ 1.9 x 10^{-4} g
CYST:	2.73 x 10^{-4} g	x	of assays	+ 2.45 x 10^{-3} g
GAP:	105 μl	x	of assays	+ 945 μl
SERUM:	35 μl			

Sufficient quantities of reagent are prepared to include amounts expended during the instrument's automatic prime cycle. The TR is operated according to normal operating instructions, with the cuvette thermostated at 37°C.

4. *Stability of Enzyme:*

There is no information in the literature regarding the stability of serum GAPDH, hence, samples of freshly drawn serum were spun in a refrigerated centrifuge at 4°C, then assayed for activity at intervals of 1,2,3 and 4 hours with sera kept at room temperature and at 4°C. Further sera were assayed within one hour, then stored at −20°C for periods up to 3 days, with assays performed at daily intervals.

Shonk and Boxer[8] examining tissue GAPDH, refer to its instability in that vehicle and suggest immediate assay.

B. *Establishing Normal Isoenzyme Activity:*

1. ***Apparatus:***

a. Beckman Model R101 Microzone Electrophoresis Cell
b. Beckman Duostat Power Supply
c. Beckman CDS-100 Computing Densitometer
d. Thelco Model 2B Incubation Oven.

2. ***Reagents:***

a. Agarose Gels – Beckman Rehydratable
b. Chemicals – as described for the total activity with addition of phenazine methosulfate (PMS) (Aldrich P-1340-1) and nitro blue tetrazolium (N.B. T.) (Aldrich N1540-5) (Aldrich Chemical Co., Inc., 940 West Saint Paul Ave., Milwaukee, Wiconsin, 53233).

3. ***Procedure:***

The agarose gels are rehydrates in distilled water for 45 minutes; then allowed to equilibrate with the barbiturate buffer at 4°C for 15 minutes. The gels are blotted dry with Whatman 42 filter paper, taking care each well is cleaned of buffer. A 10-microliter sample of serum is placed in each well, the gel placed in the electrophoretic cell containing pre-cooled buffer at 4°C, and electrophoresis performed at a constant voltage of 150V for 45 minutes at 4°C.

Incubation and staining are accomplished by a paper overlay technique. Substrate aliquots were prepared as follows: To 265 ml of 0.1 mol/l sodium pyrophosphate buffer pH 8.6 is added 2.6004×10^{-3} gm cysteine hydrochloride, 0.082 gm dithiothreitol, 3.72×10^{-1} gm EDTA and 2×10^{-4} gm magnesium chloride. One hundred aliquots were prepared and stored at –20°C. For use, after thawing, 0.002 gm NAD, 0.001 gm ADP, 0.25 ml of PDK solution and 0.005 gm NBT were added to a single aliquot; a spatula tip of PMS was added in the dark to prevent light initiated oxidation.

The overlay paper was saturated with the mixture, applied to the gel and finally 0.5 ml of 3.75×10^{-3} M GAP solution is added to the overlay. The gels were incubated for 15 minutes at 37°C, cleared in an 8% acetic acid solution for 10 minutes, and then scanned on the densitometer. Peak areas were calculated by the computerized attachment.

NORMAL
19 U (60%)
4 U (13%)
9 U (27%)
GAPDH
1
2
3
Albumin
α 1
α 2
β
γ

Fig. 1

MYOCARDIAL INFARCTION
26 U/L (32%)
41 U/L (52%)
13 U/L (16%)
GAPDH
1
2
3
Albumin
α 1
α 2
β
γ

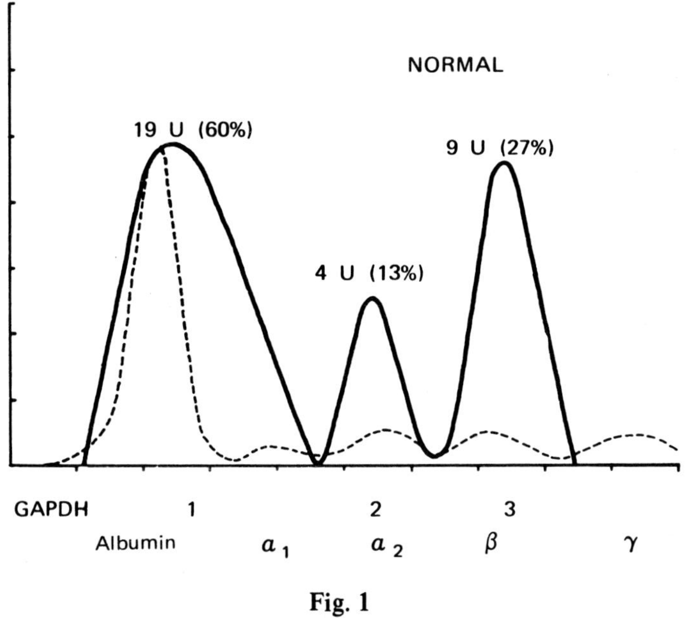

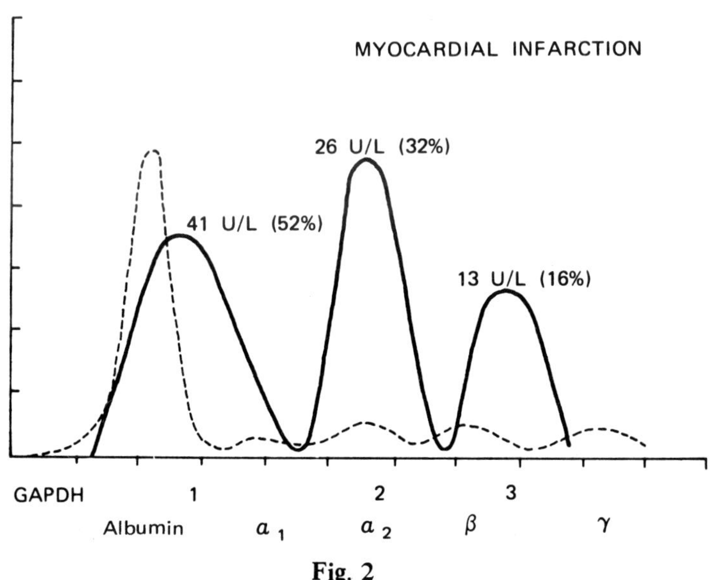

Fig. 2

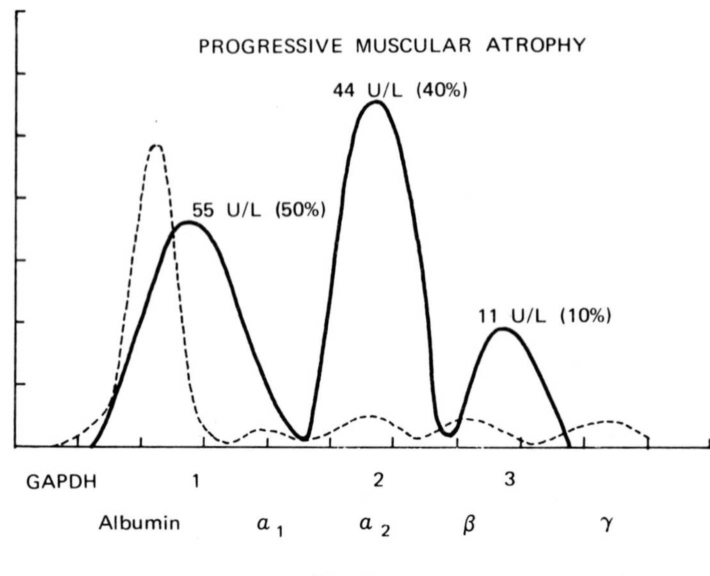

Fig. 3

4. *Tissue Activity:*

Selected tissue removed at surgical biopsies and cardiac muscle removed at an autopsy performed within two hours of death were examined together with erythrocytes. Approximately 1 gram of tissue was homogenized in 10 ml of 0.1 mol/1 tris buffer pH 7.4 in a Waring blender for one minute. The homogenate was spun at 500G for 10 minutes in a refrigerated centrifuge at 4°C, then the supernatant was spun for an additional 10 minutes at 10,000G at 4°C. The initial step removes the bulk of the cells; the second step allows a clear supernatant for enzyme assay. The red cells were lysed with ice-cold distilled water.

Initial levels of activity were determined, then diluted to approximately 40 U/liter with tris buffer for electrophoresis.

C. *Patient Selection*

Samples were taken from 26 patients who had clear clinical symptoms suggesting myocardial infarction which commenced approximately 45-60 minutes before blood was drawn. After initial venipuncture, serial hourly samples were taken for enzyme assay. No attempt was made to relate the enzyme assays to size, location of the infarct, concomitant arrythmias or the administration of drugs.

Results

A. *Normal Activity:*

This was established by examining some 200 consecutive patient samples. Normal activity for males ages 20-70 was 23-41 U/liter.

B. *Serum Isoenzyme Activity:*

Established by subjecting 50 fresh samples taken from the normal range.

Figure 1 illustrates a typical tracing from a sample of 32 U/liter. Three isoenzyme bands were found to be consistently present and were assigned in order of decreasing electrophoretic mobility, the prefix $GAPDH_1$, $GAPDH_2$ and $GAPDH_3$.

Mean values for each band were:

$GAPDH_1$	55-70 %	13-23 U/liter
$GAPDH_2$	3-21 %	4-9 U/liter
$GAPDH_3$	12-27 %	6-9 U/liter

From electrophoresis of the tissue homogenates, it is suggested that $GAPDH_1$ is derived from the erythrocyte, $GAPDH_2$ from both skeletal and cardiac muscle, and $GAPDH_3$ from the liver.

Figure 2 illustrates an electrophoretic tracing from a confirmed case of MI 5 hours after infarction, showing elevation of $GAPDH_2$ only.

Figure 3 shows an electrophoretic tracing of a patient with a progressive muscular atrophy in which other enzymes indicating skeletal muscle damage, i.e., CK, MM band and aldolase were considerably elevated. This also shows elevation of $GAPDH_2$ and is similar to Figure 3.

The hourly time course over a 6-hour period shows total activity of GAPDH and a similar time course for CK is shown in Figures 4 & 5. A comparative percentage elevation above normal between the two enzymes is shown in Table 1, indicating that total GAPDH is elevated slightly earlier than CK.

Table I. *Percentage elevation above normal of GAPDH activity in serum compared with CK**

Time 1st Examination	*Percent Above Upper Limit of Normal*	
	Elevated GAPDH	*Elevated CK*
0 hours	None	None
1	17	5
2	54	10
3	90	50
4	90	90
5	100	90
6	100	100

*Mean values for 26 patients with M.I.

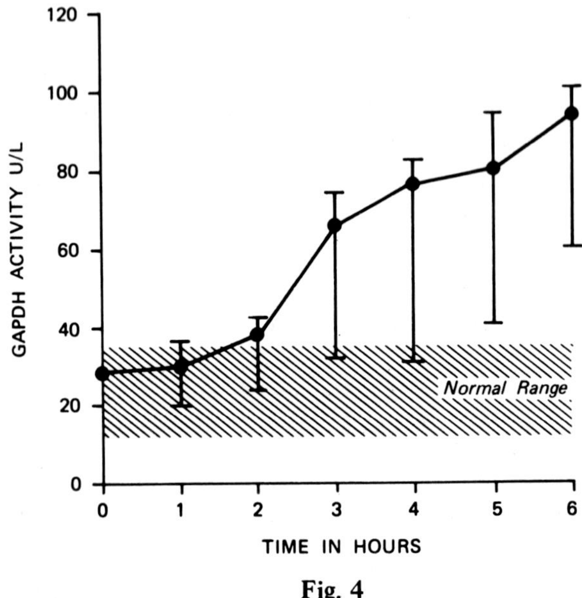

Fig. 4

Table II. *Stability of GAPDH activity in serum (Percentage change)*

	Time Following Initial Analysis			
	1h	*2h*	*3h*	*4h*
Room Temperature	N/C	+ 20	– 30	– 50
At 4°C	N/C	+ 30	– 30	– 50
	1d	*2d*	*3d*	
At –20°C	– 30	–75	– 75	

Details of the stability study of GAPDH (Table 2) indicate that at both room temperature and at 4°C, an increase in GAPDH activity 2 hours post venipuncture is evidenced, followed by an appreciable loss in total activity.

At –20°C storage, there was considerable loss of activity after 24 hours storage.

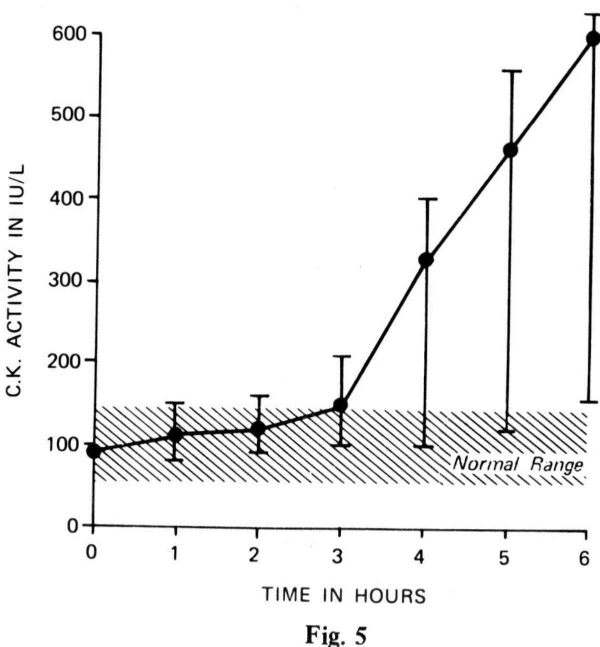

Fig. 5

Discussion

GAPDH, an enzyme in the Embden-Myerhof pathway, has a wide tissue distribution, rivaling that of lactate dehydrogenase in ubiquity[9]. In experimental animals, skeletal muscle has by far the highest tissue concentrations, showing a threefold greater activity than cardiac muscle, its closest rival. Brain, kidney and liver have roughly similar activities, however, less than muscle tissue[8]. Erythrocytes were not examined in that study, but other authors[10, 11] have noted very high concentrations, both in matrix and membrane of the red cell.

Our study confirms the report of Karliner et al.[5], who noted early release of total GAPDH in M.I. In 16 of 63 patients, GAPDH appeared in the serum before total CK, with GAPDH activity peaking in the first twelve hours. This observation was also substantiated by von Pall[6] who regarded GAPDH as one of the most sensitive parameters to measure myocardial hypoxaemia.

In the patients examined in the present study, based on total activity at three hours, 90% had shown elevation of GAPDH, whereas only 50% had elevation of total CK. This information suggests that GAPDH may offer a real advantage in the early diagnosis of M.I. and may replace CK, or at least be a useful addition to the cardiac enzyme panel.

There appear, however, definite objections to its routine use. The relative instability of the enzyme under normal conditions of storage indicate its assay must be performed within one hour. At this time, efforts to either increase stability or reactivate the enzyme were not made.

A further major defect relates to the failure by the present agarose electrophoretic technique to separate the multiple molecular forms of skeletal and cardiac muscle GAPDH. Compared with the cardiospecific nature of the CK BM band, this must present a real diagnostic disadvantage. Concurrent studies suggest that the CK BM band may appear sooner than is evident, if *isoenzyme fractionation* is examined from the first sample.

This study also indicates a specific hepatic band, $GAPDH_3$. Elevation of total GAPDH in patients with congestive cardiac failure and hepatic failure and in patients with primary hepatic disease are reported by Karliner et al.[5]. Preliminary data suggests that GAPDH may be a sensitive and early indicator of hepato-cellular damage.

The present study concludes that despite indication of early GAPDH release from hypoxic myocardium, there are major objections to adding this enzyme to the laboratory panel relating to the early diagnosis of M.I.

References

1. Porembska Z., and Kedra M.: Early Diagnosis of Myocardial Infarction by Arginase Activity Determination, Clin. Chim. Acta, *60*:355 (1975).
2. Krause E. et al.: The Assay of Glycogen Phosphorylase in Human Blood and its Application to the Diagnosis of Myocardial Infarction, Clin. Chim. Acta, *58*:145 (1975).
3. Ellis G., and Goldberg D.: Serum Guanase Activities after Myocardial Infarction, Clin. Chim. Acta, *63:*205 (1975).
4. Mikunis R., Servoka V. and Prikhod'ko V.: Activity of Blood Serum Elastase in Patients with Acute Myocardial Infarct in the Pre-hospital Period, Kardiologiia *15*(1): 132 (1975).
5. Karliner J., Gander M. and Sobel B.: Elevated Serum Glyceraldehyde Phosphate Dehydrogenase Activity Following Acute Myocardial Infarction, Chest *60*(4):318 (1971).
6. Pall, V., et al.: Neuroleptanalgesie beim frischen Herzinfarkt, Wiener klinische wochenschrift, 85, Dec. 7 (1975).
7. Duggelby R., and Dennis D.: Nicotinamide Adenine Dinucleotide-Specific Glyceraldehyde-3-Phosphate Dehydrogenase from Pisum Sativum. Journal of Biological Chemistry *249*(1):167 (1974).
8. Shonk C. and Boxer G.: Enzyme Patterns in Human Tissues I. Methods for the Determinations of Glycolytic Enzymes, Cancer Research *24*:709 (1964).
9. Schmidt E. et al.: Measurement of Enzyme Activity, In Methods of Enzymatic Analysis, Section C. H.U. Bergmeyer, Ed. Academic Press, New York, N.Y., p. 658 (1965).

10. Anderson J., and Giblett E.: Intraspecific Red Cell Variation in Pigtailed Macaque (Macaca Nemestrina) Biochemical Genetics, *13* (3/4):189 (1975).
11. Letko G. and Bohnensack R.: Investigations on the Release of Membrane Bound Glyceraldehyde-3-Phosphate Dehydrogenase, Febs Letters *39*(4):313 (1974).

EVALUATION OF A NEW TURBIDIMETER FOR SERUM AMYLASE MEASUREMENTS USING A TURBIDITY REFERENCE MATERIAL AND A STABLE PECTIN SUBSTRATE

B.E. Copeland, A. Doherty, J. Jordan and R. Parkey

Summary

The Perkin-Elmer #91 Turbidimeter with a stable turbidity reference material and a stable pectin substrate was compared with an amyloclastic method using stabilized starch substrate.

The normal human ranges were similar-47to 208 and 70 to 204.

The relative standard deviations were similar for control sera having concentrations at the upper limit of normal RSD 10% and 12% and at an elevated range 10% and 7%. The turbidity reference material offers for the first time a reproducible and stable reference point upon which serum amylase measurements may be based over an extended period of years. Both instruments variability and variability of the pectin substrate materials are eliminated.

Introduction

In my experience, the human serum amylase is of critical importance in a single disease state – acute hemorrhagic pancreatitis. In all cases of acute abdominal pain, acute hemorrhagic pancreatitis is one of the diseases which must be considered in the differential diagnosis. A serum amylase value over 500 Somogyi units will influence a surgical decision to explore or not to explore the acute abdomen. Somogyi units are the reference point for amylase measurements in the USA. Generally speaking, the Somogyi saccharogenic method is meant. Serum amylase measurements are characteristically emergency measurements. There has always been the need for an amylase measurement with an immediate turn around time within less than 10 minutes including control measurements.

Many modifications of amylase measurements have been developed. In general, the purpose has been to reduce the turn around time in the emergency situation.

In our experience, the turn around time for the saccharogenic method has been from 1-3 hours; and for the amyloclastic method, 30 minutes to one

---

New England Deaconess Hospital, William Meissner Laboratory Center Pilgrim Road, Boston, M.A., U.S.A.

hour and a half. The turbidometric method with two controls can be performed in ten minutes. Because we have several surgeons from the Lahey Clinic who specialize in pancreatic surgery and because of the short turn around time, we have evaluated the PE 91 instrument[1]. This is especially critical during the fifteen emergency evening and night hours from 5:00 p.m. to 8:00 a.m.

Method and instrument

The instrument was developed from the method of Seligson et al[2] of Yale University Medical School. This method uses a purified solution of amylopectin.

There is a cleavage action of alpha amylase on the central area of the large amylopectin molecule. As the length of the amylopectin chain is reduced by enzymatic cleavage, the turbidity of the amylopectin solution is reduced.

The rate of change in turbidity is observed in the new instrument. An automatically timed observation of the rate of change of turbidity in one minute of the reaction is converted into units which can be adjusted to approximate the usual amylase number system.

The instrument is standardized with a stable turbidity standard so that the output of incident light is reproducible.

Using an expanded nephelos-turbidity scale, the rate of turbidity change per minute is observed and is converted to a numerical value approximately equal to the saccharogenic Somogyi method[3]. This value is used to calibrate an automatically timed measurement circuit which converts decrease in turbidity into a conventional amylase number system.

This instrument consists of[1] a nephelometer which includes a dry bath for temperature control of reaction system and two separate electronic measuring circuits. The first circuit measures absolute turbidity units. The second circuit measures rate of change. The instrument uses a stable turbidity standard which is used to standardize the amount of incident light. Using the absolute nephelos-turbidity scale, the change in turbidity in one minute is measured with a precise stopwatch. This is an absolute measurement based only on the turbidity standard. This absolute turbidity change is used to calibrate the automatic rate of change scale. The second scale can be changed to fit any amylase unit scale. Seligson converted his measurements to approximate Somogyi saccharogenic units and we have followed this practice. The absolute nephelometric standard units could be used with equal justification.

The automatic measuring circuit measures the time for preincubation of two minutes and then automatically reads out the rate of reaction in one minute in the units of calibration chosen.

The sample size is twenty microliters.

Amylase units

It is appropriate to use the phrase "usual amylase number system" since in scientific areas "a convention" is something which everyone agrees to accept in order to simplify communication. Of the many different available reactions for demonstrating amylase activity, there is not as yet a theoretical basis for converting one activity unit to another activity unit except by a pragmatic experimental relationship.

Because the saccharogenic method of Somogyi (using glucose reduction as a reference point) was the first widely used amylase method in the United States, it has become a convention to express all new methods in comparison to this reference method. The reason for doing this is to make the new amylase methods more acceptable to practicing physicians who experience difficulty when unit systems change with respect to tests upon which they base critical decisions.

In the United States the usual upper limit of normal for the healthy human population is 200 units.

The usual practice in presenting a new method is to compare it with either the original saccharogenic method or with an amyloclastic method which has units related to the saccharogenic method.

Secondary standardization

Another common occurrence is the use of a lyophilized human serum sample as a secondary standard for amylase methods. Many of the kit methods contain a secondary standard of this type which bears a labelled amylase value. This value may or may not be related to a saccharogenic amylase procedure. There is considerable confusion in this regard. The recent proliferation of kit amylase methods has produced an even greater confusion of substrates and reference secondary standards.

A further complication is that the copper reduction glucose methods which were regularly used in all laboratories (1920-1960) have largely been replaced and the primary reference method, the saccharogenic amylase method, is becoming less and less available. Because the reference method is becoming less and less available, and because it is not an ideal method at best, we may consider other possible reference methods. We would like to suggest that the turbidity standard using a stable amylopectin substrate could become a stable reference point which would extend twenty to thirty years into the future.

The turbidity standard is much easier to maintain than the saccharogènic method. At present the National Bureau of Standards of the United States is considering the development of such a turbidity standard as a Standard Reference Material.

To have a completely controlled standard system, it would also be necessary to control the reproducibility of the substrate.

Difficulties with the starch substrate

During my experience over the past twenty-five years, the reproducibility of starch substrate mixtures from batch to batch has been a major problem. One starch batch will work very well and the next one will be terrible. For an essential emergency method, this is a grave defect.

Starch is a commercial product which comes from corn, potatoes, wheat and other plant sources. Some methods have used household products. Other methods have used purified products with limited success, as well, in reproducibility and stability. This has resulted in frequent problems of interlaboratory systematic bias. Starch deteriorates in solution usually due to bacterial action which is another source of variability.

Starch has a variable solubility.

Different commercial batches of starch behave differently. It is hoped that amylopectin will be a reproducible substance whose batch to batch stability will be consistently maintaned by the manufacturer. This is not an easy problem. It is essential that the manufacturer use daily quality control in his control laboratory in the same way that we use it in the medical laboratory.

Observations

Normal Values. Control Values. Patient Values – Normal Range. Patient Values – High Range.

Table 1 shows normal human values for the saccharogenic, the amyloclastic, and the turbidometric methods.

The range for the amyloclastic and the turbidometric methods are similar. Note that the saccharogenic method has a slightly lower range. Statistically, it is significantly different from the amyloclastic and the turbidometric.

From the medical point of view, this is not a significant problem.

Table I. *Normal Human Range*

	Range Male & Female	*n*	*Ave*	*SD*
Amyloclastic[4, 5]	32-206	20	98	44
Turbidometric[2] (this method)	20-188	20	104	42
Saccharogenic[3]	34-174	20	92	35

Table 2 shows a comparison of these three methods using a control serum in the normal range and another control serum in the borderline medically diagnostic range.

Table II. *Comparison of Quality Control Pool Samples*

	*Pool I**		*Pool II**		*AV*	
	SD	*RSD*	*SD*	*RSD*	*Pool I*	*Pool II*
Amyloclastic[4,5]	17	8	38	8	200	479
Turbidometric[2]	22	10	48	10	219	490
Saccharogenic[3]	9	6	24	5	154	460

*Massachusetts Society of Pathologists Regional Quality Control Program.

Between the average values there are not any differences of clinical significance. The precision for each method is similar in the normal and abnormal range. None of these methods has the precision shown in the report of Professor Guarino and associates[6] using the amido blue method with an RSD of ± 2%. However, for the intended medical use, the precision is acceptable. For the saccharogenic method, the Relative Standard Deviation (RSD) or Coefficient of Variation (CV) is ± 5%. For the turbidity and amyloclastic methods, the RSD was from ± 8-10%. Our data are regular operating values from day to day.

High amylase study

A study was conducted using both patient samples (9 values) and diluted pancreatic juices (10 values) with amylase values in the 200-2600 range.

The diluted pancreatic juices were prepared by mixing a piece of pancreatic tissue (from a patient with a negative test for Australian antigen) and normal saline. Small drops of this solution were added to different tubes containing 2 ml. of normal saline. Amylases were run on these solutions by both the turbidometric and amyloclastic methods.

Table 3 shows the results of this study using the turbidometric and the amyloclastic methods.

In the first group, from 200-800 units, there was an average difference of 153 units with the turbidity method giving higher vales. This is a statistically significant difference, and at the decision level of 500 units, could be a problem.

In the range from 800-2600 units, the average difference is 700 units. Again, the difference is statistically significant. However, at these levels, there is less medical significance.

Table III. *Amyloclastic Versus Turbidometric*

	Amylo. Mean	*Turbido. Mean*	*Observ. Av. Dif.*	*S.D. of Av. Dif.*	*Stat. Signif.*
Group I (200-800 Range)	218	371	153	33	Yes
Group II (Over 800 Range)	707	1417	710	118	Yes

It is interesting to note that Guarino et al[6] suggested that the amido bleu method gave higher values than the amyloclastic method in certain circumstances..

Conclusions

1. At present, there are many problems in the standardization of methods for alpha amylase: the units, the substrate preparations, the secondary standards, and the change in reference methods.
2. The turbidity method using the PE 91 is a method which has a turn around analysis time of ten minutes. It is comparable for clinical purposes to current reference methods.
3. In the search for a new reference point for amylase measurements, the turbidity standard and the amylopectin method should be given consideration.

References

1. Operating Directions, Perkin Elmer Model 91 Amylase-Lipase Analyzer, 91-900, D-604D, January 1975. Coleman Instruments Division, Oakbrook, Illinois.
2. Zinterhofer L., Wardlaw S., Jatlow P. and Seligson D.: Nephelometric Determination of Pnacreatic Enzymes 1. Amylase Clin. Chim. Acta, *43,* 5-12, 1973.
3. Somogyi M.: Micromethods for the Estimation of Diastase. J. Biol. Chem., *125,* 399-414 (1938).
4. Biomedix Kit Method. Biomedix Incorporated, Alexander Road, Princeton, N.J.
5. Van Loon E.J. et al.: Photometric Method for Blood Amylase by Use of Starch-Iodine Color. Am. J. Clin. Pathology, *22*: 1134-1136 (1952).
6. Guarino A. et al.: Evaluation of a New Method for the Determination of Alpha-Amylase in Biological Fluids. Lab *2,* 239-249, (1975).

EXPERIMENTAL STUDIES ON LIPASE. THE INHIBITION BY SODIUM CHLORIDE

L. Galzigna and A. Burlina

Summary

The different behaviour of lipase activity and lipoprotein lipase activity in the presence of high concentration of NaCl has been studied. The lipolytic activity has been monitored by a de-emulsification procedure with a turbidimetric assay and by determination of the free fatty acids, monolein and diolein released with thin layer chromatography.

The ability of albumin added to the incubation mixture to switch on a lipoproteinlipase-like activity has been observed with the turbidimetric procedure.

The influence of heparin has been studied under different conditions and the response of lipase to changing environmental factors investigated.

Introduction

Lipases (Triacylglycerol hydrolase, EC 3.1.1.3) are a rather peculiar class of enzymes known to be active on substrates in micellar form although the lowest degree of molecular aggregation of triglycerides compatible with lipase action is not yet defined[1].

Since 1971 we have described the micellar characteristics of a substrate[2] which was further characterized while a turbidimetric procedure[3] for lipase determination in serum was set up. Our studies on lipase were centered on the problem of the analogies between lipase and lipoproteinlipase[4,5] and the possible effect of heparin and albumin as modulators of lipase activity[6].

We have stressed the possible function of albumin as a surface-active agent and fatty acid acceptor and postulated the formation of aggregates of lipase-albumin-heparin-triglyceride. Moreover we have proved, by the turbidimetric method, the main microsomal localization of lipase in pig pancreas and liver and suggested that probably the same enzyme is able to exhibit different activities depending on the substrate and on the presence of possible modulators of activity such as albumin and heparin.

On the other hand, studies on the charge distribution of lipase have suggested the possibility that other modulators act on the orientation of lipase

Institute of Biological Chemistry, University, 35100 Padova, Center for Clinical Enzymology, Lab. Chimica Clinica, Centro Ospedaliero Borgo Trento, 37100 Verona, Italy.

at the interface of the micelles of substrate[7] and that also the size and shape of the micelles can be influenced by different agents such as temperature and concentration of electrolytes[8].

Pancreatic lipase has very low activity in the absence of added sodium chloride and the hydrolysis rate is optimal at NaCl concentration of 7 mM decreasing at higher concentrations[9]. Recently it has been shown that lipoproteinlipase is the activity inhibited by sodium chloride at high concentrations[10].

Lipoproteinlipase on the other hand is also measured as a protamine-inactivated lipase activity whereas hepatic lipase is considered to be resistant to protamine[11].

The present paper reports the results obtained with the turbidimetric determination of lipase in different species, different organs and different subcellular particles in the presence or absence of sodium chloride at the concentration considered to be inhibitory for lipoproteinlipase. The turbidimetric method was checked with a parallel determination of the products of hydrolysis by thin layer chromatography.

Materials and methods

Wistar strain male rats (180 to 210 g), male domestic rabbits (3 Kg) and male guinea-pigs (400 to 530 g) were used for the experiments. Fresh pig organs were obtained from the slaughter-house and human organs were obtained from autopsies performed within a period of 24 hrs. after death. The methods used for the fractionation of the organs and the enzyme assays have been described previously[12].

Purified pig lipase (Sigma, St. Louis) was used for the "in vitro" experiments together with albumin (Fraction V, Sigma, St. Louis) heparin (Roche, Paris), trasylol (Bayer, Milan), triolein and purified olive oil (Sigma, St. Louis).

An Eppendorf recording filter photometer was used for the kinetic measurements at 25°C and the method for the determination of lipase activity was the turbidimetric procedure at 546 nm previously described[3].

Thin layer chromatography was carried out with silica-gel 60 alluminum sheets (Merck, Darmstadt) with a layer thickness of 0.2 mm. The solvent was a mixture of petroleum benzin and ethyl ether (80:20) and the references were triolein (Rf 0.45), diolein (Rf 0.07), mono-olein (Rf 0.03) and oleic acid (Rf 0.15). All products were Sigma (St. Louis). After a chromatographic run of 15 min the plates were stained with phosphomolibdic acid (20% in ethanol) according to Stahl[13] and the quantitative determination of the spots was carried out with a Chromoscan-Joyce (Joyce-Loebl. & Co. Inc., Burlington) at 620 nm. A calibration curve was set up with oleic acid in concentration ranging from 0.5 to 7.5 nMoles/l. The A_{620} values corresponding to those concentrations ranged between 0.013 and 0.265.

Table I. *Experimental conditions utilized for the determination of serum lipase and serum lipoproteinlipase activity*

	Lipase (Burlina & Galzigna 1973)	*Lipoproteinlipase* (Krauss et al. 1974)
Triglycerides	0.45 mg/ml	8.3 mg/ml
Protein	1.75 mg/ml	16.7 mg/ml
Na^+	0.20 mM/ml	0.1 mM/ml
Protein/triglycerides ratio	3.8	2.0

Table I shows the amounts of components used in the assay medium for the turbidimetric measurement of lipase activity carried out in the present study and those used by other authors in the assay of lipoproteinlipase activity[11].

It is obvious that lipase-catalyzed de-emulsification reactions do not follow Michaelis-type kinetics although the relationship between rate of reaction and concentration of triolein is expressed by a hyperbolic function corresponding to an adsorption isotherm[14].

The relative value of the dissociation constant of the enzyme-substrate complex derived from turbidimetric measurements[3] should therefore be transformed into a parameter related to the area of the interface of the emulsion[15]. In our preceeding studies[2,3] we have related the optical absorption (A) of a triglyceride emulsion with the volume of triglyceride micelles according to the equation:

$$A = k.n.v.^2 (1 + \alpha\, t/_T) \qquad (1)$$

In the very first period (t very low) the expression $1 + \alpha\, t/_T \cong 1$ and

$$A = k.n.v.^2 \qquad (2)$$

At saturation (high concentration of substrate) A = 0.050 and n = 0.1 x 6.02 x 10^{23} according to our previous data[3] and since 50 molecules of triolein correspond to one micelle, 5 mol. l^{-1} triolein correspond to 0.1 mol. micelles l^{-1}. If the volume of one micelle is v = 5 x 10^{-5} cm^3 the value of the constant k becomes:

$$k = \frac{0.6 \times 10^{23} \times 25 \times 10^{-10}}{0.050} = 300 \times 10^{13}$$

and this allows us to transform any value of absorbance into number of micelles present in the reaction medium.

The Km can also be expressed in terms of micelle surface[7] and, with the above values, it becomes:

$$Km = 9 \times 10^{-2} \text{ mol. } l^{-1} = 1.8 \times 10^{-1} \text{ mol. micelles. } l^{-1}$$

and, if the surface of one micelle is:

$$s = 4\pi\ r^2 = 4 \times 3.14\ (1 \times 10^{-6})^2 = 12.56 \times 10^{-12}\ cm^2$$

the affinity of the enzyme for the micelle surface should be of the order of:

$$Km\ (surface) = 22.6 \times 10^{-14}\ cm^2 \cdot l^{-1}$$

The micellar behaviour of the triglycerides in the medium used for lipase determination (i.e. triolein, 2-aetoxy-ethanol and universal buffer pH 9.15) is dependent on the presence of albumin or electrolytes and the value of the so-called[16] *critical micellar concentration* (CMC) changes as shown in Figure 1. In the Figure we plotted the values of absorbance at 546 nm of different emulsions prepared with varyng concentrations of tryglycerides in normal conditions, in the presence of added albumin and in the presence of sodium chloride at 1 M final concentration. The correspondence between turbidimetric determination of lipase and fatty acid liberation due to the lipolysis is illustrated in Figure 2. Purified porcine lipase was used as an enzyme and a medium composed by triolein, 2-aetoxythanol and buffer was the substrate.

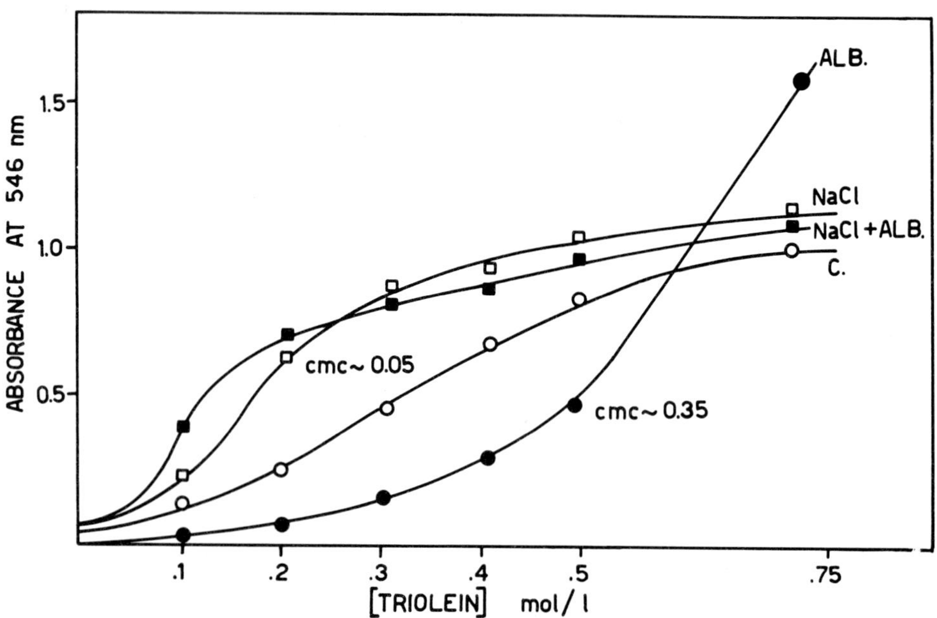

Fig. 1 – *Absorbance value as a function of triolein concentration*. Suspensions were prepared by mixing different volumes of triolein dissolved in 2-ethoxyethanol (1 ml in 99 ml) with universal buffer. The final volume was always 10 ml and the volumes of triolein/2-ethoxyethanol ranged from 0.01 to 1 ml. The absorbance values were determined 1 hour and 24 hours after mixing.

The amount of albumin (alb) added to the mixture was 0.1 mg in a final volume of 10 ml. NaCl was 1 M final concentration. The approximate values of CMC are calculated from the slope of the part of the curve which follows the initial lag-phase.

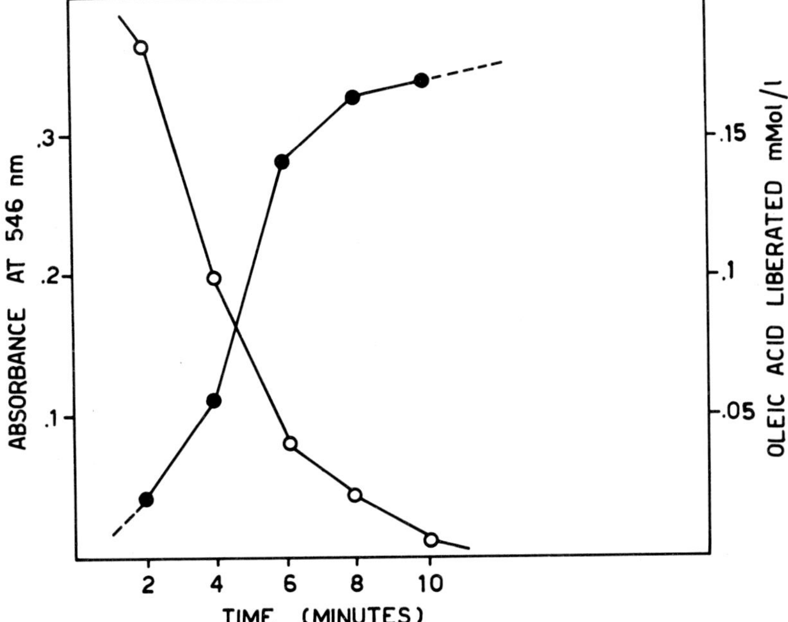

Fig. 2 – *Lipase activity followed turbidimetrically and by oleic acid liberation.* The assay mixture is that described by Burlina and Galzigna[3] with 2.2 × 10^{-3} M triolein and 0.1 μg/l lipase final concentration. Temperature was 25°C. The amount of oleic acid liberated was quantitated with a calibration curve. Chromatographic determination was carried out as described in caption to Table II.

Results

Figure 3 shows the behaviour of lipase in mitochondria and microsomes of rat liver and brain and serum lipase in rats maintained at hypoproteic and hyperproteic diet compared to controls.

The relative concentration of lipase in the microsomes of different organs of three species is shown in Figure 4.

The behaviour of purified porcine lipase in the lipolysis is shown in Figure 5 which illustrates the effect of high concentrations of NaCl in the medium in the presence and absence of albumin.

Table II summarizes the effect of different compounds added "in vitro" on the de-emulsifying activity of purified porcine lipase and shows the slight apparent activatory effect of heparin and the strong apparent inhibitory effect of trasylol.

The effect of the different compounds was also checked with the chromatographic method, but the results are not coincident with those obtained with the turbidimetric procedure.

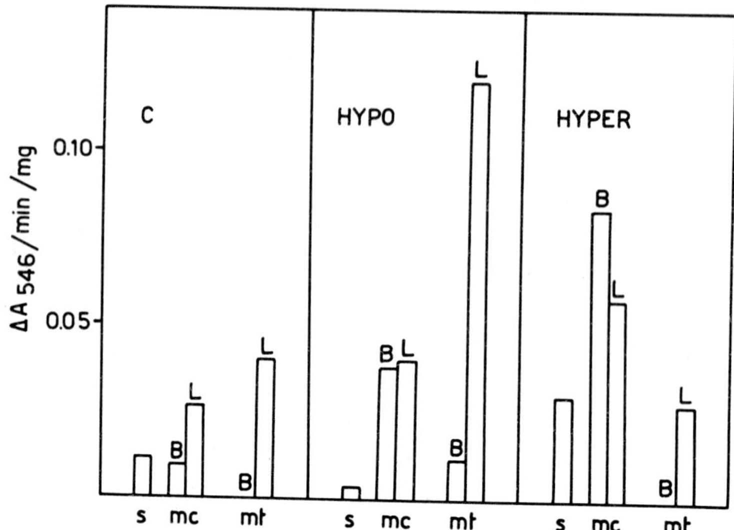

Fig. 3 – *Effect of different environmental conditions on lipase activity.* The results are average values obtained from 27 rats divided into three groups and maintained for 40 days on different diets as described previously[12]. C is the control group, HYPO is the group on hypoproteic diet and, HYPER is the group on hyperproteic diet. The symbols indicate serum (S), microsomes (mc), mitochondria (mt), in brain (B) and liver (L) respectively.

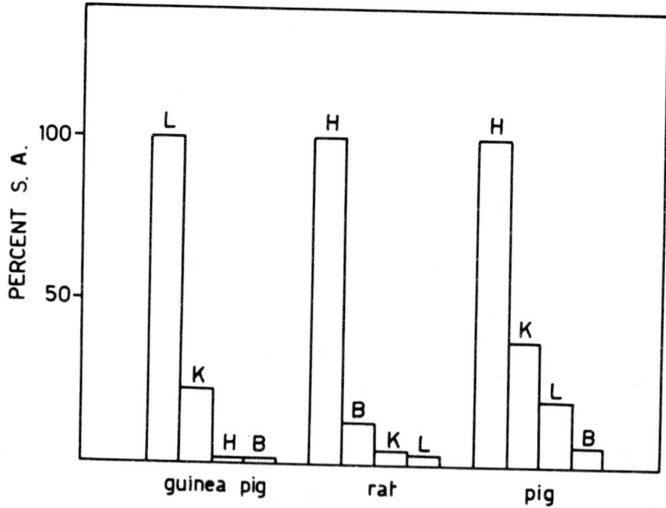

Fig. 4 – *Relative lipase activity in microsomes of different tissues in guinea-pig, rat and pig.* The specific activity (SA) is calculated as ΔA_{546}/min/mg protein and kidney is (K). A value of 100 corresponds to 0.019 ± 0.001 for guinea-pig, to 0.04 ± 0.005 for rat and to 0.8 ± 0.01 for pig.

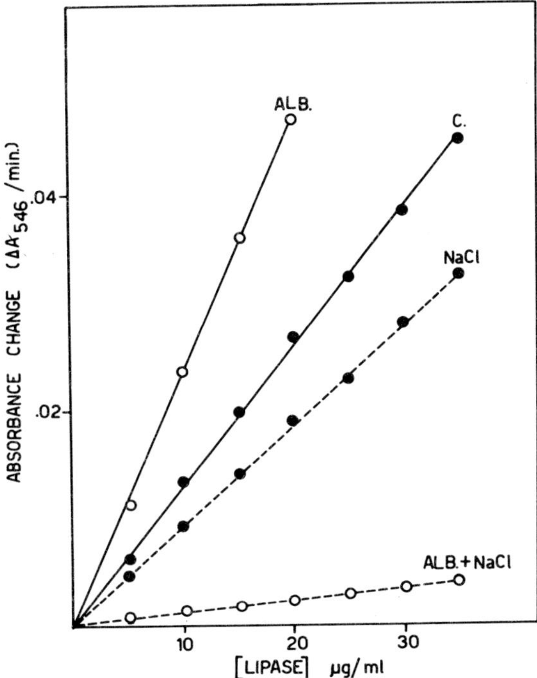

Fig. 5 – *Velocity of lipolysis as a function of lipase concentration.* Triolein was 2.2 x 10^{-3} M, NaCl was 1 M and albumin was 3 μg/l final concentrations.

Tab. II. *Effect of different compounds added in vitro on the activity of purified pig lipase*

Substrate is 2.2 x 10^{-3} M triolein and lipase is 15 ug/ml. The parallel determination by thin layer chromatography was carried out by incubating lipase and substrate for 10 min under the different conditions. After incubation the mixture was extracted with an equal volume of n-hexane (Merck, Darmstadt) and 20 μl of the supernatant obtained after 5 min shaking with Vortex and centrifugation at 500 x g in clinical centrifuge were utilized for the chromatography. The values were corrected for the amount of oleic acid trapped by albumin and are mean ± SD from 5 determination.

	De-emulsification velocity (ΔA_{546}/min)	*Oleic acid liberation* (mMol/l)
Control	0.020 ± 0.0100	0.0159 ± 0.002
+ Heparin (2.5 mg/ml)	0.021 ± 0.0120	0.0162 ± 0.003
+ Albumin (3 mg/ml)	0.036 ± 0.0080	0.0128 ± 0.002
+ Albumin + Heparin (1.25 mg/ml)	0.025 ± 0.0130	0.0160 ± 0.004
+ Trasylol (500 U/ml)	0.002 ± 0.0005	0.0220 ± 0.004
+ Trasylol + Albumin	0.021 ± 0.0110	0.0166 ± 0.003
+ NaCl (1 M final)	0.015 ± 0.0010	0.0200 ± 0.002
+ NaCl + Alb.	0.002 ± 0.0005	0.0140 ± 0.003

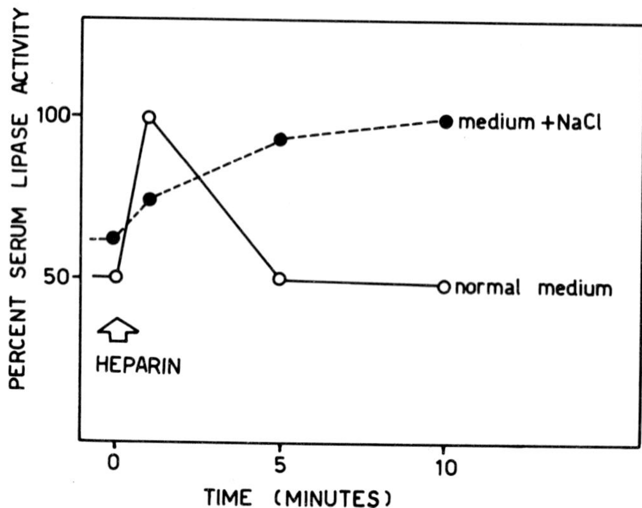

Fig. 6 – *Typical behaviour of serum lipase in rabbit after heparin administration.* 2500 U of heparin are given intravenously and serum lipase assayed with the normal medium and with the NaCl medium.

A percent value of 100 corresponds to a specific activity of 80 for lipase assayed with the normal medium and a specific activity of 16 for lipase assayed in the presence of 1 M NaCl. Specific activity is expressed as ΔA_{546} /min/l serum.

Figure 6 illustrates the result of intravenous administration of heparin on serum lipase of rabbit assayed turbidimetrically in the presence and absence of NaCl.

The localization of lipase in different organs and different subcellular fractions of rabbit measured in the presence and absence of NaCl is shown in Table III. The behaviour of subcellular fractions of liver and pancreas in the presence of NaCl in man and pig is shown in Table IV and Table V.

The effects of heparin both "in vitro" and "in vivo" are completely abolished by protamine and the effects of NaCl are mimicked by KCl when added at the same concentration. Divalent cations such as Mg exert an inhibitory effect on purified porcine lipase regardless of the presence of albumin.

The effect of temperature was checked in the interval between 20 and 50°C both on the turbidity of the suspension without any added enzyme and on the velocity of lipolysis. The Q_{10} value for lipolysis is 1.25 for the control, 1.07 in the presence of NaCl, 1.2 in the presence of albumin and 1.15 in the presence of NaCl and albumin. The turbidity of the suspension without any added enzymes decreases with the temperature and the decrease in absorbance at 546 nm for an increase of 10°C is 0.07 for the control medium, 0.04 for the medium with NaCl, 0.095 for the medium with albumin and 0.100 for the medium with NaCl and albumin.

Tab. III. *Subcellular localization of lipase determined in the presence and absence of 1 M NaCl in different organs of rabbit*

The method of assay is turbidimetric[3]. The results are average from three determinations

	Specific activity (ΔA_{546}/min/mg)					
	Mitochondria		*Microsomes*		*Cytosol*	
	–	+ NaCl	–	+ NaCl	–	+ NaCl
Heart	0	0.0021	0	0.009	0	0.048
Brain	0	0.0027	0	0	0	0.013
Pancreas	0.075	0.005	0.045	0.01	0	0.019
Kidney	0	0.032	0	0	0	0
Liver	0.002	0	0	0	0	0.002

Table IV. *Subcellular localization of lipase determined in the presence and absence of 1 M NaCl in liver and pancreas of pig.*

The method of assay is turbidimetric[3]. The results are average from three determinations

	Specific activity (Δ_{546}/min/mg)					
	Mitochondria		*Microsomes*		*Cytosol*	
	–	+ NaCl	–	+ NaCl	–	+ NaCl
Liver	0	0.011	0.025	0.016	0	0.024
Pancreas	0.345	0.111	0.590	0.336	0.094	0.026

Table V. *Subcellular localization of lipase determined in the presence and absence of 1 M NaCl in human liver and pancreas*

The method of assay is turbidimetric[3]. The results are average from three determinations

	Specific activity (ΔA_{546}/min/mg)					
	Mitochondria		*Microsomes*		*Cytosol*	
	–	+ NaCl	–	+ NaCl	–	+ NaCl
Liver	0	0.019	0	0.014	0.010	0.007
Pancreas	0.102	0.006	0.140	0.015	0.024	0

Discussion

The results presented in this paper show that albumin, when present in the assay medium, has the effect of drastically increasing the value of the critical micellar concentration of triglyceride suspensions. This indicates a higher amount of free triolein in equilibrium with the micelles. On the other hand NaCl either alone or in the presence of albumin does not influence the CMC value in a significant way.

The turbidimetric method appears to be more suitable for measuring lipase activity and the method based on the measurement of oleic acid liberation certainly is the result of an activity which is more complex than a simple lipase activity.

The chromatographic method in any case allowed us to show that oleic acid liberation is parallel to an increase of both mono- and di-olein and a decrease of triolein. As for the turbidimetric change we observed that only when triolein is hydrolyzed to oleic acid is a de-emulsification detectable where the attack of only one ester bond with formation of diolein results in a slight or barely detectable turbidimetric change.

The presence of NaCl at a 1 M final concentration causes a marked inhibition of purified porcine lipase activity measured turbidimetrically only when albumin is added to the incubation mixture, that is in conditions where the substrate is practically a lipoprotein and the activity can be considered a lipoproteinlipase activity. This result is not observed by the method of fatty acid liberation.

The injection of heparin induces two effects on the de-emulsifying activity of rabbit serum. The first effect is immediate and might be related to an activation of the circulating lipase resulting in lipoproteinlipase activity. If NaCl is added to the assay mixture a secondary effect appears which is consistent with a gradal stimulation of the endothelia and a release of lipase into the circulatory stream.

Alternatively, the result can be interpreted by postulating that two triglyceride lipase activities are released by heparin injection, one similar to the adipose tissue lipase inhibited by 1 M NaCl and the other similar to the hepatic lipase.

This does not contradict the idea that the same enzyme is able to exhibit two different activities depending on the type of substrate (triglycerides or lipoproteins) and heparin, albumin and NaCl are able to act as modulators of such an activity.

When considering lipoproteins we figure out aggregates of proteins and lipids bound through saline (i.e. electrostatic) bonds, stable within a certain pH range[17]. Heparin, strongly polar and acidic, can also control "in vivo" the attachment of lipids to proteins through formation of protein-heparin or lipoprotein-heparin complexes.

If micelles are formed when hydrophobic portions of triglycerides interact and a surface-active agent (e.g. 2-aetoxyethanol) regulates the interaction with water[2] the presence of salts at high ionic strength should influence, among others, the interaction of the monomer with the solvent by increasing the interfacial free energy. In fact we have seen that all the observed effects of NaCl are obtained when NaCl is replaced with KCl at the same concentration.

The localization of lipase in rabbit tissues with the turbidimetric method in the presence or absence of NaCl and that of lipase in liver and pancreas of pig and man gives comparable results.

We observe in fact that there is a certain complementarity between NaCl-inhibited lipase and "normal" lipase activity. In other words when NaCl is present we have a true lipase activity and in the other case we can consider the activity as a lipoproteinlipase activity. This is certainly related to the observed liberation of lipoproteinlipase from different organs after heparin administration and with the still obscure function of the co-lipase in serum[18].

The inhibitory action of trasylol on purified lipase and the protection exerted by albumin is an effect which supports the idea of the importance of the formation of molecular complexes in the regulation of lipase activity. The effect is consistent in any case with the protective action of trasylol in experimental pancreatitis[19].

Acknowledgement

The expert assistance of M. Bianchi, W. Dozza, M. Zaninotto and M.A. Galfano is gratefully acknowledged.

References

1. P. Desnuelle: The Enzymes, *7,* 575 (1972).
2. A. Burlina, L. Galzigna, G. Miconi: 3d Int. Symp. Clin. Enzymol., Conegliano, 1971.
3. A. Burlina and L. Galzigna: Clin. Chem., *19,* 384 (1973).
4. A. Burlina and L. Galzigna: Scand. J. Clin. Lab. Invest., *29,* 126 (1972).
5. A. Burlina and L. Galzigna: VIII World Congress of WASP, Munich 1972.
6. A. Burlina and L. Galzigna: 4th Int. Symp. Clin. Enzymol. Conegliano, 1972.
7. H. Brockerhoff: Chem. & Phys. of Lipids, *10,* 215 (1973).
8. J.M. Corkill, J.F. Goodman and J.R. Tate: Proc. Symp. on Hydrogen-bonded solvent systems, Taylor & Francis Ltd., London 1968.
9. G. Benzonana and P. Desnuelle: Bioch. Bioph. Acta, *164,* 47 (1968).
10. J.K. Huttunen, C. Ehriholm, P.K.J. Kinnunen and E.A. Nikkila: Clin. Chim. Acta, *63,* 335 (1975).
11. R.M. Krauss, R.J. Levy and D.S. Fredrickson: J. Clin. Invest., *54,* 1107 (1974).
12. L. Galzigna, M. Bianchi, U. Lippi and A. Burlina: 6th Int. Symp. Clin. Enzymol., Venezia, 1974.
13. E. Stahl: Thin-Layer Chromatography. A Laboratory Handbook p. 139, Springer Verlag, Berlin, New York, 1969.
14. F. Schoenheyder and L. Volqvartz: Acta Physiol. Scand., *9,* 57 (1945).

15. G. Benzonana and P. Desnuelle: Bioch. Bioph. Acta, *105,* 121 (1965).
16. G. Tettamanti, B. Cestaro, B. Venerando and A. Preti: 6th Int. Symp. Clin. Enzymol., Venezia, 1974.
17. E. Chargaff: Adv. Prot. Chem., *1*, 1 (1944).
18. C. Erlanson and B. Borgstrom: Bioch. Bioph. Acta, *271*, 400 (1972).
19. D.C. Nabach, K. Apoatolon, B.P. Kekis, S. Choren and A. Hirach: Fed. Proc. *23*, 605, (1963).

PHYSIOLOGICAL VARIATION OF ENZYME ACTIVITIES IN SERUM

B. E. Statland, P. Winkel and R. E. Cross

Summary

Attempts to estimate the physiological day-to-day variation of enzyme activities in healthy subjects have often met with difficulty in that the analytical errors (both within-run and run-to-run) have been of such magnitude that the underlyng physiologic variation cannot be estimated with certainty. In this present study we were able to minimize the analytic variation by[a)] analyzing all the specimens collected over many days on one run and[b)] using the highly precise fastcentrifugal analyzer (Rotochem) for the analyses. We drew blood specimens from 14 healthy volunteers, aged 22 to 40 years (8 male and 6 female) on six separate days over a 10 day interval. The venipuncture sessions were standardized, i.e. all specimens were obtained at the same hour of the day (8.00 a.m.), the tourniquet was applied for no longer than one minute, all subjects sat upright for at least 15 minutes before venipuncture, and all blood specimens were drawn in duplicate The sera were separated and then frozen for no longer than 10 days in the most extreme case. On one occasion all the 12 specimens from each subject (6 days x 2 replicates) were assayed for the enzymes alkaline phosphatase (AP), gamma-glutamyl transpeptidase (GGTP), lactate dehydrogenase (LDH), aspartate aminotransferase (AST), alanine aminotransferase (ALT), and creatine kinase (CK). The 12 specimens for each subject and for each enzyme test were assayed on one disc (run) of the fast centrifugal analyzer and all assays were performed during a 6 hour period of one day in the routine laboratory. The data were evaluated using a twoway ANOVA model with the factors being "subject" and "day of venipuncture". The total variation was separated into the biologic intra-individual day-to-day variation (subject-day interaction term), the biologic inter-individual variation (main effect of subject), and the within-run (within-disc) analytic variation (error term). The "ratio value" is defined as the ratio of the biologic intraindividual day-to-day standard deviation divided by the biologic inter-individual standard deviation. The ratio value for each enzyme characterizes that enzyme in terms of the relative magnitudes of the within-subject variation and the subject-to subject variation. The ratio values for the six enzymes as well as the coefficients of variation (c.v.) for the analytic and biologic components of variation are presented.

The purpose of this present study is to characterize the biologic sources of variation in healthy subjects for the activity values in serum of six enzymes.

Department of Hospital Laboratories, University of North Carolina Medical Center Chapel Hill, North Carolina. Clinical Chemistry, Department A, Rigshospitalet, Copenhagen, Denmark.

Table I. *Demographic characteristics of the 14 volunteers used in this present study*

No.	*Sex*	*Age (Years)*	*Race*	*Height (meters)*	*Weight (kg)*	*Smoking (cig/day)*
1	male	25	White	1.70	73.5	none
2	male	25	White	1.84	72.6	none
3	male	26	White	1.80	65.8	none
4	male	28	White	1.75	74.9	20/day
5	male	28	White	1.82	71.3	20/day
6	male	29	White	1.82	64.9	none
7	male	30	White	1.80	97.6	10/day
8	male	35	White	1.85	88.5	none
9	female	23	White	1.52	49.0	15/day
10	female	23	White	1.66	52.2	none
11	female	26	Black	1.63	74.9	none
12	female	29	White	1.52	46.3	none
13	female	38	Black	1.63	77.2	20/day
14	female	40	White	1.79	88.5	40/day

In order to decide whether or not the magnitude change in a result from day to day is clinically significant, we must first have an understanding of sources of variation in healthy subjects independent of change of health status.

Methods and materials

Subjects: Fourteen healthy volunteers were used in this study. Table 1 presents the demographic characteristics of each member of the group, all of whom were graduate students or employees of the routine chemistry laboratory at the North Carolina Memorial Hospital. None of the female subjects was taking oral contraceptives.

Protocol: On six days: December 2, 3, 4, 8, 9, and 11 (1975), each of the subjects came to work in a fasting state and then adopted a sitting posture for at least 15 minutes; then an experienced phlebotomist applied a tourniquet and filled a glass tube with whole blood so that approximately 15 ml of blood were obtained per venipuncture. All venipunctures were performed between 0800 and 0830 a.m.

Processing of specimens: The whole blood was allowed to clot at room temperature and then centrifuged. The serum was decanted and then separated into two equivalent aliquots with two aliquots (per venipuncture) being frozen (–20°C) until 2 days after the last venipuncture and then assayed in one batch. From each day's venipunctures we obtained 28 serum aliquots (2 per venipuncture x 14 subjects) or a total of 168 (28 x 6 days) frozen specimens at the conclusion of the study.

Chemical Methods: Each serum specimen was assayed for activity values of AP, LDH, AST, ALT, and CPK, while the serum specimens for subjects 1, 2, 3, 4, 7, and 8 were also assayed for γ-GTP. The 12 specimens from each subject (6 days x 2 duplicates) were dispensed into 12 places of the transfer disc of the ROTOCHEM fast centrifugal analyzer along with control sera. All activities were measured at 30°C. The LDH, AST, ALT, and γ-GTP activities were measured with kits from Boehringer Mannheim Corporation (LDH-L: 14314; GOT:15791; GPT:15792; and γ-GT:15794; respectively). The AP activity was measured using 15mM p-nitrophenylphosphate as substrate in 2-amino, 2-methyl, l-propanol buffer, pH 10.2. The CPK activity was measured using Statzyme CPK kit from Worthington Biochemical Corporation (CPK:7251).

Statistical Methods: The data set containing the 168 daily results (14 subjects x 6 days x 2 duplicates) for each of the enzymes was analyzed according to a two-way ANOVA model with the factors being "subject" and "day". The analytic component of the variation was based on the error term (within-batch variation). The biologic components of variation were derived from the main subject effect (inter-individual variation of mean levels) and the subject-day interaction term (average intra-individual variation)[1]. In addition, the 12 results for each subject-enzyme combination (6 days x 2 duplicates) were analyzed according to a one-way ANOVA. The within-subject's variation (day-to-day) was estimated on the basis of the main day effect. For both the analytic and biologic components of variation, the variation is presented as coefficients of variation: (standard deviation/mean) x 100%.

Results and discussion

Analytic Variation: As noted in Table 2 the within-batch analytic variation varied from 2.0% for alkaline phosphatase to 8.5% for alanine aminotransferase. Applying the two-way ANOVA model to the results on the frozen sera, three of the six enzymes demonstrated a statistically significant main effect of day: ALT ($p < .05$), LDH ($p < .01$), and AST ($p < .01$). This indicates that for ALT, AST, and LDH there was either a common pre-instrumental procedural factor and/or a common biologic factor affecting these three assays. A very likely source would be the differential effects of freezing the samples. Had we elected to assay the sera on six separate days, we would have eliminated the variation due to variable freezing time and thawing; however, we would have added the long-term (batch-to-batch) analytic variation which often is the largest component of variation in the analytic procedure.

Biologic Variation: Figure 1 illustrates the relative sizes of the mean biologic intra-individual day-to-day variation and the biologic inter-individual variation for the six enzymes studied. The actual coefficients of variation

Table II. *Components of Variation for activity values of six enzymes in sera of healthy subjects expressed as percent coefficient of variation*

Enzyme Assay	*Within-batch analytic variation (%c.v.)*	*Biologic mean intra-individual variation (%c.v.)*	*Biologic inter-individual variation (%c.v.)*	*Ratio of intra-individual to inter-ind. variation (%c.v.)*
Alkaline Phosphatase	2.0	3.5	24.8	0.14
γ-Glutamyl Transpeptidase	3.7	3.7	23.8	0.16
Lactate Dehydrogenase	6.0	5.5	12.6	0.44
Aspartate Aminotransferase	6.2	10.9	12.1	0.90
Alanine Aminotransferase	8.5	13.2	29.6	0.45
Creatine Kinase	3.0	25.7	46.0	0.56

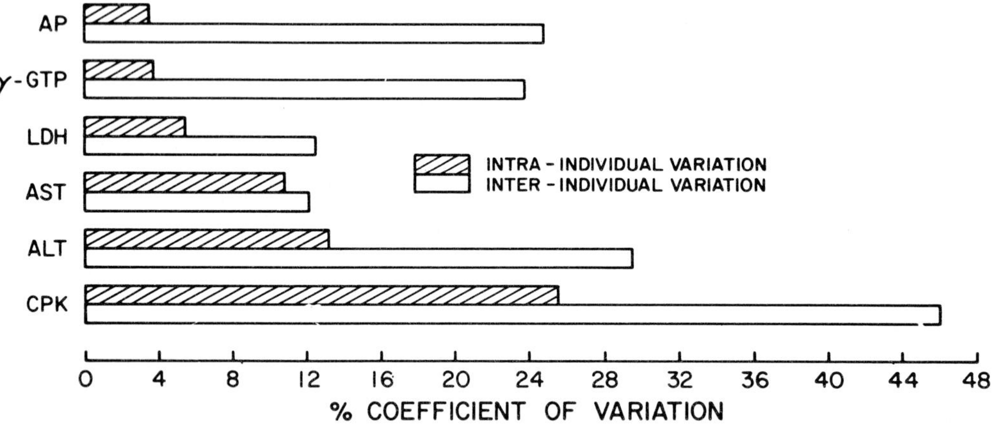

Fig. 1 – Mean biologic intra-individual variation and inter-individual variation for AP, γ-GTP, LDH, AST, ALT, and CPK in 14 healthy subjects expressed as percent coefficients of variation.

and the ratio values (intra/inter) are presented in Table 2. For all six enzymes, the intra-individual variation (subject-day interaction term of the two-way ANOVA) was statistically significant ($p < .01$).

The magnitude of the 14 within-subject day-to-day variations varied dramatically (Table 3). The stimated intra-individual day-to-day variation (in terms of percent coefficient of variation) varied from 1.4% to 9.2% for alkaline phosphatase; from 2.6% to 10.7% for γ-glutamyl transpeptidase; from 5.4% to 15.7% for lactate dehydrogenase; from 4.6% to 31.6% for aspartate

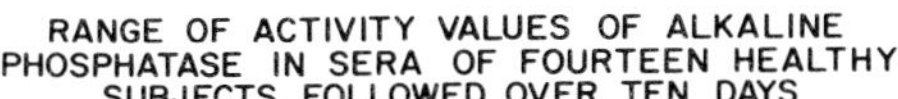

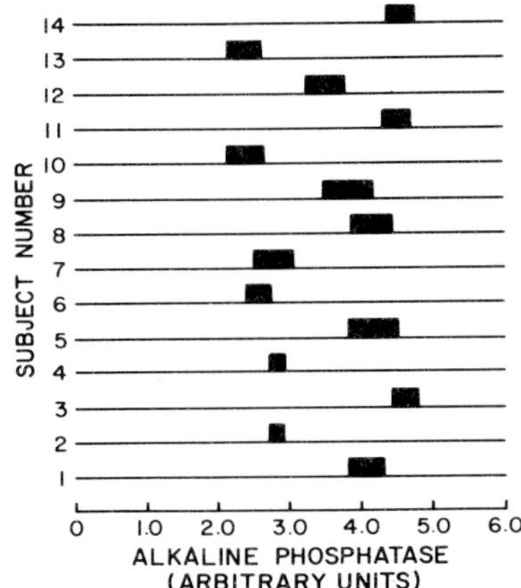

Fig. 2 – Range of activity values of alkaline phosphatase in sera of 14 healthy subjects followed over ten days.

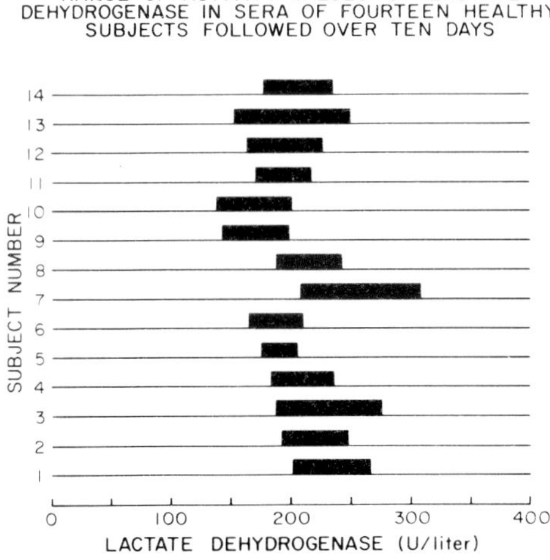

Fig. 3 – Range of activity values of lactate dehydrogenase in sera of 14 healthy subjects followed over ten days.

aminotransferase; from 7.6% to 29.1% for alanine aminotransferase; and from 4.6% to 53.3% for creatine kinase.

Figures 2 and 3 graphically illustrate the ranges of the activity values of AP and LDH, respectively, in each of the 14 subjects. For AP and γ-GTP the

Table III. *Means and intra-individual day-to-day variation expressed as percent coefficient of variation for each of 14 healthy subjects for the activity values in serum of six enzymes*

Subject number	*Alkaline phosphatase*		*Gamma-glutamyl transpeptidase*		*Lactate dehydrogenase*		*Aspartate aminotransferase*		*Alanine aminotransferase*		*Creatine kinase*	
	mean (arb. units)	*c.v. (%)*	*mean (U/ liter)*	*c.v. (%)*	*mean (U/ liter)*	*c.v. (%)*	*mean (arb. units)*	*c.v. (%)*	*mean (arb. units)*	*c.v. (%)*	*mean (arb. units)*	*c.v. (%)*
1	4,12	5.2	11.3	4.3	230.8	12.2	38.9	27.7	39.5	29.1	13.6	13.5
2	2.87	1.4	7.7	10.7	215.3	6.6	29.4	9.1	18.2	8.9	10.1	6.1
3	4.65	3.4	10.6	6.8	238.3	14.5	39.3	13.6	26.5	12.0	19.5	53.3
4	2.87	2.2	15.7	2.6	211.2	9.7	35.7	9.1	31.9	7.6	9.77	38.7
5	4.17	7.4	–	–	190.1	5.7	39.2	22.0	30.6	25.4	10.3	7.8
6	2.53	5.0	–	–	184.8	8.2	30.6	11.6	24.7	18.6	5.36	11.2
7	2.83	5.7	13.9	3.6	245.8	14.8	38.2	31.6	45.0	28.2	8.46	23.7
8	4.14	5.5	11.0	7.3	211.4	10.3	31.1	10.6	28.0	10.8	9.30	11.3
9	3.77	6.5	–	–	163.5	11.1	27.4	12.3	17.2	9.2	5.15	32.6
10	2.50	9.2	–	–	155.5	15.6	29.2	22.5	20.3	19.6	4.96	4.6
11	4.54	2.5	–	–	197.7	5.4	34.3	15.4	25.7	13.3	8.30	6.1
12	3.57	5.2	–	–	184.3	13.3	35.4	4.6	19.4	7.6	6.24	29.0
13	2.36	6.6	–	–	202.8	15.7	28.3	13.7	24.0	8.2	4.05	21.7
14	4.67	2.3	–	–	198.5	10.7	31.9	18.7	21.6	25.2	6.44	16.8

ratio value of (mean intra-individual variation)/(inter-individual variation) is less than 0.2 and for these enzymes, it may be desirable to rely on subject-specific reference intervals. However, for LDH and CPK, the ratio values are sufficiently large as to make us doubt the usefulness of subject-specific reference intervals as compared to the more conventionally-used group-specific reference intervals.

Acknowledgements:

We wish to acknowledge the technical assistance of Ken Hill and Linda Woodard in performing the enzyme assays. In addition we appreciate the support and encouragement of John Savory in carrying out this study.

References

1. Winkel P., Bokelund H. and Statland B.E.: "Factors contributing to intraindividual variation of serum constituents: 5. Within-hour and short-term day-to-day variation of serum constituents in healthy subjects". Clin. Chem., *20,* 1520 (1974).

SUBSTRATE DETERMINATIONS ON DISCONTINUOUS AUTOMATED SYSTEMS

A. Hagen, J. Ziegenhorn and H.U. Bergmeyer

Summary

Today automated systems have firmly established themselves in clinical chemistry laboratories.

At the beginning of their development they were utilized solely to simulate manual procedures; thus some of their obvious advantages were neglected. Only recently new approaches have been made: an automated system may surpass manual performance with respect to uniformity of processing and precision. This ability can be used for substrate determinations on a kinetic basis. It offers the advantage of a high sample frequency and reduced susceptibility to interferences, without negative effects on accuracy and precision.

Recently developed kinetic methods for the enzymatic determination of urea and glucose on a centrifugal fast analyzer will be presented as examples. It will be demonstrated that such estimations can be performed with high precision during short analysis times. Furthermore, it will be shown that with kinetic methods the influence of interfering compounds is often much smaller than with comparable endpoint procedures.

The basic theory of substrate determinations by reaction rate measurements will be outlined. Mathematical treatment will indicate that the use of automated systems will permit a broader application of kinetic methods in the clinical laboratory than was origin nally anticipated.

Automation has established itself in clinical chemistry. Indeed, it seems difficult to imagine what large and medium sized laboratories would do without automated systems today. Such instruments have reached a degree of perfection that allows their economical and reliable use for routine purposes. We have had much experience over the years with automated systems and a more generalized overview of their use is the contribution I would like to make today.

The era of laboratory automation began near the end of the fifties, when Technicon introduced the AutoAnalyzer. It was ten years later, however, before other types of instruments appeared in significant numbers. A most im-

Boehringer Mannheim GmbH, D-8132 Tutzing, West-Germany.

pressive exhibition of instruments could be found at the VIIth International Congress of Clinical Chemistry in Geneva 1969 where approximately 16 different companies presented their automated systems, as is shown in Table I.

Table I. *Automated systems presented during the VIIth International Congress of Clinical Chemistry, Geneva 1969*

Beckman DSA 560	Bioanalyst
Eppendorf 5030	Clinomac
Eppendorf 5010	Chem-o-lab
LKB 8600	AutoAnalyzer
Braun SysteMatik	Greiner GSA
Perkin Elmer C 4	Dupont ACA
Mecolab	Centrifichem
Quickfit 617	Gemsaec Fast Analyzer
Autolab	

Most of these instruments are similar to each other in that they simply mechanize the manual procedures. The only recent innovation in automated systems since Technicon's AutoAnalyzer was the centrifugal analyzer, which afforded Norman G. Anderson the award "Biochemische Analytik" award nich in 1972. It seems noteworthy that even this instrument was not utilized advantageously in the beginning. Initially one generally tried to simulate the existing manual procedure, thus neglecting to use the instrument's true advantages of complete automation.

Time has passed and much has been learnt by the instrument manufacturer, by the reagent producer and especially by the clinical chemist, who has become more confident today and is more critical in comparing analytical results arising from different procedures. This is true mainly for substrate determinations, which naturally can be quantified more readily than enzyme activity determinations.

The modern concept of automated procedures is not analogous to manual procedures. The criteria which make a procedure acceptable are convenience and accuracy and they have to be demonstrated for every single method. In the beginning of automation in clinical chemistry the true advantage of the machine was not realized. We know that the machine surpasses man in its precision of performance of every working step and its exactness in the whole analytical run. This power opens new approaches which cannot be simulated by manual procedures.

In this respect the method of kinetic substrate determination is an example for us to discuss in more detail.

The kinetic substrate determination offers two main advantages that are listed in Table II.

Table II.

- High sample frequency
- Reduced susceptibility to interferences without negative effects on accuracy and precision

These are high sample frequency and reduced susceptibility to interferences, without negative effects on accuracy and precision. A high degree of synergism between instrument and reagent is a most important condition. Both have to be optimized in regard to time and sample volume. The key criterion is accuracy which must hold true even under the presence of interfering substances.

For practical purposes the so–called fixed–time kinetic procedure is often used. I want to demonstrate this technique with the example of the

Table III

1.

1. $NH_2{-}CO{-}NH_2 + H_2O \xrightarrow{\text{Urease}} 2NH_3 + CO_2$

2. $2NH_4^+ + 2NADH + 2\ \text{Oxoglutarate} \xrightarrow{\text{Glutamate dehydrogenase}}$
$2NAD^+ + 2\text{-L-glutamate} + 2H_2O$

600 μl	Triaethanol amine buffer (142 mmol/liter, pH 8.0)	
	2-oxoglutarate	(15.1 mmol/liter)
	β-NADH	(453 μmol/liter)
	Urease	($\geq 4.7 \cdot 10^2$ U/liter)
	Glutamate dehydrogenase	($\geq 2.1 \cdot 10^4$ U/liter)
100 μl	Saline	
5 μl	Serum	

Reaction temperature	25° C
Wavelength	365 nm
First reading	160 sec after start
Second reading	260 sec after start

urea determination on the GEMSAEC FAST ANALYZER of Electro-Nucleonics. The reaction sequence is shown in Table III.

Below I have noted the concentrations of the components and the instrument settings. The reaction is performed at 25°C. At 365 nm two readings are taken at 60 and 240 seconds after mixing the components in the cuvette. The optical readings are related to a standard and calculations are performed by the on-line PDP 8 Computer. The results obtained with this method look excellent, as Figure 1 indicates.

The linearity of the assay was determined by analyzing appropriate dilutions of a human serum which contained 35 m mole urea/l. Figure 1 demonstrates that a direct proportionality between fixed time absorbance change and concentration exists over the whole concentration range.

Furthermore the figure shows that the sensitivity of the assay at normal urea values is sufficiently high. Data for the within-run precision are provided in Table IV.

You will notice that the coefficient for variation lies in the range of 1%. Day to day precision was studied for 5 days. The CV was 3.1% for the control serum Precilip which contained 3.9 mmol urea/l. Results obtained by the kinetic method are compared to those obtained by the manual endpoint procedure (Fig. 2).

The calculated linear regression coefficient indicates excellent agreement between the two methods. No interferences have been observed from lipemia in the kinetic procedure. Furthermore the addition of bilirubin to sera did not influence the accuracy of the assay.

The method is impressive because of convenience and accuracy. However it has to be mentioned that reagents used for kinetic methods have high demands on quality. In this case it is a pre-condition to use a urease of good stability and a well defined activity in every assay. Only then is it possible to use it as a routine method, having the same instrument settings every day.

With regard to the theory behind the experiments, the time course of the kinetic urea determination is shown in Figure 3.

In our so-called kinetic fixed-time procedure, the absorbance change between two absorbance readings at definite times is measured. As already shown, this absorbance change ΔA is proportional to the initial urea concentration to be determined. How can one explain this observation?

To answer this question, we have to take a look at the basic theory of kinetic substrate determinations:

Theoretical considerations of this type of analysis were discussed in detail by Ingle and Crouch in 1971. The authors pointed out that for these analytical procedures reactions following first or pseudo-first order kinetics are of unique importance, and that for the reasons listed in Table V.

In a first order or pseudo-first order reaction the reaction rate is directly proportional to the present concentration of the reactant (equation 1). If

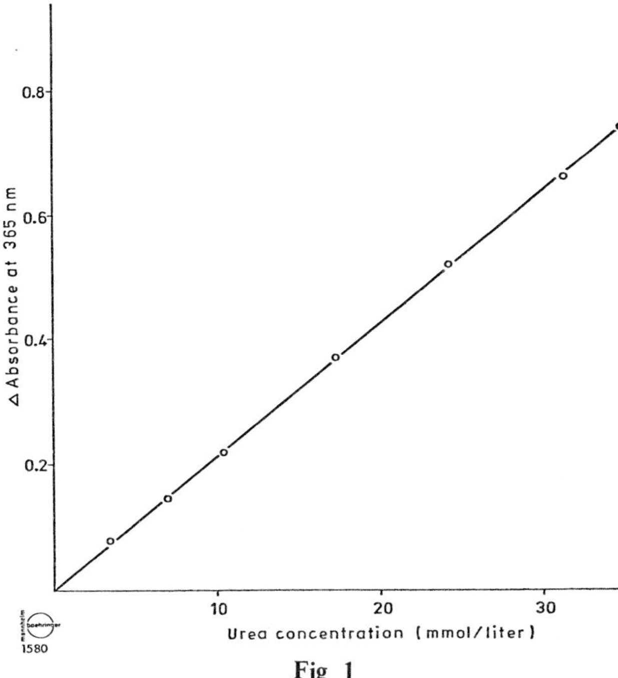

Fig 1

Table IV. *Within-run precision of the kinetic assay for urea*

Sample	*No tests*	*Mean mmol/liter*	*SD mmol/liter*	*CV*
Standard	14	6.66	0.053	0.8
Serum 1	14	6.92	0.059	1.0
Serum 2	14	15.8	0.16	1.0

you integrate that equation and, furthermore, introduce Lambert-Beer's law, then you obtain the basic equation shown below in the Table (equation 2). You can see that there is a linear relation between the initial substance concentration c_0 to be determined and the change of absorbance ΔA, if the reading times t_1 and t_2 are held constant. Both requirements can easily be achieved on automated instruments. Therefore, the fixed-time

GEMSAEC kinetic method (mmol/liter)
30
20
10
0
10
20
30
Manual endpoint method (mmol/liter)
n = 70
y = 1.00 X + 0.08
r = 0.999
p{0.99 < β < 1.01} = 0.995
mannheim
boehringer
1581

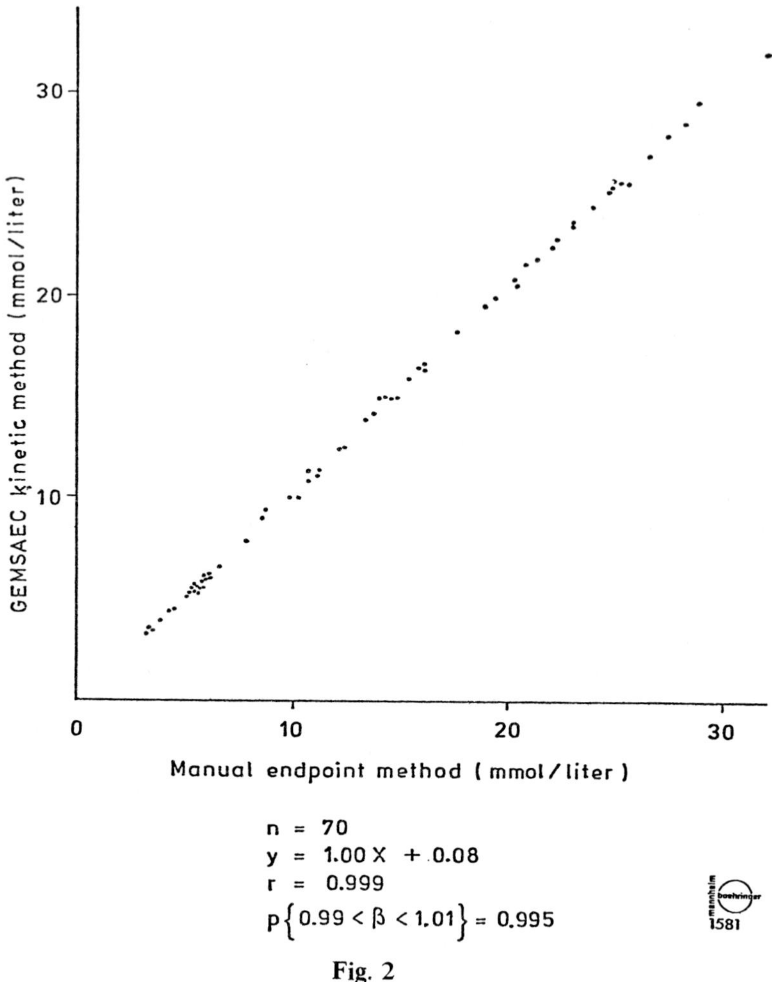

Fig. 2

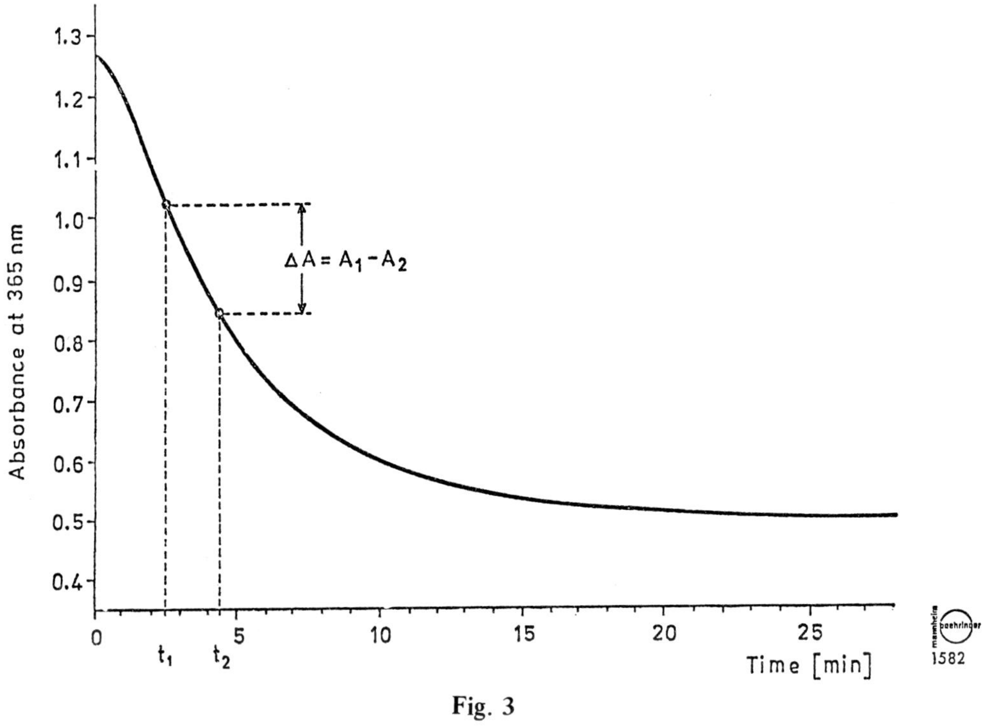

Fig. 3

Table V

1. $-\frac{dc}{dt} = k \cdot c$

2. $c_0 = \frac{1}{d \cdot \epsilon (e^{-k \cdot t_1} - e^{-k \cdot t_2})} \cdot \Delta A$

c = Substrate concentration; c_0 = Initial substrate concentration; t = Time
k = Velocity constant of first order reaction; d = Pathlength of cuvette; ϵ = Molar absorptivity;
ΔA = Absorbance difference

Table VI

$$v = \frac{-dS}{dt} = \frac{v_{max} \cdot S}{S + K_m}$$

$$s = K_m$$

$$v = \frac{-dS}{dt} = \frac{V_{max} \cdot s}{K_m} = k \cdot S$$

v = rate; s = substrate concentration; v_{max} = maximum rate; K_m = Michaelis constant.

approach in combination with pseudo-first order reactions is especially suited for this type of instrument.

This basic equation is valid for both enzymatic or non-enzymatic reactions. In the case of enzymatic reactions, the desired pseudo-first order kinetics are obtained, if the Michaelis constant K_M of the enzyme in the rate determining reaction is much greater than the initial substrate concentration. These facts are demonstrated in Table VI.

The equation shown follows from the theory of Michaelis and Menten. If K_M is much greater than the substrate concentration S, then the term S in the denominator of the equation can be neglected. Thus, you again obtain the typical rate equation of a first order reaction.

Results from different laboratories indicate that the substrate to K_M ratio normally must be below about 0.05. This restricts the use of the kinetic enzymatic methods to enzymes with high K_M-values. The K_M-value of urease with respect to urea is about 4×10^{-3} mol/liter. This fundamentally allows the quantitative analysis of urea by reaction rate measurements, if appropriate test conditions are used.

Figure 4 supplies the evidence for this statement.

The figure shows the time course of the overall reaction of our urea assay using a semilogarithmic plot. As you may know, in a plot of that kind, first order behaviour is indicated by a straight line. You can see in the figure, that under the conditions of our test procedure, the reaction passes through an induction period after which it follows first order kinetics for a long time. Therefore, the fixed-time approach can be used successfully for this assay system, if the optical readings are taken in the linear part of the curves.

There is a number of other substrate assays for which enzymes with sufficiently high Michaelis constants are available. However, in some cases, such enzymes do not exist, or, at least could not have been isolated so far. Then one can apply a special trick, which was first demonstrated by Mueller-Mathesius for the enzyme uricase: If the K_M-value of an enzyme is too low it can apparently be increased by the addition of a competitive inhibitor (Table VII).

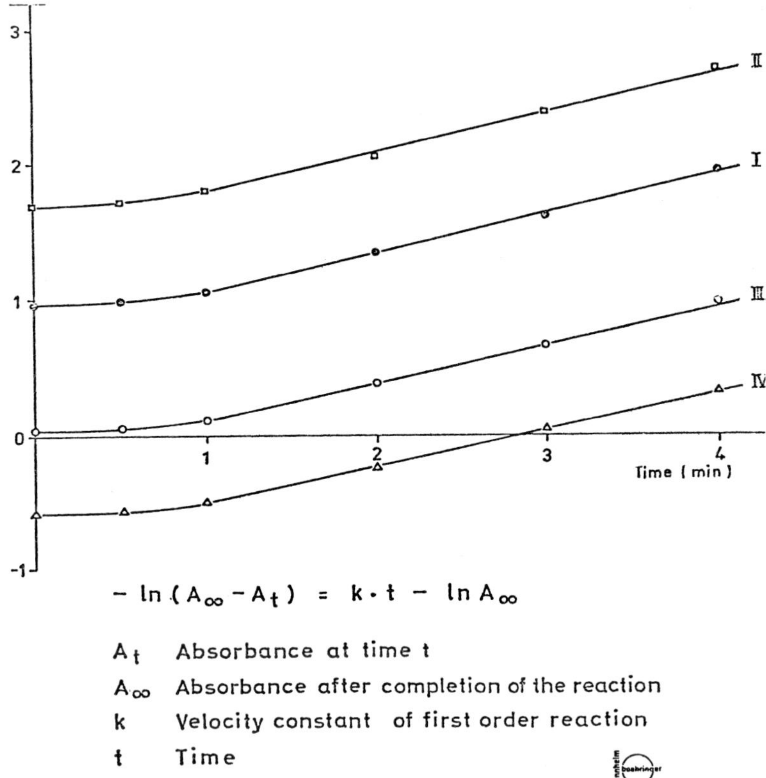

$$-\ln(A_\infty - A_t) = k \cdot t - \ln A_\infty$$

A_t Absorbance at time t

A_∞ Absorbance after completion of the reaction

k Velocity constant of first order reaction

t Time

mannheim boehringer 1584

Fig. 4

Table VII

$$V = \frac{-dS}{} = \frac{V_{max}\, S}{S + K_m\,(1 + \frac{I}{K_I})}$$

$$S \ll K_m\,(1 + \frac{I}{K_I})$$

$$V = \frac{-dS}{dt} = \frac{V_{max} S}{K_m\,(1 + \frac{I}{K_I}}$$

V = rate; S = substrate concentration; I = inhibitor concentration; V_{max} = maximum rate; K_m = Michaelis constant; K_I = inhibitor constant.

In the simplest case of a one-substrate reaction, the competitive inhibition can be described by the expression shown in that Table. Again, this equation follows from the theory of Michaelis and Menten. If S is much smaller than the term K_M $(1 + \frac{I}{K_I})$, it can be neglected and the equation represents a pseudo-first order reaction, which again can be easily measured on automated instruments by use of fixed-time methods. We successfully applied this competitive inhibition effect on a complex assay system, the enzymatic determination of serum triglycerides. The reaction sequence is well known (Table VIII).

Table VIII.

$$\text{Triglycerides} \xrightarrow{\text{Lipase, esterase}} \text{fatty acids} + \text{glycerol}$$

$$\text{Glycerol} + \text{ATP} \xleftrightarrow{\text{glycerol kinase}} \text{glycerol-3-phosphate} + \text{ADP}$$

$$\text{ADP} + \text{phosphoenolpyruvate} \xrightarrow{\text{pyruvate kinase}} \text{ATP} + \text{pyruvate}$$

$$\text{Pyruvate} + \text{NADH} + H^+ \xleftrightarrow{\text{Lactate dehydrogenase}} \text{lactate} + \text{NAD}^+$$

The Michaelis constants of the enzymes involved in these reactions are of the order of 10^{-5} to 10^{-4} mol/liter, thus limiting the applicability of the kinetic assay to relatively low glycerol or triglyceride concentrations. We observed that the range of the workable concentration could be considerably extended, if the amount of ATP in the reaction mixture was increased. This finding can be explained by the fact that ATP competitively inhibits the pyruvate kinase in the rate determining third reaction, which gives rise to an apparent increase of the K_M.

As a consequence, the overall reaction follows pseudo-first order kinetics over a wide range of glycerol concentration. This can be utilized for the kinetic determination of glycerol through a fixed-time procedure. By that means, one obtains an assay system which allows the fast determination of triglyceride concentrations up to 9 g/liter of serum, without pre-dilution of the sample, and without running a sample blank.

I would like to mention one further example of a kinetic procedure, the enzymatic cholesterol determination. Its reactions are outlined in Table IX.

Without going into details, I only want to stress one fact which might be of general importance. Plotting the logarithm of absorbance against time, we observed an unacceptably long induction period, in this case of almost five minutes.

Table IX. *Schema der enzymatischen Bestimmung des Gesamt-Cholesterins*

$$\text{Cholesterinester} + H_2O \xrightleftharpoons{\text{Cholesterin-Esterase}} \text{Cholesterin} + \text{Fettsäure}$$

$$\text{Cholesterin} + O_2 \xrightarrow{\text{Cholesterin-Oxydase}} \Delta^4\text{-Cholestenon} + H_2O_2$$

$$H_2O_2 + CH_3OH \xrightarrow{\text{Katalase}} H_2CO + 2\,H_2O$$

$$H_2CO + 2\,CH_3COCH_2COCH_3 + NH_3 \longrightarrow [\text{3,5-diacetyl-2,6-dimethyl-dihydropyridinium}] + 3\,H_2O$$

$$\lambda_{max} = 412\text{ nm}$$

Table X

1: $\frac{dc}{dt} = k \cdot C$

2. $\frac{dc}{dt} = f(t) \cdot c$

3. $c_0 = \frac{1}{\epsilon \cdot d \cdot [e^{-F(t_1)} - e^{-F(t_2)}]} \cdot \Delta A$

4. $F(t) = \int f(t)\, dt$

The question arises, whether fixed time kinetics could be performed if the first reading is taken in the nonlinear part of the curve. Let us look at the fundamental equation once more (Table X).

We may substitute the constant k in equation (1) by a hypothetical time-function f (t). This will lead to equation (2). An analogous derivation as before leads to equation (3) where F (t) represents the integral from equation (4). It is a mathematical consequence, that the multiplying factor of ΔA in equation (3) is a constant, if t_1 and t_2 are fixed. Then a direct proportionality between absorbance difference and initial concentration exists. Using the former plot we have to expect the lines intersecting the concentration-curves to be parallel (Fig. 5).

This parallelism is to be demanded strictly on the premises of the theory, therefore it follows that, irrespective of the degree of curvature, the fixed

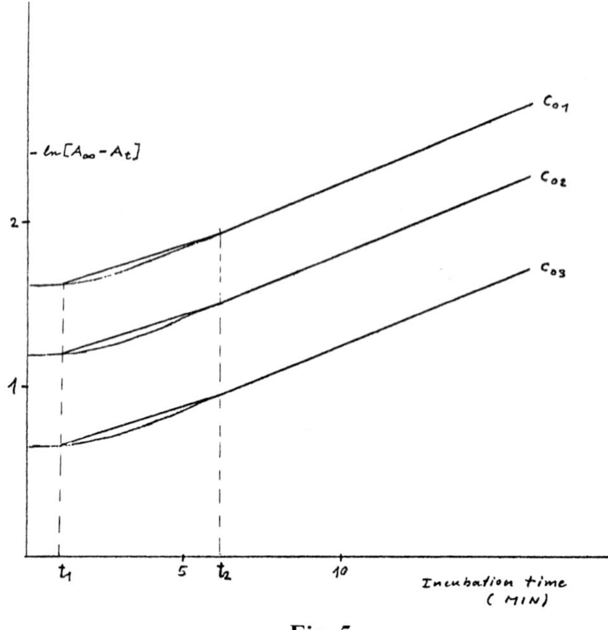

Fig. 5

time kinetic determination can be used if parallel lines are observed. Exactly this is true for the cholesterol experiment, from which the present data originate.

For what reason do we obtain a time function with such properties? Several reasons can be proposed: a change of temperature during the induction period of the reaction; or a structural alteration of the enzyme relating to changes in its activity. We have studied some theoretical cases and were able to demonstrate that under certain assumptions curves can be calculated that are congruent with those observed in the experiment. Unfortunately there is not enough time to discuss this theory in detail.

It can easily be understood that it is only because of the strictly repetitive behaviour of the machine, that any time dependency changes are exactly the same from one analysis to the other. It would be impossible to simulate this behaviour in a manual procedure.

Kinetic procedures very often have the great advantage of being less sensitive to interfering compounds than endpoint procedures. A good example is the Glucose GOD-PAP-method originally described by others. Figure 6 compared the fixed-time kinetic procedure performed at the Gemsaec analyzer with the AutoAnalyzer method and the manual method in respect to ascorbic acid interference. The superiority of the kinetic method is obvious. Differences between the velocities of the competing reactions must be the reason. It has to be made clear, however, that in regard to ascorbic acid interference endpoint procedures are applicable too, since such high ascorbic acid concentrations normally do not exist.

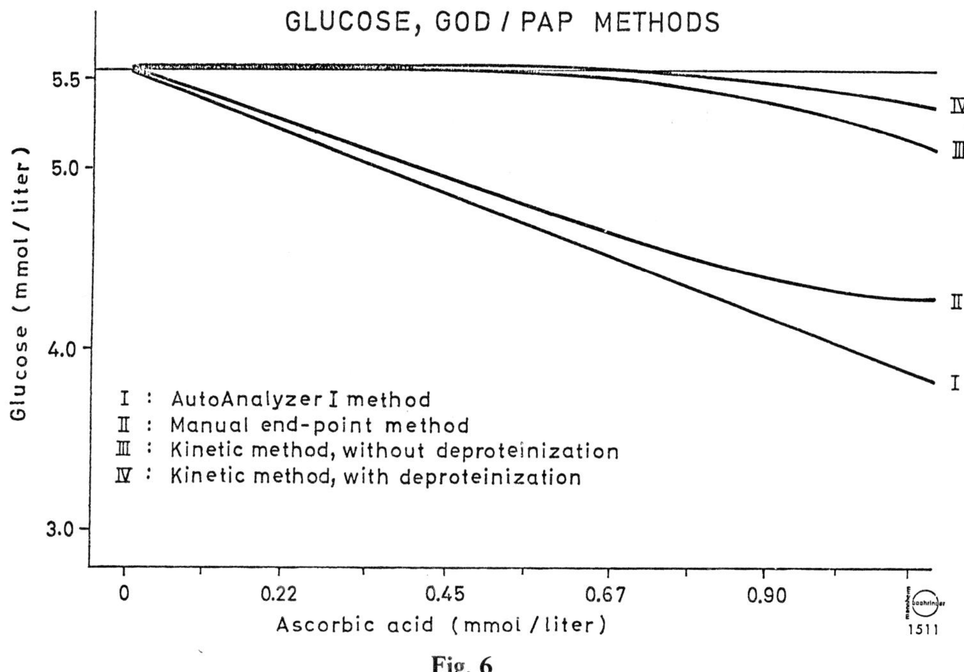

Fig. 6

In conclusion, I should like to summarise: today we feel free to develop the optimal automated method which may be quite different from the manual procedure. Kinetic procedures for substrate determinations show special advantages which can be carried out with automated systems, where manual performance will fail. We may anticipate that in the near future suitable methods will be developed for most of the interesting substrates. The theoretical penetration of the problems will lead us to the best method for every special case.

References

1. I.D. Ingle, jr., and S.R. Crouch, Analytical Chemistry *43*, 697-701 (1971)
2. G.I. Buffone, I. Savory and R.E. Cross, Clin. Chem. *20*, 1320-1323 (1974).

EVALUATION OF ENZYME DETERMINATIONS IN MYOCARDIAL INFARCTION: MEASUREMENTS OF S-CREATINE KINASE, S-ASPARTATE AMINOTRANSFERASE, AND S-LD CATALYTIC ACTIVITIES WITH THE SCANDINAVIAN (S.C.E.) METHODS SUPPLEMENTED WITH QUALITATIVE ESTIMATION OF S-CREATINE KINASE AND S-LACTATE DEHYDROGENASE ISOENZYMES

W. Gerhardt and S. Hofvendahl

Summary

Data are presented on the creatine kinase (CK, EC 2.7.3.2) reaction system according to the recommendation of the Scandinavian Committee on Enzymes (S.C.E.). By a simple modification of the LKB 8600 measurements with this reagent were carried out during the equilibrium phase 120-180 seconds after start of the reaction at a throughput of 50 samples per hour. Qualitative fluorescence estimations of S-CK MB and MM isoenzymes were made with the S.C.E. reagent on agarose film electrophoresis.

S-Aspartate aminotransferase (S-ASAT, EC 2.6.1.1) catalytic activities in patient sera were measured in the presence and absence of pyridoxal phosphate with the IFCC and S.C.E. S-ASAT reagents, respectively.

In a group of patients from a coronary care unit samples were taken every 6 hours and the following enzyme activities were measured with the S.C.E. reaction systems: S-CK, S-ASAT, S-ALAT and S-LD, supplemented with qualitative estimations of both S-CK and S-LD isoenzymes.

The value of these enzyme measurements including the possible benefit of S-CK MB estimations in the diagnosis and management of patients with suspected acute myocardial infarction is critically evaluated.

I. SCE* CK reaction system

A. *Reaction*

1. Creatine phosphate + ADP $\xrightarrow{\text{EC 2.7.3.2. (CK)}}$ creatine + ATP

2. ATP + glucose $\xrightarrow{\text{EC 2.7.1.1.}}$ glucose-6-phosphate + ATP

Dept. of Clin. Chem. and the Coronary Care Unit. Lasarettet, S-251 87 Helsinborg, Sweden.

*SCE: Scandinavian Committee on Enzymes of the Scandinavian Society for Clinical chemistry and Clinical Physiology.

Members: J. Stromme (chairman) and L. Theodorsen, Norway; M. Horder, Denmark; M. Härkonen and E. Pitkänen, Finland; N. Tryding, J. Waldenström and W. Gerhardt, Sweden.

3. Glucose-6-phosphate + NADP $\xrightleftharpoons{\text{EC 1.1.1.49}}$ gluconate-6-phosphate + NADH + H^+

Reaction 1 can be measured in both directions but since creatine phosphate has a significantly higher free energy than ATP the equilibrium is shifted to the "backward" direction. At pH optimum the rate of turn-over is about four times higher in the "backward" direction as compared to the forward.

The proposed SCE CK method (fig. 1) was worked out by team-

Fig. 1 – Composition of proposed SCE CK reaction medium.

S.C.E. reaction system (37°C)	
ATP: creatine-N-phosphotransferase EC 2.7.3.2. creatine kinase (CK)	
Imidazole acetate buffer	100 mmol/l
pH 6.5 (37°C) pH 6.7 (25°C)	
Creatine phosphate	30 mmol/l
N-acetyl cysteine	20 mmol/l
ADP	2 mmol/l
AMP	5 mmol/l
Ap5A	10 μmol/l
D-glucose	20 mmol/l
Magnesium acetate	10 mmol/l
NADP	2 mmol/l
HK (EC 2.7.1.1)	3600 U/l*
G-6-PDH (EC 1.1.1.49)	2000 U/l*

Reduction of –S–S– groups: 5 min
Lag phase: 2 min
Continuous monitoring of reaction rate
at steady state conditions
Volume fraction of serum 0.044

*Measured in CK reagent at 37°C as specified by S.C.E.

work within the Scandinavian Committee on Enzymes. From the early stages, experimental data have been exchanged and discussed with members of the German Enzyme Committee. It is the common objective of both committees that the final CK method be multinational rather than national.

One basic problem has been the differences in the selected reaction temperatures: 37°C in Scandinavia and certain other countries, 25°C in Germany, and 30°C in some parts of the world. As will be shown in the following, it has been possible to select components and concentrations that function satisfactorily at both 25°C and 30°C.

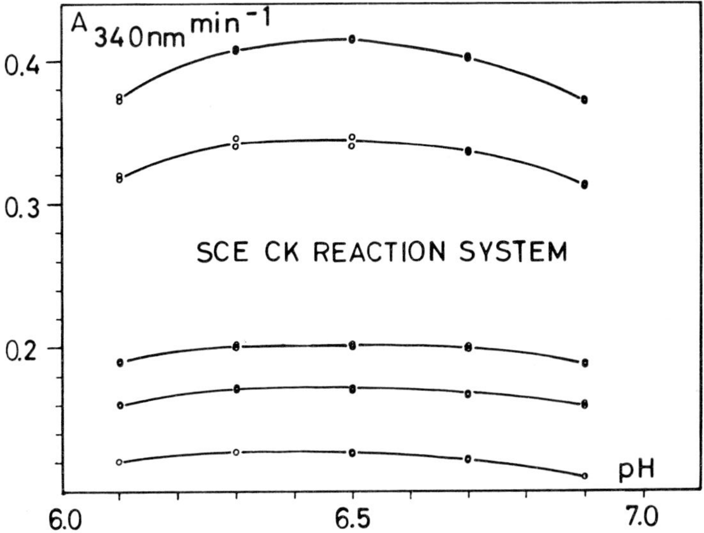

Fig. 2 – pH optimum of CK in the proposed medium. Reaction plotted as a function of pH (37° C).

The data given below are from experiments carried out at 37°C with human sera from cases of myocardial infarction except where specified otherwise.

Final concentrations of all components were found to agree well with those given at 25°C by Szasz[15, 16].

Reaction rates were measured under steady state conditions in the time interval 120-180 seconds after start of the reaction with creatine phosphate.

1. *pH optimum.* (37°C) Fig. 2 shows a flat pH optimum from pH 6.4 to pH 6.6 within 98 per cent of V max. A pH of 6.5 (37°C) was selected.
2. *Buffer.* Imidazole has a pK of 7.1. Imidazole – acetate buffer has a temperature coefficient corresponding to a decrease of pH from 6.7 (25°C) to 6.5 (37°C). The reaction pH of 6.7 (25°C) selected by the German Enzyme Committee[16] will thus be 6.5 at 37°C, which corresponds to the pH optimum at 37°C as shown in fig. 2.
   A quality requirement for the Imidazole is that 340 nm absorbance of the buffer shall be less than 0.050.
3. *Creatine phosphate* (fig. 3). With a Km value for creatine phosphate of 1.3 mmol/l a concentration of 30 mmol/l = 23 x Km was selected.
4. *ADP.* (fig. 4) 2 mmol/l was selected.
5. *NADP*$^+$. As has been shown by Szasz[15, 16] NADPH produced will inhibit reaction rate. The extent of this inhibition depends on the respective NADP$^+$ – NADPH concentrations. Increase of NADP$^+$ concentration to 2

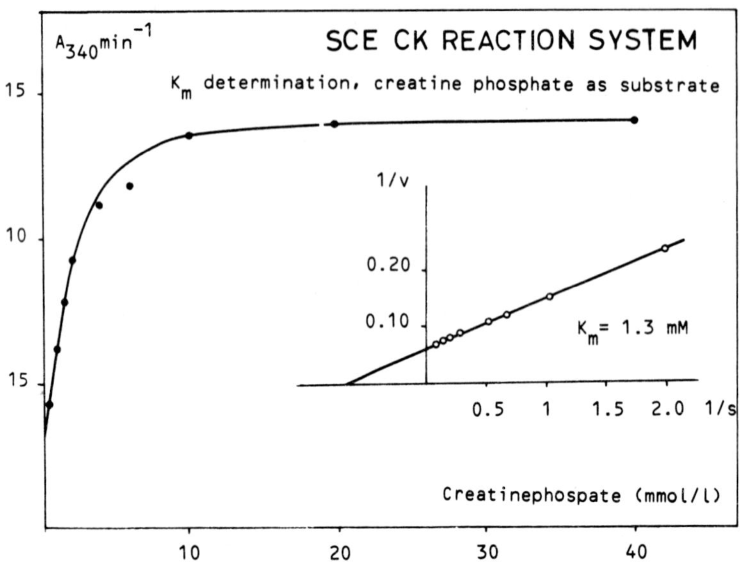

Fig. 3 – Saturation curve of CK for creatine phosphate. Reaction rate plotted as a function of creatine phosphate concentration in the proposed CK medium.

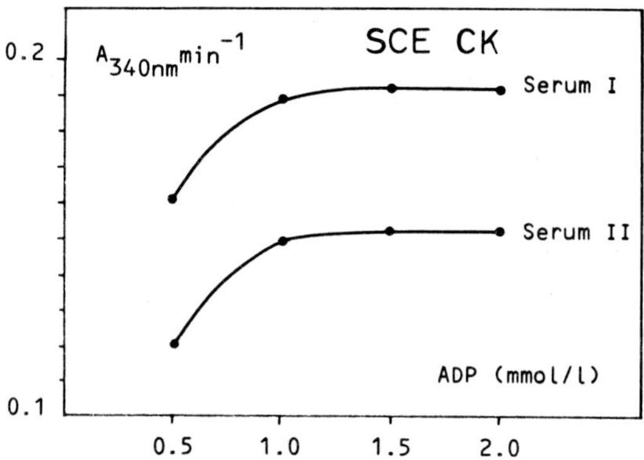

Fig. 4 – ADP saturation curves.

mmol/l extends the period of constant reaction rates (fig. 5) and also permits direct measurements of high CK activities.

6. *Specificity*. Adenylate kinase (EC 2.7.4.3, AK) of erythrocytes, muscle and liver[7, 13, 15, 16, 17] may interfere with CK determinations appearing as increased, unspecific activity. As shown by Szasz, S-AK is normally up to 50 U/l (25°C,[15, 16]).

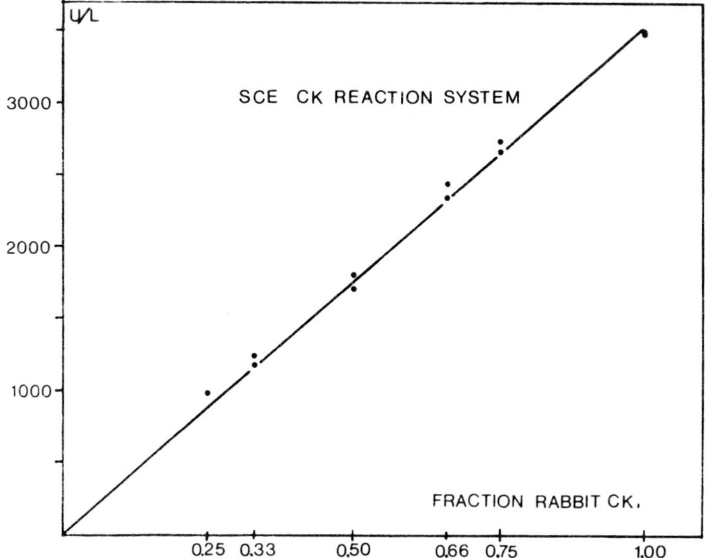

Fig. 5 – Linearity of SCE CK reaction system. Reaction rate as a function of rabbit CK diluted in imidazole-acetate buffer, pH 6.7 (25°C). Volume fraction of sample kept constant. Reaction rates read as soon as steady state conditions were observed.

AK may be inhibited by the competitive inhibitor AMP. The AMP concentration required depends on the ADP concentration. However, AMP also inhibits CK. As in the German method, a compromise concentration of 5 mmol/l AMP was selected. At this concentration of AMP, however, 10 per cent to 20 per cent of the erythrocyte or muscle AK is not inhibited. This may be seen as relatively large sample blank reaction rates of up to 15 U/l in the absence of creatine phosphate, which are increased even further if some hemolysis is present in the sample.
Addition of yet another competitive AK inhibitor. p^1, p^5-di (adenosine-5¹) pentaphosphate ($AP_5 A_1$[7, 13]) 50 μmol/l causes nearly complete inhibition of AK and none of CK. Inclusion of AP_5 A 50 μmol/l reduces sample blank rates as high as 15 U/l in the presence of 5 mmol/l AMP to less than 1 U/l.

7. *Reactivation of CK sulfhydryl groups.* Among several possibilities Dithiothreitol (DTT) and N-acetylcysteine (NAC) were closely investigated. Details of these experiments will be given elsewhere. DTT was finally excluded due to a variable sample blank turbidity that increased with time.
8. *Auxiliary enzymes.* The required Hexokinase (EC 2.7.1.1, HK) and Glucose-6-phosphate dehydrogenase (EC 1.1.1.49, G-6-PDH) activities have been varied through many experiments. Details will be given with the final version of the method. The enzyme concentrations are the same as in the

German CK method[15, 16] but are measured in the CK medium at 37°C. An important point is that auxiliary enzymes must be free of ammonium sulphate which inhibits the apparent CK reaction rate up to 20 per cent.
9. Linearity of method is shown in fig. 5.

The complete method will be published by the SCE in the Scand. I. clin. Lab. Invest. 1976.

Acknowledgement

The author wishes to thank the Segerfalska stiftelsen, Helsingborg, for financial support for this project. He also wants to thank Professor Gabor Szasz, University clinic of Giessen and Dr. Wolfgang Gruber, Boehringer Tutzing for valuable discussions and advice in the elaboration of the CK method and the research group of Boehringer Tutzing for generous gifts of chemicals and enzymes for the CK method.

References

1. The complete list is given at the end of the following part.

II. EVALUATION OF THE DIFFERENTIAL DIAGNOSTIC INFORMATION OF S-CK, S-ASAT, S-LD AND S-CK MB AND S-LD H_4 ISOENZYME DETERMINATIONS IN MYOCARDIAL INFARCTION

W. Gerhardt, S. Hofvendahl, J. Börgesson and B. Hedenäs

II. Enzyme patterns in serum of patients with suspected acute myocardial infarction

Material

47 patients taken into a coronary care unit on suspicion of an acute myocardial infarction. Samples were taken on admission and subsequently at six hour intervals until normalization of the enzyme values.

Sampling time for the first samples varied between 1 and 24 hours with an average of 6 hours after onset of the acute illness.

Methods

S-ASAT, S-ALAT and S-LD were determined by the SCE methods[12, 4]. The upper reference limits (URL) used in this study were 40 U/l for both S-ASAT and S-ALAT[10] and 450 U/l for S-LD.

In a few cases S-ASAT determinations were carried out in the presence of pyridoxal phosphate, 100 μmol/l.

S-CL was determined by the proposed SCE CK method, using a preliminary URL of 150 U/l. As described elsewhere[5] a modified LKB was used in which the CK reaction was initiated by injection of the starting reagent, creatine phosphate, through an extra nozzle 2 minutes before samples reached the measuring position. Even without mixing, the concentrations of intermediates start to build up during these 2 minutes so as to approximate steady state conditions. After mixing by the instrument reaction rates were monitored at steady state conditions in the time interval 120-180 seconds after start of the reaction. Throughput with this procedure is 50 samples per

---

Unofficial abbreviations: S-ASAT = S-alanine aminotransferase, EC 2.6.1.1; S-ALAT = S-alanine aminotransferase, EC 2.6.1.2.; S-LD = S-lactate: NAD-oxido-reductase, EC 1.1.1.27.

hour. S-CK and S-LD isoenzyme activities were qualitatively estimated by agarose electrophoresis at peak total activity if this was above the upper reference limit.

10 μl samples were applied to 1.2 per cent agarose gels in 750 mmol/l diemal buffer, pH 8.6. Electrophoresis was carried out under continuous buffer circulation and water cooling. S-CK isoenzymes were evaluated by NADPH fluorescence. After electrophoresis the agarose slides were incubated with the SCE CK reaction medium concentrated by a factor of two so that final concentrations within the agarose gel would approximate those of the proposed SCE CK reaction system. A filter paper soaked with the concentrated CK reagent was placed on the agarose slide. After incubation for 60 minutes in a humid chamber at 37°C this filter paper was removed. A new, dry filter paper was placed in close contact with the wet agarose slide, soaking up NADPH produced. Both filter papers were then dried at about 50-60°C with a hair drier and viewed in UV light. After drying, the fluorescence of NADPH formed at the sites of the CK MM and MB bands on the filter paper appeared considerably intensified as compared to the fluorescence seen on a wet slide.

The detection limit of this CK isoenzyme procedure was 25-30 U/l estimated by electrophoresis of dilutions of known activity from a nearly pure CK MB preparation from the human heart.

S-LD isoenzymes were estimated on the same electrophoresis system by the Nitro-Blue Tetrazolium technique[11]. Increase of LD H_4 to an intensity greater than that of LD H_3M was considered positive for myocardial infarction[1,6]

The enzyme data were expressed as relative values and plotted as a function of time after onset of the acute symptoms as shown in figs. 6-9.

Results

In order to illustrate the basis for the subsequent grouping of the material some examples of enzyme patterns are shown in figs. 6-9.

All these cases had had chest pain suggestive of acute myocardial infarction. Fig. 6 represents the enzyme patterns in a 47-year-old male. The first samples were taken 9 hours after onset of acute symptoms. Relative to the upper reference limit (URL), CK, ASAT and LD were increased by factors of 20, 9, and 5, respectively. On the other hand an initial ALAT value of 2 URL remained relatively constant. At peak total activities CK MB, and LD H_4 isoenzymes were increased.

In this report such an enzyme pattern is considered diagnostic for an acute myocardial infarction even in the absence of conclusive ECG changes.

In this case, however, also the ECG was also diagnostic of an AMI.

On the other hand, increase of total CK alone and even the combination of highly increased CK and ASAT activities may be found in the absence of AMI.

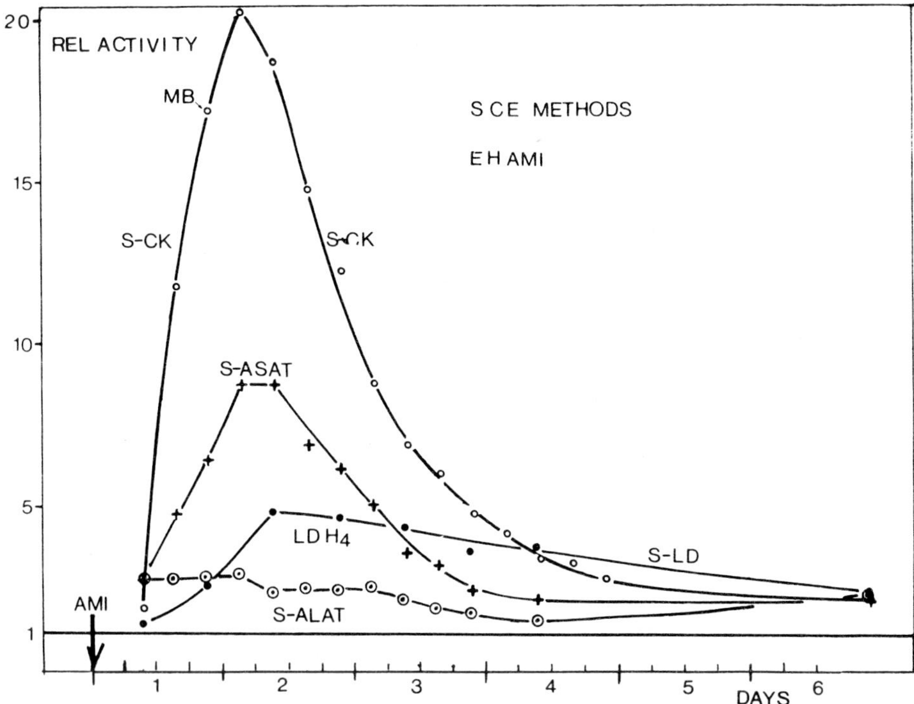

Fig. 6 – Serum enzyme pattern in a 49-year-old male with AMI. Total enzyme activities in serum were plotted as a function of time after acute onset of illness (arrow marked AMI). The enzyme activities are shown as relative values with respect to the upper reference limit. The reference range is shown as the area below the horizontal line in the graph.

Increase of S-CK MB and S-LD H_4 isoenzymes are marked with arrows at the appropriate point on the curves that represent the corresponding total activities.

ECG diagnostic for AMI.

Fig. 7 demonstrates this well-known fact. A 46-year-old male was admitted with fainting followed by chest pain. The first sample was taken 3 hours after onset of symptoms. Within 24 hours a peak CK of 7 URL, and 12 hours later a peak ASAT of > 5 URL were found. In contrast, LD remained constantly close to the URL. An initially high ALAT is not characteristic of AMI, but indicates a previous liver affection. This was confirmed by the demonstration of increased LD M_4 isoenzymes throughout the observation period. The failure to demonstrate CK MB or increased LD H_4-values at peak total activities excluded the presence of AMI. It was subsequently revealed that the patient was a heavy drinker and that the acute illness had been preceded by a heavy drinking bout.

ECG showed changes corresponding to an older infarction with no signs of new acute myocardial infarction.

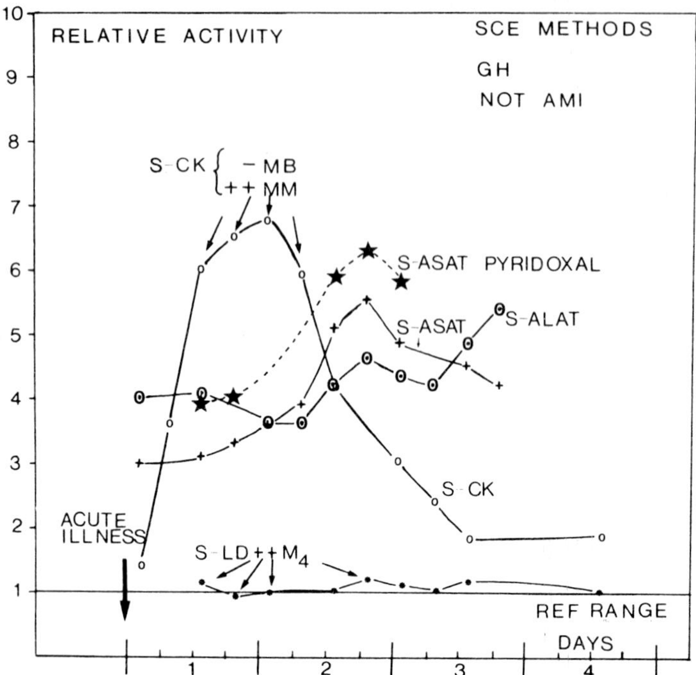

Fig. 7 – Serum enzyme pattern in a 46-year-old male with an acute episode in chronic alcoholism. Observe similarity of the enzyme pattern to fig. 6. S-CK MB and increase of S-LD H_4 isoenzymes could not be demonstrated.

ECG negative. Pt was classified as not AMI. Note parallel course of S-ASAT ± pyridoxal phosphate.

Fig. 8 shows essentially the same type of enzyme changes as in fig. 6 but on a smaller scale. This case was a 68-year-old female. The first sample was drawn 1 hour after the onset of chest pain. However, as discussed above, (fig. 7) only the demonstration of increased CK MB and LD H_4 isoenzymes can be considered conclusive for the enzymological diagnosis of AMI. In this case the ECG-findings were non-conclusive.

The last example in this series (fig. 9) is a case in which the acute increases of ASAT and LD activities were suggestive of an AMI. However, the absence of increased CK values and the atypically high increase of ALAT do not support a diagnosis of AMI.

The patient was a 52-year-old female with acute cholelithiasis. The hepatic origin of the increased enzyme values was confirmed by the elevated LD M_4 and ALAT values in the absence of increased LD H_4 value.

As shown by numerous other cases typified by the examples given above even complete total enzyme activity patterns do not constitute conclusive

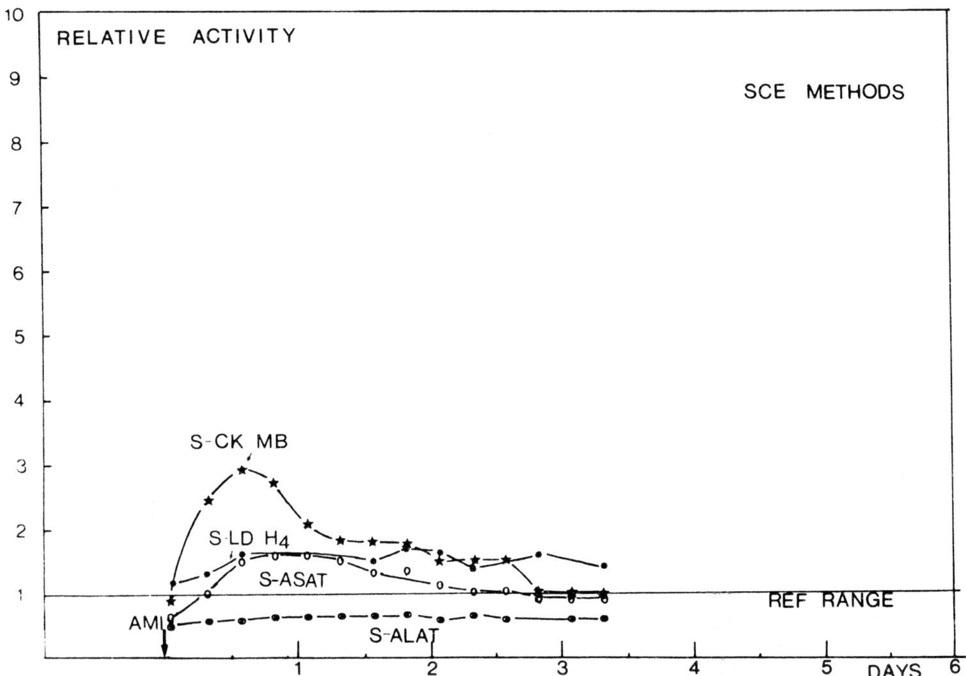

Fig. 8 – Serum enzyme pattern in a 68 year old female. Note the smaller scale as compared with fig. 6. Both S-CK MB and increased S-LD H_4 isoenzymes were demonstrated at the arrows. ECG was non-conclusive. Classified as AMI.

enzymological diagnostic evidence for AMI. For this purpose CK MB and/or LD H_4 isoenzyme determinations are necessary.

Many years of experience in several laboratories have established the increase of LD H_4 activity as one of the most reliable enzymological criteria for AMI provided that certain interference such as hemolysis can be excluded. In the light of this evidence we decided to separate our total material into two groups according to the presence or absence of increased LD H_4 isoenzymes values determined at peak total LD activity. Within each group the frequency of increased or normal values of each of the other tests were tabulated (tab. 1-2). Tests were considered as possible indicators of AMI according to the following preliminary set of criteria:

CK > 2 URL within 12 hours
CK MB Demonstrable (> 25 U/l)
ASAT > 1.5 URL within 24 hours in combination with a lower ALAT value within the first 36 hours.

In the group of 23 patients (table 1) with increased S-LD H_4 activity all

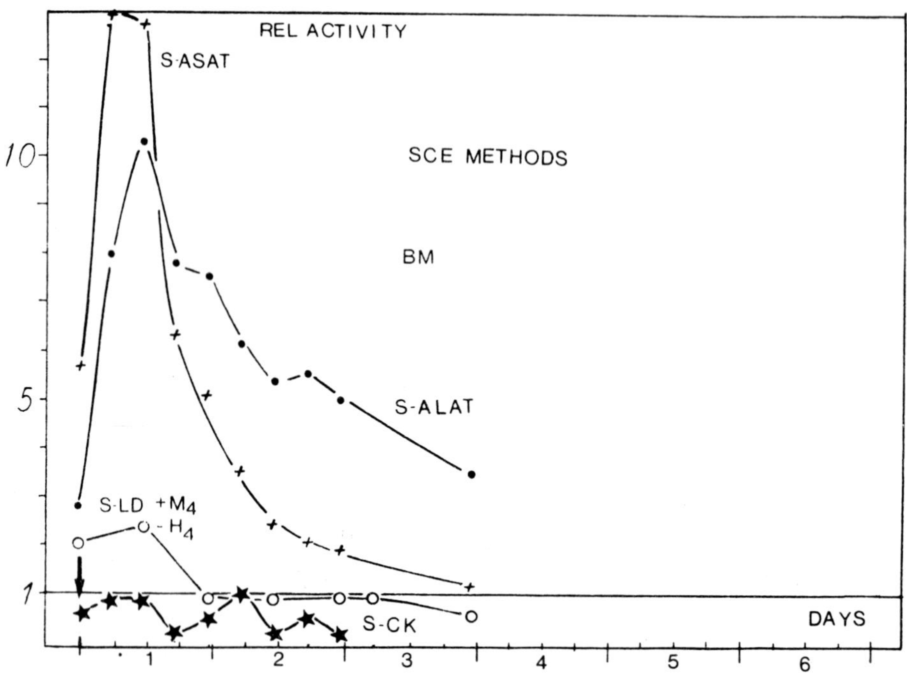

Fig. 9 – Serum enzyme patterns in a 52-year-old female. Increased total S-CK, S-CK MB, and S-LD H_4 isoenzymes could not be demonstrated.

S-LD M_4 was found to be increased at arrow. Pt was classified as not AMI and found to have had an acute attack of cholelithiasis.

Table I. *Frequency of positive S-CK, S-CK MB, and S-ASAT values in the group with increased S-LD H_4.*

Patients with increased S-LD H_4		
Total number	*23*	*Per cent*
S-CK > 2 · URL	23	100
S-CK MB +	22	96
S-ASAT > 1.5 URL/24 hours with S-ALAT < S-ASAT/36 hours	23	100
ECG conclusive for ami	15	65
Non conclusive	8	35

had increased S-CK and S-ASAT values according to the criteria set out above. CK MB isoenzyme was demonstrable in all but one of these patients.

Table II. *Frequency of positive S-CK, S-CK MB, and S-ASAT values in the group with no increase of S-LD H_4 at peak total S-LD activity.*

Patients with no increase of S-LD H_4 at total S-LD maximum		
Total number	*24*	*Per cent*
S-CK > 2 · URL	3	13
S-CK MB⁺	0	0
S-ASAT > 1.5 · URL/24 hours with S-alat < S-asat/36 hours	6	25
ECG conclusive for ami	0	0

In this case the first sample was taken 24 hours after the acute onset of pain and no CK MB was found. ECG changes conclusive for AMI were found in 65 per cent of this group.

Table 2 contains the group of 24 patients without increased LD H_4. Three of these patients had peak CK values between 2 and 22 times the upper reference limit. Six more showed CK values of 1.1 to 1.5 URL. One of these was a case of pericarditis, two were alcoholics.

One of the three cases with high total CK has been discussed above. CK MB could not be demonstrated in any of these 9 cases at peak CK activity.

Increase of ASAT in this group was related to liver cell damage as indicated by increased S-LD M_4 and/or S-ALAT in 5 cases. None of the patients in this group had ECG changes conclusive for AMI.

Discussion

Myocardial *infarction* (MI), is a *clinical* entity. It is a complex clinical and laboratory diagnosis which is based on numerous relative criteria including the patient's history reviewed against the physician's experience, electrocardiographical findings, and laboratory enzymological data. As experience has accumulated during several years with improvement of methodology more and more emphasis has been put on the enzymological data. For classification of our material an increase of the LD H_4 isoenzyme in serum was considered the best investigated single enzymological criterion for acute myocardial infarction[1,6]. This was confirmed by the finding of increased total CK and ASAT in serum from all patients in table 1, and CK MB isoenzyme in all but one of these patients with increased serum LD H_4.

The finding of ECG changes conclusive for AMI in 65 per cent of a series of AMI cases is in good agreement with cardiologist's experience. Four of the 5 tests applied were thus positive for each patient in this group.

Consequently, the group in table 1 were classified as positive with respect to AMI.

The other group contained the patients with no increase of LD H_4 at peak total serum LD activity. In none of these could either CK MB or conclusive ECG changes be demonstrated. No patient in this group had 4 positive tests of the 5 tests applied. Consequently, this group was classified as negative with respect to AMI.

Accepting this classification more than 10 per cent of the AMI negative group showed false positive total CK increases of more than 2 URL. This figure is in good agreement with other reports[2, 9, 14].

The incidence of false positive ASAT values was 25 per cent. All of these cases had liver affections as verified by an increased LD M_4 isoenzyme or other criteria.

The patients in this report constitute a selected material. All were admitted to a coronary care unit on suspicion of AMI. The incidence of false positive CK or ASAT values could be higher in patients from general surgical or medical wards in which incidence of muscle lesions or e.g. hepato-biliary conditions is higher.

Conclusion

Our data indicate that even serial determinations of total CK, ASAT, and LD activities in serum do not provide the differential diagnostic information required for a conclusive enzymological diagnosis of AMI in the *individual* case.

In contrast, increase of either CK MB or LD H_4 in serum determined as described can be considered highly specific indicators of acute myocardial infarction. The two tests appear to be equally informative. The advantage of CK MB lies in the earlier appearance of this enzyme in serum.

Acknowledgement

The authors want to thank the Segerfalska stiftelsen, Helsingborg, for the generous financial support that made this project possible.

References

1. Auvinen S.: Evaluation of Serum Enzyme Tests in the Diagnosis of Acute Myocardial Infarction. Acta Med. Scand. Suppl., 539, 1972.
2. Blomberg D.J., Kember W.D. ans Burke M.D.: Creatine Kinase Isoenzymes. Predictive Value in the Early Diagnosis of Acute Myocardial Infarction. Am. J. Med. *59,* 464-469, 1975.
3. Friman G.: Serum Creatine Phosphokinase in Epidemic Influenza. Scand. J. Infect. Dis. *8,* 13-20, 1976.
4. Gerhardt W.: Evaluation of some Determination Materials used in the Scandinavian

Recommended Methods (S.C.E.) for Determination of four Enzymes in Blood and the Performance of Control Materials with these Methods. P 45 in: Quality Control in Clinical Chemistry. (Ed: G. Anido, S.B. Rosalki, E.J.U. Kampen, M. Rubin, W. de Gruyter). N.Y. 1975.

5. Gerhardt W., Kofoed B.: Modification of LKB 8600 for 1-minute Determination of Creatine Kinase Activity in Serum. Scand. J. Clin. Lab. Invest. *35,* 381-383, 1975.
6. Konttinen A., Auvinen S. and Somer H.: Reliability of Serum LD Determinations in the Diagnosis of Myocardial Infarction. Clin. Chim. Acta, *52,* 245-247, 1974.
7. Lienhard G.E., Secemski I.I.: P^1, P^5-Di (Adenosine-5^1) Pentaphosphate, a Potent Multisubstrate Inhibitor of Adenylate Kinase. J. Biol. Chem. *248,* 1121-1123, 1973.
8. Moss D.: Reactivation of the Apoenzyme of Aspartate Aminotransferase in Serum. Clin. Chim. Acta, *67,* 169-174, 1976.
9. Nygren A.: Serum Creatine Phosphokinase Activity in Chronic Alcoholism in Connection with Acute Alcoholic Intoxication. Acta Med. Scand. *179,* 623-629, 1966.
10. Penttila I.M., Jokela H.A., Viitala A.J., Heikkinen E., Nummi S., Pystynen P. and Saastamoinen J.: Activities of Aspartate and Alanine Aminotransferases and Alkaline Phosphatases in Sera of Healthy Subjects. Scand. J. Clin. Lab. Invest. *35,* 276-284, 1975.
11. Rosalki S.B.: Scand. Committee on Enzymes of the Scand. Soc. Clin. – Chem. and Physiol. Recommended Methods for the Determination of Four Enzymes in Blood. Scand. J. Clin. Lab. Invest. *33,* 291-306, 1974.
12. Sachsenheimer W., Goody R.S. and Schirmer R.H.: Elimination und Exkretion von Adenylatkinase noch Zellschädigungen. Klin. Wschr. *53,* 617-622, 1975.
13. Somer H.: Determination and Clinical Significance of Creatine Kinase Isoenzymes in the Serum. Acad. Dissert., Wikuri Res. Inst. and Dept. Medicine University of Helsinki, 1975.
14. Szasz G.: Laboratory Measurement of Creatine Kinase Activity. Proc. 2nd Int. Symp. Clin. Enzymol. Chicago, 1975.
15. Szasz G., Gruber W. and Bernt E.: Creatine Kinase in Serum. I. Optimum Reaction Conditions. Clin. Chem. (in press).
16. Szasz G., Gruber W. and Bernt E.: Interference of Adenylate Kinase with Creatine Kinase Activity Measurement. Proc. 28 Nat. Meeting of A.A.C.C., Huston, 1976.

BIOCHEMICAL PROBLEMS OF AMINOTRANSFERASES ASSAYS

A. Castelli

Summary

In 1954 Wroblewsky, La Due and Karmen started the modern age of clinical enzymology, introducing the measurement of transaminase activity in serum. In 1955 Karmen devised a spectrophotometric method in which the transamination reaction is coupled to the reduction of oxalacetate to malate in the presence of "indicator enzyme".

In the last few years the most important parameters for the standardization of AST (E.C. 2.6.1.1.) and ALT (E.C. 2.6.1.2) determinations have been discussed by almost every clinical enzymologist.

As far as the reaction temperature. one of the most important factors in the optimization of enzyme assay, is concerned, different Enzyme Commissions recommended as standard temperatures 25°C, 30°C, 37°C, showing little agreement among their experts.

The optimization of the physical and chemical conditions of enzyme assays deserves further discussion for in many cases there is still no correlation between biochemical data and clinical features.

Nevertheless Aminotrasferases are the enzymes most frequently determined in a Clinical Chemistry Laboratory.

AST and ALT isoenzymes have been separated from different human tissues and their serum activities have been found to be differently elevated in the course of various diseases.

Furthermore evident differences in kinetic behaviour between mitochondrial or cytosol isoenzymes from different cells have been described.

In the light of possible applications to diagnosis of the different kinetic behaviour of AST and ALT isoenzymes, a critical revision of recommended standard methodology would be opportune, in particular with regard to the kinetics of reversible reactions, to the "indicator enzymes" and to any other physicochemical parameters.

New optimized procedures are then needed if we want to research further differential diagnostic criteria based on isoenzyme activity from several sources.

Among the last recommendations of IFCC methods for the measurement of catalytic concentration of enzymes, Bowers et al.[1], there are the following considerations (1:1:1): "the diagnostic value of determining the

---

Istituto di Chimica Biologica, Università Cattolica S. Cuore - Roma, Italy.

catalytic activity of an enzyme in serum which has made its importance in medicine should be maintained or enhanced. The effect on total activity of alterations in the proportions of multiple molecular forms of a particular enzyme found in human serum must also be considered".

The IFCC recommendations were published 20 years after transamination reactions were first applied in the diagnosis of heart infarction by La Due et al. in 1954[2] and of acute hepatitis by de Ritis et al. in 1955[3].

During the same period a considerable growth in the utilization of enzyme reactions for clinical diagnosis, among which transaminase determinations are still the more commonly performed, has been reported (Castelli, A., unpublished data)[4].

A review of the literature shows that some problems of prevailing interest can be grouped together:

1. Problems concerning the method
2. Problems concerning the reagents concentration
3. Problems concerning the reaction temperature
4. Problems concerning the instrumentation.

1.1. *Choice of method*

The techniques employed for aminotransferase assays are based mainly on two fundamental measurement procedures:

1. The first type of method, introduced by Karmen et al. 1955[5], measures indirectly point to point the reaction rate by coupling the system with a NAD-dependent redox reaction in which NADP oxidation can be followed by means of U.V. spectrophotometry.
2. The second type, introduced by Reitman and Frankel in 1956[6] and subsequently modified, directly measures the enzyme reaction product by means of an end point *colorimetric reaction* with 2.4. dinitrophenylhydrazone.

For a series of historical reasons the Reitman method had been prevalently used by clinical chemists in the period 1955-1970, while from 1970 on the Karmen approach has been preferred.

The various methods chosen, anyway, have not yet reached the result of reproducibility desired by many researchers, even though in the last five years many standardization and optimization procedures have been recommended by competent national and international authorities among which I may briefly mention the German Society "Empfehlunghen etc." in 1972[7], Bergmeyer et al. in 1970[8], The English Clinical Biochemistry Association and the Royal College of Pathology, Moss et al. in 1971[9], Wilkinson et al. in 1972[10], IFCC Draft N. 8, Cooper et al. in 1973[11], the already mentioned IFCC 1975 recommendations, Bowers et al. in 1975 and the one issued by the Scandinavian Society in 1973[12].

A very important role as a discussion forum has been performed by the International Symposia of Clinical Enzymology where outstanding contributions on the subject were presented by Szasz in 1972[13], Holtz in 1973[14], Sandifort in 1975[15], Rich and Lang in 1973[16], and Bergmeyer in 1975[17].

In the kinetic determinations of aminotransferases, in which we deal with two substrates reactions, the relative concentrations of six variables should be taken care of, i.e., having prefixed the temperature: substrates concentration, pH, ionic strength, buffer, cofactor concentration.

Complete studies have been published by many authors: Bergmeyer in 1972[18], Bergmeyer et al. in 1970[8], Wilkinson et al. in 1972[10], Bowers et al. in 1975[1], Sandifort et al. in 1973[19]. A substantial agreement toward method standardization is emerging from their results. Some differences still exist in commercial preparations, both in the buffers used and in substrates concentration.

Temperature standardization

The IFCC draft by Bowers et al. in 1975[1] recommends the reaction temperature of 30.00°C.

Up to this date the opinions have been very controversial among national and international societies' committees on the choice of different reaction temperatures.

The problem has been approached from the optimal reaction temperature view-point by Bergmeyer in 1973[20] and by Szasz in 1974[21], from the enzyme protein denaturation view-point by Michie et al. in 1969[22], Thompson in 1969[23] and again Szasz in 1972[24], and from the enzyme stability at various temperatures view-point by Carter et al. in 1971[25].

The different experimental conditions make it difficult for the analysts to express their values of serum aminotransferase activity in an unified manniter even for the physiological range. On this particular subject papers have been published by several German and English researchers, Bergmeyer et al. in 1971[26], Wilkinson et al. in 1971[27] and recently Siest et al. in 1975[28].

Actually the problem of determination of serum aminotransferases in the Clinical Chemistry laboratory cannot be completely solved only by means of reagent and reaction temperature standardization.

The instruments, too, particularly the continuous and discrete automatic ones, influence the output quality in some cases, by modifying the dilution ratio and preincubation time in mixing the reagents.

It could then be hypothesized that by standardising methods and instruments, reliable results of aminotransferase assays could be obtained.

As a matter of fact, the problem seems more complex when a short review of actual biochemical knowledge on aminotransferases is taken into consideration.

They were first shown by Braunstein in 1937[29], and their 40 years of history have been enriched by continuous observations which clarified the role of vitamin B_6, Snell in 1945[30], and defined the first isoenzymes in 1968.

Up to today, at least 50 types of transaminating reactions are known in the various organs and tissues and the list is likely to be extended.

The fields of study of pre-eminent interest in the last 20 years, contemporarily to aminotransferases activity diagnostic application, have concerned:

- separation, purification and identification of monomers, with aminoacid sequence analysis.
- clarification of active site structure and chemical topography, and molecular mechanism of the reaction.
- macro-molecular geometry (conformational and quaternary structure) with localization and identification of the role of pyridoxal-phosphate.
- multiple aspects of isoenzyme kinetics from various sources.

As far as the structural organization is concerned, the enzyme was shown to be a dimer formed by two equal subunits, by Fasella in 1968[31].

Several monomers in each subunit were demonstrated electrophoretically by Fasella et al. in 1967[32].

The Enzyme is known to behave as a pH indicator: at a low pH the active form with an absorbance peak at 362 nm is transformed into the inactive form with absorbance peak at 426 nm. The spectral variation is due to the ionization of the phenolic group of the coenzyme as PLM (pyridoxal phosphate); an AST enzyme was isolated bound to PMP (pyridoxamine phosphate) with absorbance peak at 330 nm independently from pH.

Molecular weight has been recently estimated at 90.000, Michuda et al. in 1969[33], Polyanovsky et al. in 1970[34], by means of x-ray diffraction technics.

One AST mole binds two coenzyme moles at the lysine residues of the protein moiety.

The primary structure of heart cytosol AST was identified by Polyanavsky et al. in 1973[35]. Each subunit is composed by 412 aminoacids with 46.344 M.W. (93.147 m.W. for the holoenzyme).

From these observations it can be inferred that the protein moiety clearly plays a determinant role for the enzyme catalytic activity and specificity and that it is important to study the isoenzyme conformation to explain the differences in substrate and coenzyme binding and transformation behaviour.

Role of the coenzyme

It is known that the aminotransferases coenzyme is in the form of a Schiff base with the aminic group of the lysine residue in the active site. This fact promotes the rapid formation of aminoacid and coenzyme binding at the active site.

The principle of reaction mechanism developed by Braunstein et al. in

1953[36] and Snell in 1958[37] has been confirmed by recent reports.

Using enzymes purified from red blood cells and human liver, the factors influencing in vitro the association of vitamin B_6 with aminotransferases have been examined.

Phosphate buffer slows down this binding to a great extent in comparison to tris buffer.

This inhibition was observed for mitochondrial AST too by Nissl-baum in 1968[38] and it could be partially reversed by vitamin B_6 preincubation in the assay mixture, Rosalki et al. in 1975[39].

In human serum the pyridoxal-phosphate level is largely variable, 12 ± 15 U/L, Hamfelt in 1967[40], and this could determine the variation in the kinetics of aminotransferases isoenzyme subforms, Cheung and Brigss in 1974[41].

The deviation of aminotransferase activity from the straight Arrhenius plot at higher temperatures is prevented by the addition of pyridoxal-phosphate Jung and Egger in 1975[42], indicating the important role played by coenzyme concentration and enzyme-coenzyme affinity variation together with the effect of temperature variation.

Isoenzymes and multiple sub-forms

Both AST and ALT contain two principal protein fractions, both showing aminotransferase activity.

They differ in subcellular localization in electrophoretic mobility and in some physico-chemical characteristics.

The fraction from animal cell supernatant is also called anionic by way of its electrophoretic mobility while the mitochondrial fraction has a cationic behaviour.

They are true isoenzymes differentiated genetically nd with independent metabolic functions and regulatory mechanisms, Katanuma et al. in 1968[43], Nisselbaum et al.[44].

At their turn, isoenzymes present numerous subforms with different catalytic properties, Fasella et al. in 1968[45], Martinez-Carrion et al. in 1967[46].

Heart AST shows 4 principal electrophoretic fractions, called α β γ δ among which q presents most catalytic activity, Martinez-Carrion 1967[46].

These subforms do not differ in their primary structure but in coenzyme binding properties[46], and possess a characteristic content in carbohydrates (> 10%)[47].

The behaviour of serum isoenzymes is in relation to their source: heart isoenzymes seem to be more activated by vitamin B_6 than the liver ones[48].

Studies on transamination reaction kinetics have not yet yielded conclusive results.

The mechanism is generally classified as "Ping-Pong Bi-Bi" in the Cleland nomenclature[49].

An α -aminoacid is bound to the enzyme and the corresponding α -oxo-acid is released, then a second oxo-acid is bound and a second α -amino-acid is released.

The experimental values of the parameters measuring the kinetics of the reaction for the various substrates differ remarkably among the researchers[50, 51].

Variations are ascribed principally to differing isoenzyme combinations, to the type and concentration of anions in the incubation media, or to the presence of mitochondrial isoenzymes[52, 53]

It is therefore clear that the diagnostic application in Clinical Chemistry did not adequately consider the results achieved in aminotransferase biochemistry.

The attempts to improve aminotransferase determinations have dealt mostly with the standardization and optimization of the measurement conditions, without being strongly influenced by the chemical and biochemical aspects of the enzyme activity.

Serum GOT and GPT simultaneously determined do not always supply valid information for a differential diagnosis in liver and heart lesions.

Furthermore, the determination methods do not show the possible varying compositions in isoenzymes of the amino-transferases present in serum in pathological conditions.

On the basis of such considerations we decided to:

1. Evaluate results variability as a function of methods and instruments.
2. Determine the liver and heart isoenzymes kinetics in a laboratory animal at different temperatures.
3. Control stability of subcellular fraction iso-enzymes stored at different temperatures.
4. Examine the possibility of a different isoenzyme behaviour in human cells.

Discussion

For a statistical evaluation of serum ALT and AST determinations a comparison has been made between Du Pont ACA, LKB 8600 and Beckman TR at 37°C, with Du Pont own Reagents for ACA and with Biochemia optimized reagents for LKB and BTR.

Serum samples from in and out patients were analyzed simultaneously on two or all three instruments.

Results were statistically evaluated with the aid of an Olivetti System P 652, according to the methods of Westgard et al.[54].

It was found that statistical tests can furnish error-parameters in good agreement over different ranges of enzyme activity (Fig. 1), thus allowing some corrective procedure in instrument calibration. As this is not always the case (Fig. 2), we considered the possibility of selecting the situations

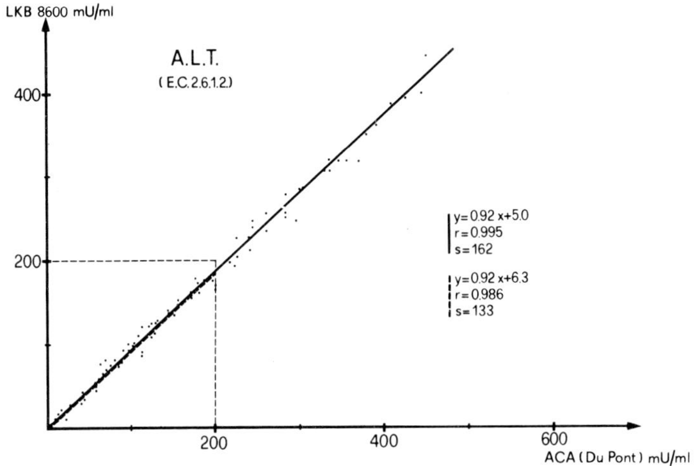

Fig. 1 – ALT results compared: LKB 8600 vs Du Pont ACA. Regression lines show the same proportional and constant errors over different ranges (55).

where interchangeability of instruments is acceptable[55].

The relation between aminotransferases activity and temperature had been one of the main concerns of clinical chemists.

Experts' agreements did not yield satisfying results, for in several instances temperatures varying from 25°C to 37°C were chosen.

Results were discussed from the points of view of general kinetic considerations and of the Arrhenius equation on the assumption that the behaviour of a well-defined enzyme structure was examined.

As a matter of fact in the case of each aminotransferase we are dealing with at least two isoenzymes showing different kinetic properties, each one of them composed by four separable and differently active sub-froms.

To contribute to the study of isoenzymes kinetic behaviour we examined the V_{max} and K_m of rat liver and heart AST and ALT activities at two different reaction temperatures 25 ÷ 0.1°C and 37°C ÷ 0.1°C.

Experiments were carried out in triplicate on subcellular fractions of livers and hearts of fasting Wistar rats, weighing around 150 g., from the same breeding lot kept on a standard laboratory diet.

The experimental results show that liver and heart mitochondrial and

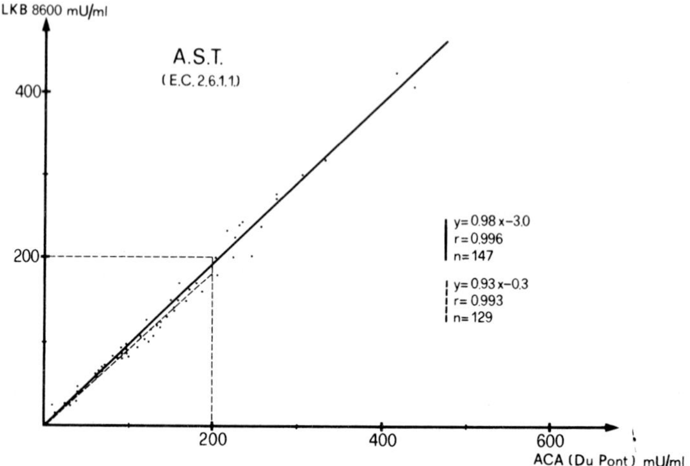

Fig. 2 – AST results compared: LKB 8600 vs Du Pont ACA. Regression lines show different proportional and constant errors over different ranges (55).

cytosol isoenzymes present different kinetics and vary differently by changing the reaction temperature[56].

Corresponding K_m values were calculated: remarkable differences can be shown between fractions and between temperatures. This is shown by the graphical representation in which the activities are confronted at the two temperatures for each fraction.

Liver cytosol AST increases by 720%, heart cytosol AST by 520%, heart mithocondrial AST 1000%. Practically no variation is shown by liver mitochondrial AST (Fig. 3).

Heart cytosol ALT increases by 790%, heart mitochondria ALT by 400%, no significant variation is shown by liver fractions (Fig. 4).

To study the problems inherent in sample activity conservation and enzymatic protein structure, liver fractions were stored at three different temperatures up to 72 hours.

ALT activity generally decays independently from storage temperature, while AST shows a paradoxical behaviour particularly evident in liver homogenate (Fig. 5).

This fact is further shown by the following figures.

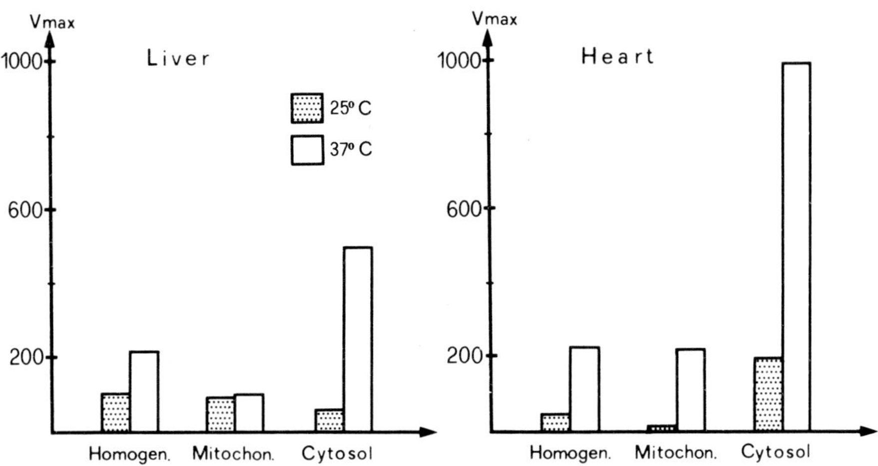

Fig. 3 – AST activity of subcellular fractions of rat liver and heart at two different temperatures (56).

Liver homogenate: (Fig. 6): there are significant differences between AST and ALT activities after storage at +4°C, –20°C, –40°C. AST in fact display a paradoxical behaviour, affected by thermic variation during storage and determination times, showing an increase in the enzyme catalytic activity.

ALT, on the other hand, display a more "regular" behaviour, presenting a clear decrease in activity already after 24 hrs., then keeping stable or decreasing further after 72 hrs.

This decrease is more gradual when storage temperature is –20°C.

Mitochondrial fraction (Fig. 7): AST: as in the homogenate. ALT: as in the homogenate, but the loss in activity is more "gradual" at the three storage temperatures.

Cytosol fraction (Fig. 8): AST: after a transient small decrease after the 24th hrs., the activity increases after the 72nd hour. ALT: loss of activity already after 24 hrs. The process increases after 72 hrs.

The "paradoxical" behaviour of AST found in liver homogenate, mito-

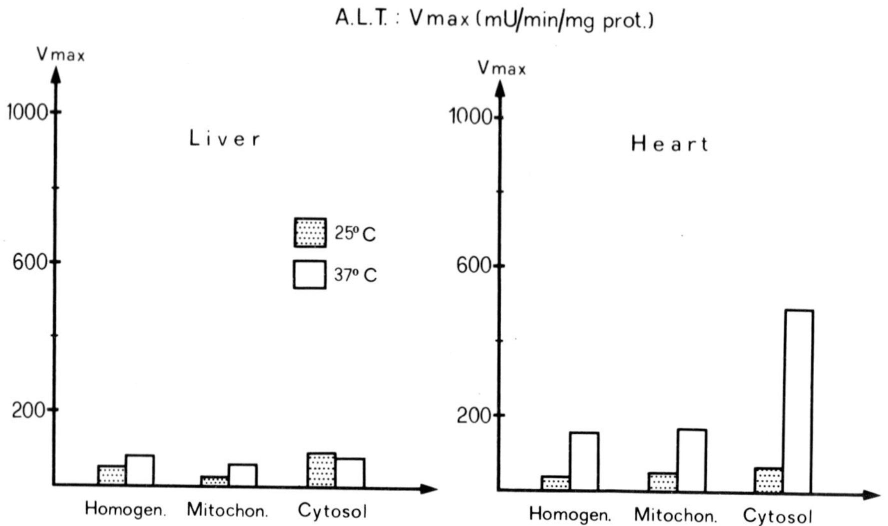

Fig. 4 – ALT activity of subcellular fractions of rat liver and heart at two different temperatures (56).

chondrial and cytosol fractions, seems to be ascribed to the thermic stress which induces a positive conformational modification of the molecule structure, hence an increase in catalytic activity, resembling an enzyme protein "renaturation" effect.

The cytosol AST isoenzyme had been already described by Giartosio in 1972[57] to present two ways of binding pyridoxal-phosphate: a kinetically inactive, although thermodynamically more stable, and thus prevalent way, and a second way, less stable from thermodynamic standpoint, but more active as regards kinetics.

It can then be assumed that temperature modifies the protein secondary, tertiary and quaternary structure in such a way that the new molecular conformation is more active kinetically.

As a concluding remark, we can say that the temperature problem should be examined experimentally for the single isoenzymes for Arrhenius equa-

Fig. 5 – Stability of liver AST and ALT stored at various temperatures (values expressed in percent of activity at 0' time)

	AST		*ALT*		*Storage Temperature*
	24 h.	*72 h.*	*24 h.*	*72 h.*	
Homogenate	131	153	35	32	+ 4° C
	127	150	64	27	— 20° C
	136	156	47	53	– 40° C
Mitochondria	113	126	102	28	+ 4° C
	140	120	71	28	– 20° C
	124	132	63	28	– 40° C
Cytosol	75	82	43	30	+ 4° C
	88	107	46	44	– 20° C
	85	91	41	40	– 40° C

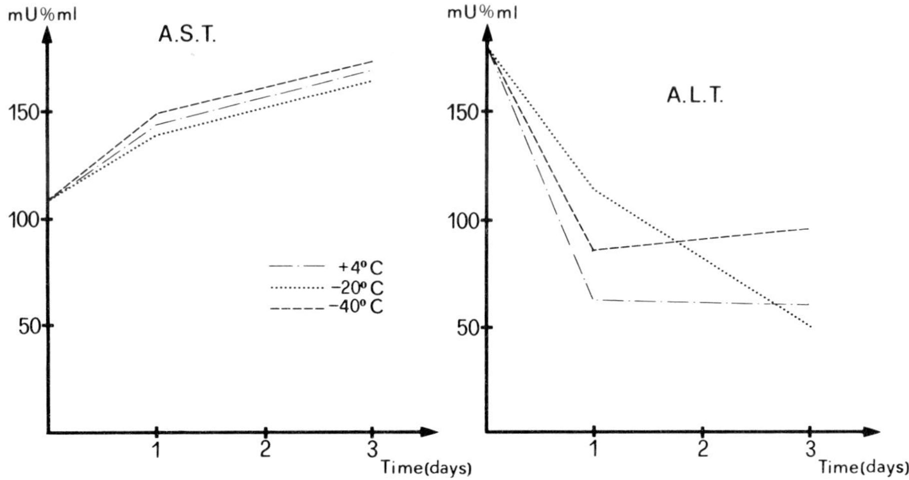

Fig. 6 – Stability of A.S.T.-A.L.T. (liver homogenate) stored at various temperatures

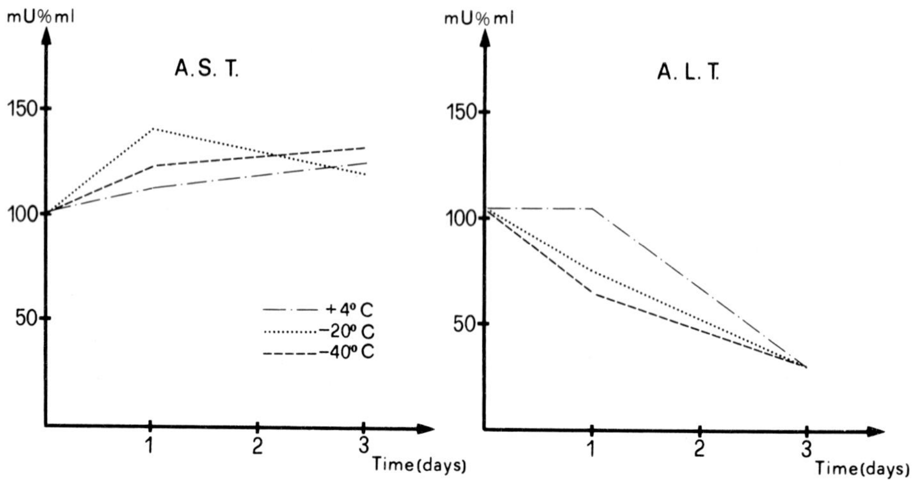

Fig. 7 – Stability of A.S.T.-A.L.T. (liver mitochondria) stored at various temperatures

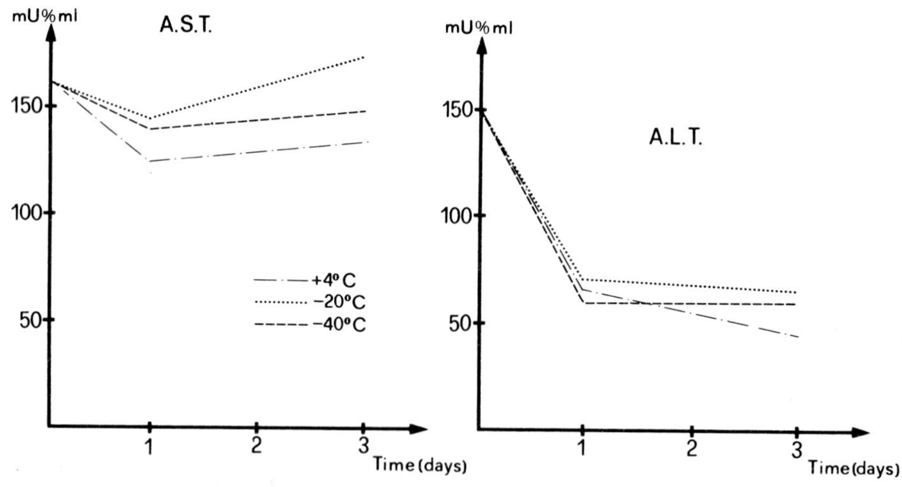

Fig. 8 – Stability of A.S.T.-A.L.T. (liver cytosol) stored at various temperatures

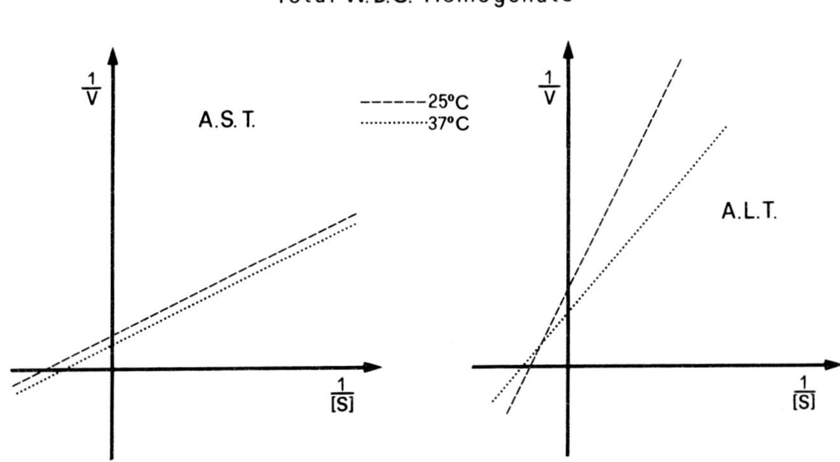

Fig. 9 – Lineweawer-Burk plots for W.B.C. homogenate aminotransferases.

tion application is limited by the temperature dependent variability of single isoenzyme activity.

Do results obtained from laboratory animals have significance in the understanding of human diagnostic problems?

Because of the different biochemical characteristics of aminotransferases we have taken the opportunity to examine AST and ALT isoenzymes behaviour in human cells too. We have then chosen to use white blood cells, which are readily available for quantitative research with the usual biochemical methodologies.

Blood was drawn from healthy donors and lymphomonocytes (MN) and granulocytes (PMN) were separated and then homogenized employing several tecniques.

The cell homogenates underwent differential ultracentrifugation and AST and ALT were determined in the sub-cellular fractions at 25°C and 37°C.

The results on aminotransferase specific activity (Tab. 1) are in good agreement with Belfiore[58] who found that a higher AST activity was present in lymphocytes than in granulocytes (although not separating them by means of mathematical calculations), and that ALT activity was very low and about the same in the lymphocytes as in the granulocytes, and with Dioguardi et al.[59, 60], who showed that mitochondrial AST represented 68% and cytosol AST 32% of total activity in lymphocytes while PMN mitochondrial AST was 35% and cytosol AST 65% of the isoenzymatic fractions isolated by means of electrophoresis and chromatography.

The experimental data indicate that also in human cells AST and ALT isoenzymes seem to present different kinetics and are differently influenced by temperature variation[61].

Lineweaver-Burk plots show that in total WBC homogenates AST K_m values are lower at 25°C than at 37°C, while ALT ones are lower at 37°C than at 25°C (Fig. 9).

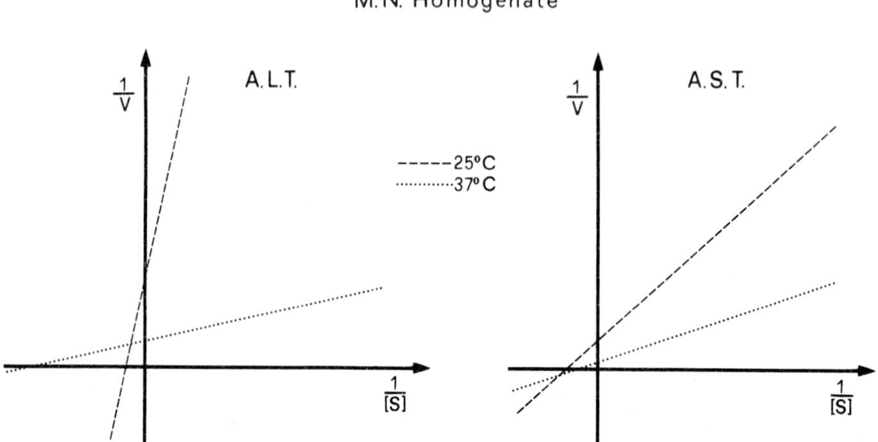

Fig. 10 – Lineweawer-Burk plots for M.N. homogenate aminotransferases (61).

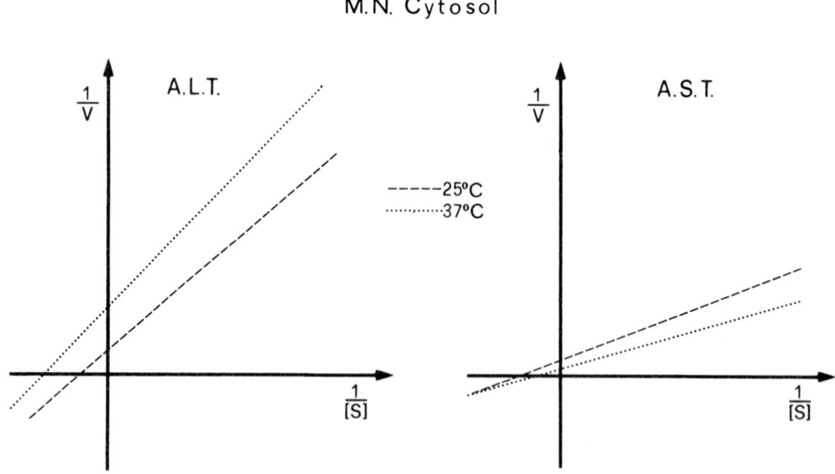

Fig 11 – Lineweawer-Burk plots for M.N. cytosol aminotransferases (61).

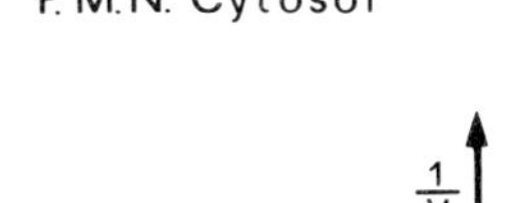

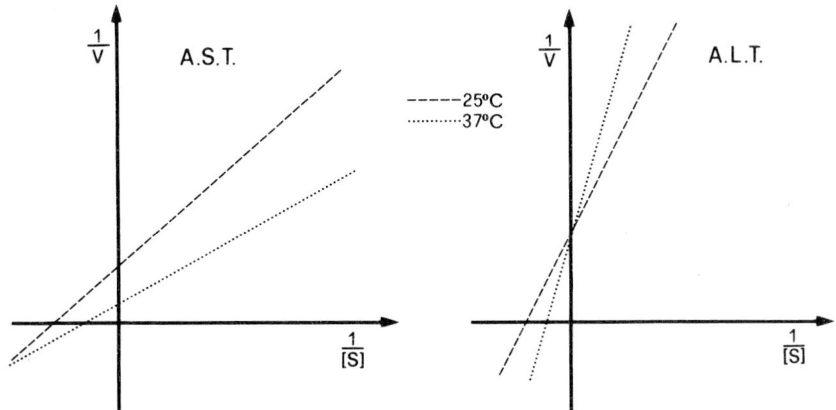

Fig. 12 – Lineweawer-Burk plots for P.M.N. cytosol aminotransferases (61).

The same behaviour for example is also shown by MN homogenate: i.e., ALT K_m values are lower at 37°C, AST K_m values lower at 25°C (Fig. 10) and by MN cytosol (Fig. 11).

On the contrary, PMN cytosol ALT shows an inversion: i.e., a higher K_m at 37°C (Fig. 12) and MN mitochondrial AST a lower K_m value at 37°C (Fig. 13).

Table I. *Intracellular distribution of Aminotransferase activities in Human WBC*[61]

	Specific activity (mU/mg prot.)			
	AST		*ALT*	
	25°C	*37°C*	*25°C*	*37°C*
WBC homogenate	24.3	29.5	9.8	11.5
WBC mitochondria	14.2	17.8	6.6	7.8
WBC cytosol	26.0	53.0	14.5	6.0
MN homogenate	23.1	93.2	5.0	31.4
MN mitochondria	74.8	57.5	9.5	29.7
MN cytosol	37.5	68.0	18.5	10.0
PMN homogenate	16.0	24.5	4.4	8.3
PMN mitochondria	6.6	21.2	5.4	14.9
PMN cytosol	13.5	31.0	9.4	7.0

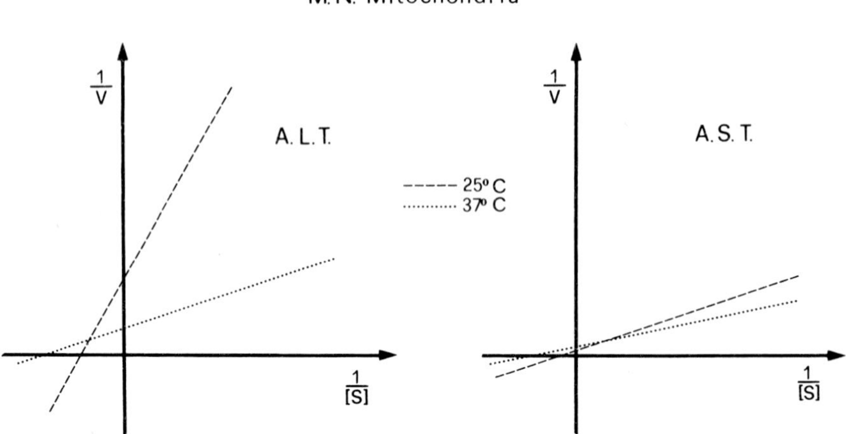

Fig. 13 – Lineweawer-Burk plots for M.N. mitochondria aminotransferases (61).

Conclusions

The results obtained on WBC are not comparable to existing data on human liver and heart and it is not clear so far if WBC metabolism is closer to one of these two cellular types or to the other.

It might however be possible to use these parameters in future for a differential diagnosis purpose, based on the metabolic role of isoenzymes at a subcellular level.

We must also consider the results of other types of research on aminotransferases modulation induced by diet protein content, vitamin supplementation and steroid therapy.

It is known that rat liver aminotransferases are a good experimental model to check hormonal effects on specific protein biosynthesis[62].

Mitochondrial matrix ALT activity varies easily and is influenced by glucocorticoid hormones, being different in this respect from the cytosol fraction[63].

In the rat liver, mitochondrial ALT activity increases very significantly by 5-10 fold after prednisolone administration[64], not because of a K_m variation but by way of an enhanced enzyme protein biosynthesis[65, 66, 67, 68].

ALT induction is affected by adrenectomy and is partially inhibited by Actinomycin D[69].

It has been proved furthermore by some researchers that protein and B_6 content in the diet affect aminotransferase activity[70, 71].

The diet protein content leads to a different quantitative effect on mitochondrial isoenzymes from liver and heart[72].

Finally ascorbic acid or its oxidation products can induce in pig heart multiple forms of AST and modify their kinetics[73, 74].

By no means do serum AST and ALT determinations maintain their diagnostic value in the case of acute heart or liver lesions for their great elevation in serum.

These results, being so elevated (5 to 10 times the normal values), do not usually require any accurate and precise method or instrumentation standardization.

The problem looks quite different in sub-acute or chronic cases in which even the most exquisite method and instruments standardization has been found unfit to reduce results variability and to explain many diagnostic questions.

In such pathologic conditions, the modest serum aminotransferases variations (no more than twice the normal range) could be most easily ascribed to:

- variation in the diet protein content;
- vitamin supplementation (B_6, ascorbic acid);
- therapeutic administration of steroid hormones.

On the basis of experimental results it might be possible in future to substitute the two serum transaminases coupled determinations, of uncertain diagnostic value, with single ALT or AST mitochondrial or cytosol isoenzyme determination, using of course methods optimized for temperature, substrates and buffers, but also specific for the single protein.

References

1. Bowers, Jr., G.N., Bergmeyer H.U., Moss D.W.: Clin. Chim. Acta *61,* F11-F24, (1975).
2. La Due, J.S., Wroblewski F. and Karmen A.: Science *120,* 497 (1954).
3. De Ritis F., Coltorti M. and Giusti G.: Minerva Med. *46,* 1207 (1955).
4. Castelli A., (unpublished data).
5. Karmen A., Wroblewski F. and Ladue J.S.: J. Clin. Invest., *34,* 131-133 (1955).
6. Reitman S. and Frankel S.: Am. J. Clin. Pathol., *28,* 56, (1956).
7. Bergmeyer H.U., Kreutz F.H., Pilz W., Schmidt F.W., Buttner H., Lang H., Rick W., Stamm D., Hillmann G., Laue D., Schmidt E. and Szasz G.: Z. Klin. Chem. u. Klin. Biochem., *10,* 182-192, (1972).
8. Bergmeyer H.U., Rich W., Büttner H., Hillmann G., Schmidt E., Schmidt F.W., Kreutz F.H., Stamm D., Lang H., Szasz G. and Lanz D.: Z. Klin Chem. u. Klin. Biochem., *8,* 658-660, (1970).
9. Moss D.W., Baron D.N. Walker P.G., Wilkinson J.H.: J. Clin. Path., *24* (1971), 740-743.

10. Wilkinson J.H., Baron, D.N., Moss D.W. and Walker P.G.: J. Clin. Path., *25,* (1972), 940-944.
11. Cooper G.R., Draft n. 8 della IFCC: Clin. Chem., *19,* N. 2 (1973), 269-275.
12. cfr. Sandifort C.R.J.: Proc. 5th Int. Symp. on Clin. Enz., *1,* (1973), 193-210. Ed. Burlina, A., Published Edizioni Lab.
13. Szasz G., Quad. Sclavo Diagnost., *9,* N. 1 (1973), 23-46.
14. Holtz A.H.: Proc. 5th Int. Symp. on Clin. Enz., *1,* (1973), 149-158. Ed. Burlina, A., Published Edizioni Lab.
15. Sandifort, C.R.J., Proc. 5th Symp. on Clin. Enz., *1,* (1973), 193-210. Ed. Burlina, A., Published Edizioni Lab.
16. Lang, H. and Rick W.: Proc. 5th Int. Symp. on Clin. Enz., *1,* (1973), 185-192, Ed. Burlina, A., Published Edizioni Lab.
17. Bergmeyer H.U., Bernt E. and Scheibe P.: Proc. 6th Int. Symp. on Clin. Enz.
18. Bergmeyer H.U.: Clin. Chem., *18*, (1972), 1305-1311.
19. Degeller K., Sandifort C.R.: Clin. Chim. Acta, *43,* (1973), 13-22
20. Bergmeyer H.U.: Z. Klin. Chem. Klin. Biochem., *11*, (1973), 39-45.
21. Szasz G.: Z. Klin. Chem. Klin. Biochem., *12,* (1974), 166-170.
22. Michie D.D., Booth R.W., Conley M., MT (ASCP), Mc Guire H.J.: Am. J. Clin. Pathol., *52,* N. 3, (1969), 329-333.
23. Thompson W.H.S.: Clin. Chim. Acta, *23,* (1969), 105-120.
24. Szasz G.: Scand. J. Clin. Lab. Invest., *29,* (1972), 126.
25. Carter J.E., Graham E.F., Lillehei R.C., Blackshear P.L.: Cryobiology; *8,* (1971), 524-534.
26. Gruber W., Bergmeyer H.U.: British Medical Journal, *IV,* october-december, (1971), 749-750.
27. Baron D.N., Wilkinson J.H., Levin G.E.: British Medical Journal, *III,* juli-september, (1971), 583.
28. Siest G., Schiele F., Galteau M.M., Panek E., Steinmetez J., Fagnani F., Gueguen R.: Clin. Chem., *21,* N. 8 (1975), 1077-1087.
29. Braunstein A.E., Kritzmann M.G.: Enzymologia, *2,* (1937), 129.
30. Snell E.E.: J. Am. Chem. Soc., *67,* (1945), 194.
31. Fasella P., Snell E.E., Braunstein A.E., Severin E.S., Torchinsky Yu.M.: Eds. Interscience Publishers, 1968, p. 1.
32. Martinez-Carrion M., Turano C., Chiancone E., Bozza F., Giartosio A., Riva F., Fasella P.: J. Biol. Chem., *242*, (1967), 2397.
33. Michuda C., Martinez-Carrion M.: J. Biol. Chem., *244,* (1969), 5920.
34. Polyanovsky O.L., Zagyansky Yu.M., Tumerman L.A.: Mol. Biol. (USSR), *4,* (1970), 458.
35. Ovchinnikov Yu.A., Egorov C.A., Aldanova N.A., Feigina M.Yu., Lipkin V.M., Abdulaev N.G., Grishin E.V., Kiselev A.P., Modyanov N.N., Braunstein A.E., Polyanovsky O.L., Nosikov V.V., Febs Letters, *29,* N. 1, (1973), 31-34.
36. Braunstein A.E., Shemyakin M.M.: Biokhimiya, *18,* (1953), 393.
37. Snell E.E.: Vitam. Horm. (New York), *16,* (1958), 77.
38. Nisselbaum J.S.: Anal. Biochem., *23,* (1968), 173.
39. Rosalki S.B., Bayoumi R.A.: Clin. Chim. Acta, *59,* (1975), 357.
40. Hamfelt A.: Clin. Chim. Acta, *16,* (1967), 19.
41. Cheung T., Briggs M.H.: Clin. Chim. Acta, *54,* (1974), 127-129
42. Jung K., Egger E.: Clin. Chem., Acta, *64,* (1975), 329-331.

43. Katunuma N., Okada M., Katsunuma T., Fujino A., Matsuzawa T. in "Pyridoxal Catalys: Enzymes and Model Systems" (E.E. Snell et al., eds.), p. 255 Wiley (Interscience), New York, 1968.
44. Nisselbaum J.S. et al.: (vol. 10, p. 275) in "Advances in Enzyme Regulation", ed. Weber, G.
45. Fasella P.: in "Pyridoxal Catalysis: Enzymes and Model Systems" (E.E. Snell et al., eds.), p. 1, Wiley (Interscience) New York, 1968.
46. Martinez-Carrion M., Turano C., Chiancone E., Bossa F., Giartosio A., Riva F., Fasella P.: J. Biol. Chem., *242,* (1967), 2397.
47. Bossa F., John R.A., Barra D., Fasella P.: Febs Lett., *2,* (1969), 115.
48. Rosalki S.B., Bayoumi R.A.: Clin. Chim. Acta, *59,* (1975), 357-360.
49. Henson, C.P., Cleland W.W., Biochemistry, *3,* (1964), 338.
50. Wada H., Kagamiyama H., Watanabe T.: in "Pyridoxal Catalysis: Enzymes and Model Systems" (E.E. Snell et al., eds.) p. 111. (Interscience) New York, 1968.
51. Nisselbaum J.S., Bodansky O.: J. Biol. Chem. *239,* (1964), 4332; idem *241,* (1966), 2661.
52. Nisselbaum J.S.: Anal. Biochem., *23,* (1968), 173.
53. Cheng S., Michuda-Kozak C., Martinez-Carrion M.,: J. Biol. Chem., *246*, (1971), 3623.
54. Westgard J.O., Hunt M.R.: Clin. Chem., *19,* (1973), 49.
55. Giocoli G., Barbaresi G., Castelli A.: Proc. of the 7th Int. Symp. on Clinical Enzimology, Venezia, (1976).
56. Barbaresi G., Miggiano G., Castelli A.: Proc. of the 7th Int. Symp. on Clinical Enzymology, Venezia, (1976).
57. Giartosio A., Politi V., Fasella P.: Febs Letters, *31,* 3 (1973), 339.
58. Belfiore F.: Boll. Soc. Ital. Biol. Sper., *34,* N. 12, (1963), 709-715.
59. Ideo G., Mannucci P.M., Musu E., Fiorelli G., Dioguardi N.: Boll. Soc. Ital. Biol. Sper., *41,* (1965), 942-945.
60. Mannucci P.M., Ideo G., Congiu P., Fiorelli G., Dioguardi N.: Boll. Soc. Ital. Biol. Sper., *41,* N. 2, (1965), 954-957.
61. Martorana G.E., Gozzo M.L., Barbaresi G., Castelli A.: Proc. of the 7th Int. Symp. on Clinical Enzymology, Venezia (1976).
62. Monder C., Coufalik A.: Enzyme, *20,* (1975), 111-116.
63. De Rosa G., Swick R.W.: J. Biol. Chem., *250,* N. 20, (1975), 7961-7967.
64. De Rosa G., Swick R.W.: Fed. Proc., *32,* (1973), 897.
65. Schimke R.T.: J. Biol. Chem., *237,* (1962), 459.
66. Kenney F.T.: J. Biol. Chem., *237,* (1962), 1610.
67. Segal H.L., Rosso R.G., Hopper S., Weber M.M.: J. Biol. Chem., *237*, (1962), PC 3303.
68. Feigelson P., Greengard O.: J. Biol. Chem., *237,* (1962), 3714.
69. Patnaik S.K., Kanungo M.S.: Biochem. Biophys. Res. Comm., *56,* N. 4 (1974), 845-850.
70. Rosen F., Roberts N.R., Nichol C.A.: J. Biol. Chem., *34,* (1959), 476.
71. Cheney M.C., Curry D.M., Beaton G.H.: Can. J. Physiol. Pharmac., *43,* (1965), 579.
72. Lalitha K., Radhakrishnamurty R.: Indian J. Biochem. Biophys., *11,* (1974), 309-313.
73. Bossa F., Giartosio A., Petruzzelli R., Fasella P.: Febs Letters, *44,* N. 1 (1974), 31-33.
74. Robinson A.B., Irving K., McCrea M.: Proc. Natl. Acad. Sci. U.S., *70,* (1973), 2122.

CLINICAL AND METHODOLOGICAL STUDIES ON SERUM GLUTATHIONE REDUCTASE ACTIVITY

D.M. Goldberg and R.J. Spooner

Summary

Optimal conditions were developed for the assay of Glutathione Reductase activity (GR; NAD(P)H: glutathione oxido reductase; EC 1.6.4.2) in human serum at 37°C using an LKB Reaction Rate Analyser. Final conditions incorporated phosphate buffer, 100 mmol/liter at pH 7.2; EDTA 0.5 mmol/liter; NADPH 0.17 mmol/liter; and GSSG 2.2 mmol/liter. The normal range determined using 63 fasting laboratory staff (mean ± 2SD) was 47-79 U/liter.

A survey of 68 cancer patients revealed 58% who exceeded the upper reference limit. Comparable data for aspartate aminotransferase, lactate dehydrogenase and alkaline phosphatase were 10%, 31%, and 19% respectively. Raised GR activity was just as common among patients lacking clinical evidence of metastases as in those with clear evidence of spread. It is of no value as an index of prognosis in cancer, but may be useful as a screening procedure.

Further studies on samples with high GR activity demonstrated that sera incubated at 37°C spontaneously lost activity. This inactivation could be slowed in sera from patients with cancer, anaemia, and myocardial infarction, and reversed in patients with liver diseases when the sample was diluted and allowed to stand prior to assay. These differences were sufficiently pronounced to render identification of hepatic GR in serum relatively simple. Tissue-specific differences in response to chromate were also observed. In addition to dilution, pre-incubation with dithiothreitol or acetate buffer, pH 6.4, protected against inactivation; this is most probably due to formation of inactive molecular aggregates through disulphide bridges and hydrogen bonding.

Glutathione reductase (GR; NAD(P)H: glutathione oxido reductase E.C.1.6.4.2) is an important enzyme involved in the maintenance of a reduced intracellular environment. It catalyses the following reaction:

$$\mathrm{GSSG} + \mathrm{NADPH} + \mathrm{H}^+ \rightleftarrows 2\,\mathrm{GSH} + \mathrm{NADP}^+$$

The equilibrium constant strongly favours the formation of reduced glutathione. The prosthetic group of GR is the riboflavin metabolite FAD.

Department of Biochemistry, Hospital for Sick Children, Toronto, Ontario, Canada, and Department of Chemical Pathology, Royal Hospital, Sheffield, U.K.

Correspondence to Dr. D.M. Goldberg, Hospital for Sick Children, 555 University Avenue, Toronto, Ontario, Canada.

Clinical interest in GR has taken several forms. Screening for hereditary erythrocyte GR deficiency has been advocated (Beutler, 1966), and the extent of activation of red cell GR by exogenous FAD has been used as a test for riboflavin deficiency (Glatzle et al., 1968; Beutler, 1969; Sauberlich, 1972, Heller et al., 1974).

Early work showed that the serum enzyme was elevated in many liver diseases (particularly infective hepatitis), pernicious anaemia, and often to a very high level in malignant disease especially where liver metastases were present (Manso and Wroblewski, 1958; Jerppola et al., 1959; West et al., 1961; Horn et al., 1962). Indeed, approximately 33 % of the patients with cancer in these studies had elevated serum GR levels.

The present study was initiated to develop an automated kinetic determination of GR activity employing optimal analytical conditions, and to determine whether the assay was of value in cancer assessment.

Patients

All subjects admitted to a cancer treatment centre had blood taken before therapy. Aspartate transaminase (AsT), Lactate dehydrogenase (LDH) and Alkaline phosphatase (ALP) activities were assayed using the SMA 12/60 (Technicon Instruments Ltd.). [Reference ranges for this laboratory are 10-50 IU/l, 100-225 IU/l and 28-100 IU/l respectively.]

Careful clinical assessment of the subjects was carried out and the salient features of their disease are summarised in Table I. Only one subject had overt hepatic metastases when first examined. [Sites of metastases were superficial glands[6], peritoneal nodes[5], pelvis[4], bones[3], brain[3], bladder[1], lungs[1], widespread[1].]

Table I. *Classification of cancer patients at time serum GR was first assayed*

Location of Cancer	*Males (per cent)*	*Mean Age*	*No. with*	
			Metastases	*No Detectable Metastases*
Respiratory Tract	95	60	4	17
Urogenital Tract	56	52	6	10
Breast	0	59	1	9
Digestive Tract	62	62	3	5
Reticuloendothelial	72	49	5	2
Others	33	46	3	3
Total			22	46

Tissue preparation

Purified preparations of human erythrocyte GR were made using the method of Scott et al. (1963). The source of human liver GR was a high speed supernate (100,000 x g for 1 hr) of fresh *post mortem* liver homogenised 1:3 in ice cold sucrose (0.26 mol/l).

Results

Choice of buffer

Initial studies showed that phosphate was superior to triethanolamine and PIPES (piperazine-N, N'-bis (2-ethane-sulphonic acid)), and that varying the molarity between 0.1 and 0.3 mol/l was without effect.

The inclusion of EDTA in the reaction mixture at a final concentration of 0.5 mmol/l enhanced enzyme activity and raised the pH optimum (fig. 1).

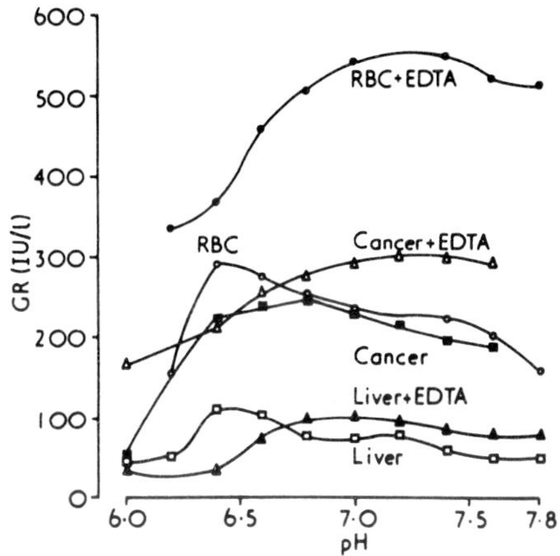

Fig. 1 – Influence of EDTA at a final concentration of 0.5 mmol/l on activity of GR under the conditions of the proposed method upon serum from a patient with cancer and one with acute liver disease, and upon a red blood cell haemolysate (RBC).

This shift in pH optimum occurring when EDTA is included in the reaction mixture allows GR from all sources to be analysed at pH 7.2 and this minimises blank reactions due to spontaneous hydrolysis of NADPH which becomes pronounced below pH 7.0.

Choice of substrate concentration

NADPH was optimal at final concentrations above 0.12 mmol/l (fig. 2). GSSG was optimal at final concentrations above 0.54 mmol/l (fig. 3).

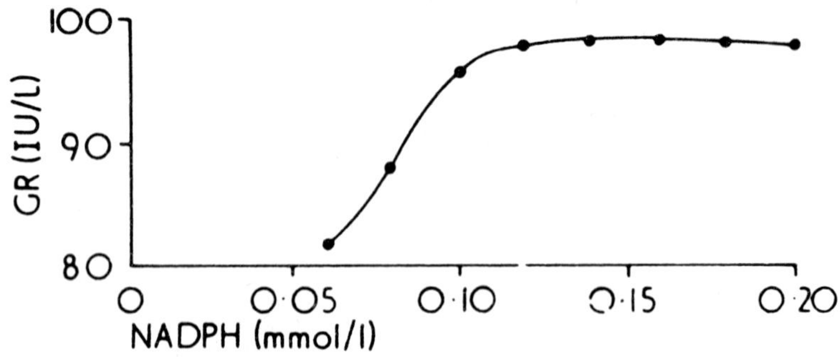

Fig. 2 – Relationship of NADPH concentration to serum GR activity, with all other conditions as in the proposed method.

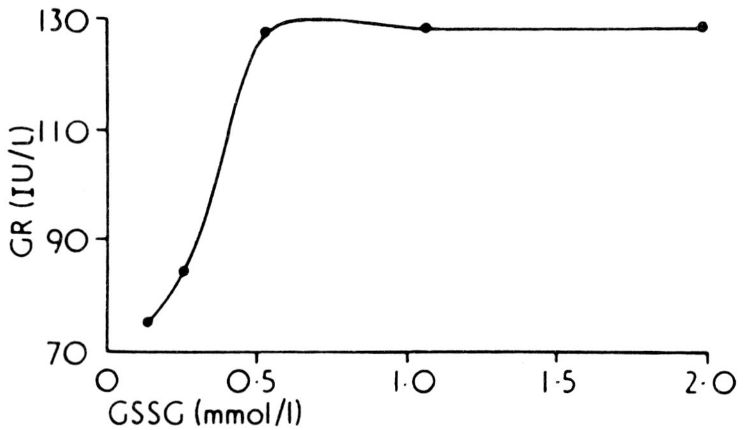

Fig. 3 – Relationship of GSSG concentration to serum GR activity, with all other conditions as in the proposed method.

Effect of enzyme activators

Dithioerythritol (DTE) was included in the reaction mixture to a final concentration of 2 mmol/l and had no significant effect on the activity of the serum enzyme.

Inclusion of FAD in the reaction mixture up to a final concentration of 10 μmol/l (concentrations above this were slightly inhibitory) had no effect on 37 of 38 randomly selected sera. The single exception was a patient with prostatic carcinoma, and riboflavin deficiency could not be ruled out.

Characteristics of the proposed method

Enzyme activity is linearly related to the amount of serum taken to an activity of 1,500 IU/l. Data for within batch precision are given in Table II for three levels of activity.

Table II. *Within batch precision of GR assay*

No. of Samples	*Mean Activity (IU/l)*	*S.D.*	*C.V. (per cent)*
18	52.7	3.72	7.1
15	140.0	6.6	4.7
15	252.5	6.3	2.5

The normal range was determined using 63 fasting laboratory staff. No sex or age dependance was observed and the results were combined to give a reference range (Mean ± 2SD) of 47-79 IU/l.

Enzyme Stability

A bovine based quality control serum was used to standardise the assay but fell in activity by 65% over a 2 h period. The reagents were perfectly stable during this time. To examine this phenomenon more closely, sera were made up in the incubation mixture and maintained at 37°C prior to initiation of the reaction with oxidised glutathione (GSSG) for varying periods of time. The results obtained are presented in Table III. Only the serum enzyme from patients with a liver pathology was activated; sera from subjects with perni-

Table III. *Effect of extended predilution on GR activity **

Patients	*Hepato-biliary disease*	*Cancer*	*Myocardial infarction*	*Pernicious anaemia*	*Red cell hemoly-sate*
Total no.	110	53	24	11	11
Increase	69	0	0	0	0
Decreased	17	49	17	9	6
No change**	24	3	7	2	5
Mean percent change in activity after 2h	+ 36	– 19	– 13	– 13	– 7

*All data refer to serum apart from final vertical column.
**No change was assigned when the percent change in activity was $<$2xCV at that level of activity.

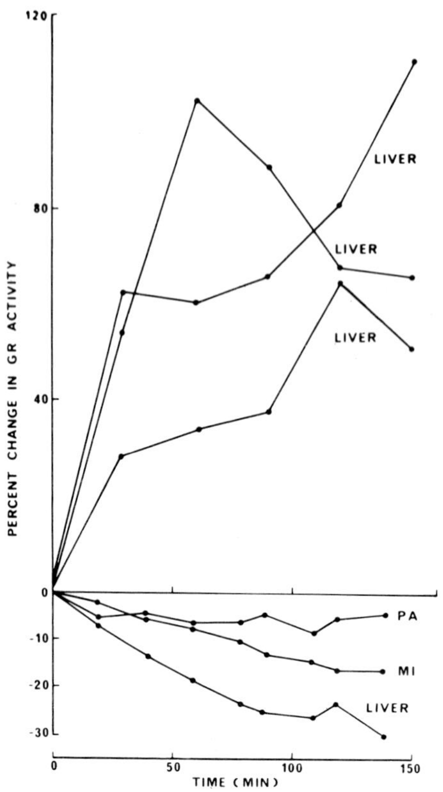

Fig. 4 – Time course of the activation and inactivation processes occurring when pathological sera are incubated in the reaction mixture at 37°C.

cious anaemia, cancer, and myocardial infarction showed a decrease or no change in activity. Figure 4 shows some typical progress curves.

In any one serum, these effects were the same irrespective of whether the reaction was initiated by adding NADPH or GSSG, and they occurred even when neither was present during the incubation process. Although dilution appeared to be the mechanism for these changes, we observed that incubation of *all* sera at 37°C without dilution led to a dramatic fall in GR activity (Table IV). In other words, dilution in phosphate buffer slowed the spontaneous inactivation of the enzyme when incubated at this temperature, and actually reversed this process in a high proportion of sera from cases with liver disease (fig. 5). Since varying dilutions of the same serum led to a strictly linear relationship between concentration and observed activity when the assay was performed without delay, it is unlikely that dissociable inhibitors of GR are present in human serum.

Table IV. *Effect of incubating serum and prediluted serum on GR activity for 2 hours at 37°C*

Nature of samples	*No. of samples*	*Mean percent fall in GR activity*	
		Undiluted	*Prediluted*
Serum (cancer cases)	10	77	17
Serum (liver cases)	10	61	0
Serum (myocardial infarction)	5	54	11
Serum (pernicious anaemia)	4	85	16
Red cell hemolysate	4	0	5

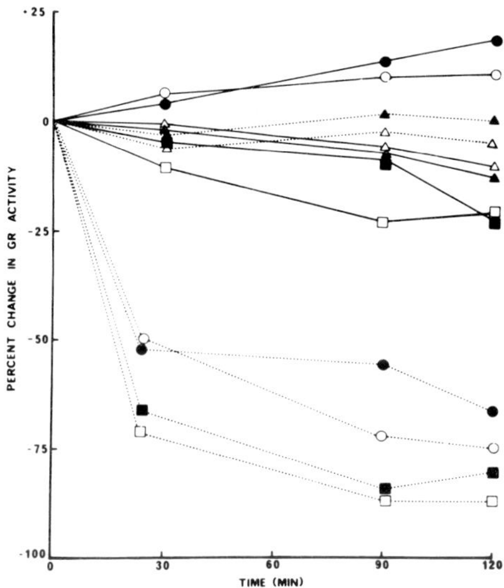

Fig. 5 – Time course of the activation and inactivation processes occurring when the same sera are incubated with (solid lines) and without (broken lines) predilution. Circles represent GR activity of sera from patients with acute liver disease, squares represent GR activity from patients with cancer and triangles represent the GR activity of red cell hemolysates.

It was not possible to demonstrate any pH dependance of these processes as at an acid pH there was excessive destruction of NADPH giving high blanks combined with low activity, and at a more alkaline pH the activities were too low to allow detection of any consistent effect.

Modification of Enzyme Stability

A number of substances previously shown by us to have no effect upon enzyme activity as such were tested to see if they could prevent these time-dependent changes in activity. Inclusion of FAD, at a final concentration of 10 μmol/liter, in the reaction mixture had no effect on the activity of samples allowed to stand after dilution in the mixture. Incubation of randomly selected sera with 20 mmol/liter DTE, in the ratio of 9 parts serum to 1 part DTE, for 24 h at 4°C caused an apparent activation of enzyme activity in 23 out of 30 samples when compared with saline-incubated controls (Table V).

Table V. *Mean percent increase in serum GR activity on incubation with activators*

No. of samples	*Activator*		
	*DTE 2 mmol/liter**	*Acetate buffer pH 6.4, 0.5 mmol/l.***	*DTE Acetate buffer*
30	69	–	–
26	–	63	–
14**	59	48	62

*Final concentration.
**These samples were incubated with DTE and with acetate buffer independently and together.

The addition of DTE had no effect on the serum pH which was approximately 8.0. Only one serum, from a patient with pernicious anaemia, showed a small decrease in activity. Three hemolysates were activated. The activation increased with DTE concentration (Table VI). Samples from cases with liver disease when incubated with DTE for 24 h showed no further activation when diluted in the reaction mixture and incubated for 2 h at 37°C, even although such incubation in the absence of DTE caused increased activity. The effects of DTE and dilution were therefore not additive. [Increasing con-

Table VI. *Effect of increasing final concentration of DTE upon the percent increase in serum GR activity from patients with liver disease. Detailed procedure in text*

No. of samples	*DTE concentration (mmol/liter)*		
	1	*2*	*4*
8	78	133	175

centrations of DTE partially protected the enzyme from inactivation at ambient temperature in that 14 sera were inactivated by a mean of 44% when incubated with saline for 24 h, as above, and then allowed to stand for 2 h at 25°C; in the presence of 1, 2 and 4 mmol/liter DTE the mean inactivations were respectively 38%, 33% and 24% over the same 2 h period.]

Incubation of sera for 24 h at 4°C with 5 mol/liter Acetate buffer over a pH range of 5.8-7.0 in the ratio of 9 parts serum to 1 part buffer also caused an activation of serum GR compared with saline-incubated controls (Table V). The buffer had no effect on the final pH of the reaction mixture. This activation was maximal at pH 6.4 at which pH the effect was slightly less than that of 2 mmol/liter DTE. [The protection against inactivation at ambient temperature was again maximal at pH 6.4. The same conditions applied as for the DTE experiments except that the incubations were with saline and with buffers, and the inactivations were 35% (saline), 31% (pH 5.8), 24% (pH 6.4) and 49% (pH 7.0).] In the presence of 2 mmol/liter DTE, acetate buffer pH 6.4 caused no further activation of serum GR. Incubation of sera with 5 mol/liter bicarbonate buffer pH 9.0 or 1% (v/v) Triton X-100 as in the experiments with DTE had no effect on the enzyme.

The effects of dilution, a reducing environment, and a pH approximating 6.4 in minimizing loss of enzyme activity are hard to interpret in the absence of detailed knowledge of the structure of human GR. It seems probable that a less ordered conformation, reduction of disulfide bridges, and an ionisable group with a pK around 6.4 confer this stability, but it is not apparent why these factors are non-additive. Loss of activity may be due to spontaneous folding of the molecule in such a way that the active site becomes buried. Any of the 3 mechanisms previously cited might then, in different ways, lead to re-exposure of active sites. Alternatively, loss of activity may be due to formation of molecular aggregates through hydrogen bonding or disulfide bonds, reversed by dilution or a pH change or sulphydryl groups.

It has recently been reported that human erythrocyte glutathione reductase is a dimer of two peptide subunits, and that in the absence of thiols aggregation to form tetramers and larger aggregates occurs (Worthington and Rosemeyer, 1975). Data on the relative enzymatic activities of these molecular species have not as yet been reported.

Effect of Chromate

Chromate has been shown to produce differential inhibition of GR in bovine thyroid slices and human erythrocytes in a concentration of 0.5 mmol/ liver over 45 min (Yawata and Tanaka, 1973; Chiraseveenuprapund and Rosenberg, 1974). We could not demosntrate inhibition of serum GR under these conditions. Only when we raised the chromate concentrations 10-fold and extended the incubation period to 24 h could we demonstrate inactivation in hemolysates and sera from patients with pernicious anaemia. Sera from patients with liver disease were totally resistant to chromate under these circumstances. The response to chromate thus represents another means of distinguishing the origin of GR activity in human serum.

Serum GR in Cancer Patients

Of the 68 patients examined, 39 (58%) had a serum GR activity above the reference range established for normal subjects. Comparable data for the other enzymes were AsT 7 (10%), LDH 22 (21%) and ALP 13 (19%). The distribution of enzyme values in all subjects is shown in fig. 6. Of the subjects with raised serum GR activity, 15% had an elevated AsT, 51% an elevated LDH and 20% an elevated ALP. Of the 29 with a normal serum GR activity only one had an elevated AsT, two an elevated LDH and five an elevated ALP. These results demonstrate the superiority of serum GR estimations over LDH in the diagnosis of malignant disease; these two enzymes are much more likely to be raised than either AsT or ALP.

Table VII shows the outcome of the disease processes six months after the initial assessment. An enzyme elevation indicates a poorer prognosis; in particular all seven patients with an AsT elevation died within six months of the analysis. Analysis of the individual raised serum GR values showed them to be randomly distributed throughout the various categories of malignancy. It could not be used as an index of possible metastases.

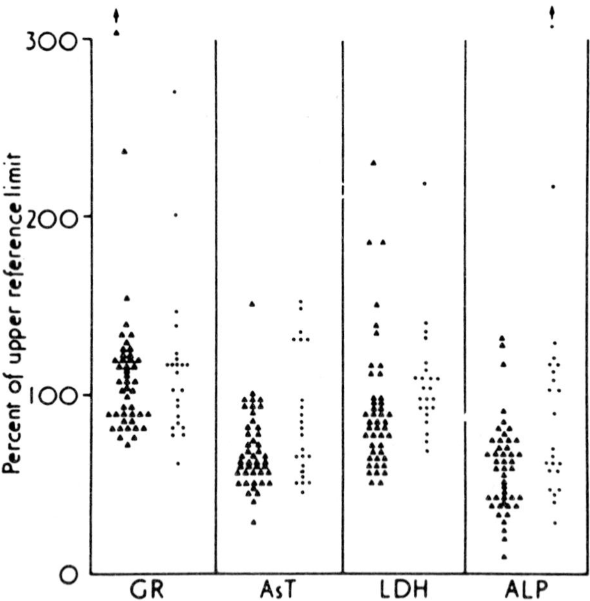

Fig. 6 – Distribution of enzyme activity in the cancer patients studied. Patients with no detectable metastases are indicated ▲ and those with metastase ●.

Table VII. *Relationship between abnormal enzymes and prognosis after six months*

	Percentage with Raised Serum Enzymes					
	GR	*AsT*	*LDH*	*ALP*	*GR & LDH*	*None*
Dead (36)*	57	20	48	29	19	29
Alive (32)*	47	0	18	9	19	47

*No. of cases in parenthesis.

The level of serum GR activity was of no value in assessing the eventual outcome and in fact GR was a worse prognostic index than any of the other enzymes assayed. AsT was the most promising in this regard since all the patients who had an elevated AsT died within six months of the sample being taken.

In view of previous emphasis on LDH as an indicator of possible malignant disease (Ticktin and Trujillo, 1970), a comparison was made of the data for this enzyme and for GR in the same patients (fig. 7). In addition, analysis was conducted on the regression and correlation between the two estimations in subjects grouped according to presence or absence of clinically detectable metastases (Table VIII). This showed that, on the whole, there was not a close relationship between the two. The factors leading to their elevation in cancer subjects therefore appear to be different, and this implies that more information can be obtained by assaying both than one alone. The

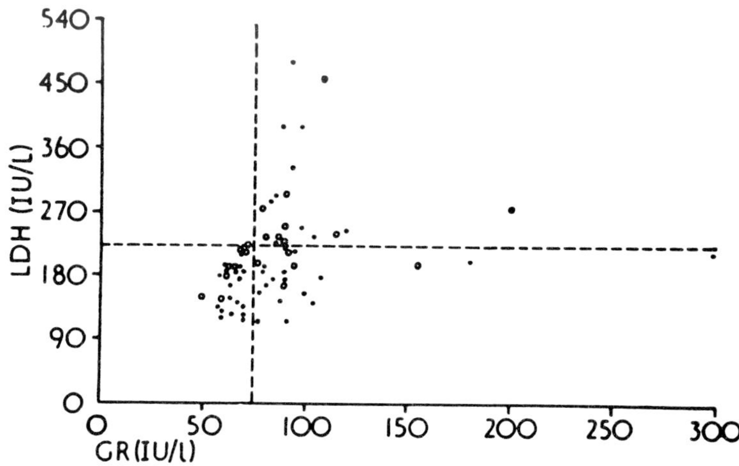

Fig. 7 – Plot of serum LDH and GR activities in the cancer patients studied. Patients with no detectable metastases are indicated • and those with metastases ■. Regression and correlation data are given in Table VIII.

Table VIII. *Data for regression analysis of LDH on GR*

Population	*Number*	*Slope*	*Intercept*	*Correlation coefficient*	*Level of significance(P)*
Without metastases	45	0.41	109.5	0.204	> 0.05
With metastases	23	0.72	137.6	0.376	> 0.05
Combined	68	0.50	119.0	0.247	> 0.05

high incidence of raised serum GR among cancer patients demonstrated in this study suggests that it may be of value in population screening. Since it is raised in liver disease and blood dyscrasias (Manso and Wroblewski, 1958; Kerppola et al., 1959; West et al., 1961; Horn et al., 1962), it cannot be regarded as a cancer-specific test, but it deserves to replace LDH in multiphasic biochemical profiling.

References

1. Beutler E.: A series of new screening procedures for pyruvate kinase deficiency, glucose-6-phosphate dehydrogenase deficiency and glutathione reductase deficiency. Blood, *28,* (1966), 553-562.
2. Beutler E.: The effect of flavin compounds on glutathione reductase activity: in Vivo and in Vitro studies. J. Clin. Invest., *48,* (1969), 1957-1966.
3. Chiraseveenuprapund P., Rosenberg I.N.: Observations on the effects of chromate on the thyroid. Endocrinology, *94,* (1974), 1714-1722.
4. Glatzle D., Weber F., Wiss O.: Enzymatic test for the detection of a riboflavin deficiency. NADPH-dependant glutathione reductase of red blood cells and its activation by FAD in vitro. Experientia, *24,* (1968), 1122.
5. Horn H.D., Mainzer K., Langrehr D.: On glutathione reductase in the serum of man. A contribution to the differential diagnosis of liver disease. Klin. Woshchr., *40,* (1962), 985-995.
6. Heller S., Salkeld R.M., Korner W.F.: Riboflavin status in pregnancy. Amer. J. Clin. Nutr., *27,* (1974), 1225-1230.
7. Kerppola W., Nikkila E.A., Pitkanen E.: Serum TPN-linked enzymes. Glucose-6-phosphate dehydrogenase, isocitric dehydrogenase and glutathione reductase activities in health and in various disease states. Acta Med. Scand., *164,* (1959), 357-365.
8. Manso C., Wroblewski F.: Glutathione reductase activity in blood and body fluids. J. Clin. Invest., *37,* (1958), 214-218.
9. Sauberlich H.E., Judd J.H., Nichoalds G.E., Broquist H.P., and Darby W.J.: Application of the erythrocyte-glutathione reductase assay in evaluating riboflavin nutritional status in a high school student population. Amer. J. Clin. Nutr., *25,* (1972), 756-762.
10. Scott E.M., Duncan I.W., Ekstrand V.: Purification and properties of glutathione reductase of human erythrocytes. J. Biol. Chem., *238,* (1963), 3928-3933.
11. Ticktin H.E., Trujillo N.P.: Enzymes in neoplastic and surgical diseases, in *Diagnostic Enzymology,* edited by E.L. Goodley, pp. 205-222. Lea and Febiger, Philadelphia (1970).

12. West M., Berger C., Rony H., Zimmerman H.J.: Serum enzymes in disease. VI. Glutathione reductase in sera of normal subjects and of patients with various diseases. J. Lab. Clin. Med., *57,* (1961), 946-954.
13. Worthington D.J., Rosemeyer M.A.: Glutathione reductase from human erythrocytes. Molecular weight, subunit composition and aggregation properties. Europ. J. Biochem., *60*, (1975), 459-466.
14. Yawata Y., Tanaka K.R.: Red cell glutathione reductase: mechanism of action of inhibitors. Biochem. Biophys. Acta, *321,* (1973), 72-83.

NEW SUBSTRATES FOR DETERMINATION OF γGT ACTIVITY IN SERUM

J.P. Persijn, W. van der Slik, G. Szasz and W. Gruber

Summary

In measuring serum γ-glutamyltranspeptidase activity (γGT), L-γ-glutamyl-p-nitranilide is generally used. Considerations such as poor solubility of L-glutamyl-p-nitroanilide prompted us to search for better substrates. We first tested a few dipeptides such as L-γ-glutamyl-L-glutamic acid and L-γ-glutamyl-D-alanine as a substrate for γ-GT.

Unfortunately, the monitoring reaction was ratelimiting. A description of these experiments will be given. We next investigated new synthetic substrates derived from Glupa. A number of substrates derived from Glupa by introduction of hydrophilic groups or alteration of the benzene ring were tested for solubility and suitability as substrates for measuring γGT. The results of these tests will be presented. The most favorable results were obtained with the substrate L-γ-glutamyl-3-carboxy-3-nitranilide which offers the advantage of producing its own chromogen directly. In our study we established optimal conditions with this substrate.

The substrate is highly soluble and the method can therefore easily be adapted to any equipment for automated assay of γGT.

The activities measured are slightly higher than with the Szasz method in which L-glutamyl-p-nitroanilide is used as substrate.

During the period 1960-1970 several synthetic substrates for determination of γ-glutamyltranspeptidase (γ-GT) activity were described, mostly by Polish investigators. Of these substrates, only γ-glutamyl-p-nitroanilide has so far established itself. The reasons are the following:

1. the cleavage product p-NO_2 aniline is coloured; the molar extinction coefficient at 405 nm is sufficient for accurate determinations in the area of transition from normal to pathological;
2. this substrate has been used by several investigators (Jacobs, Persijn, Rosalki, Schmidt, van der Slik, Szasz and their co-workers), who studied the clinical significance of γ-GT;
3. in 1969 Szasz[1] described a clinical method in a report which gave a more detailed account than other publications of various factors which can

Department of Clinical Chemistry, Antoni van Leeuwenhoekziekenhuis, Amsterdam, and the Central Laboratory of Clinical Chemistry State University, Leiden, The Netherlands. Institute of Clinical Chemistry, University Medical School, Klinikstrasse 32b, 6300 Giessen and Biochemica Werk Tutzing, Boehringer Mannheim GmbH, 8132 Tutzing, West Germany.

influence the reaction speed. As an example, we may mention the effect of glycyl-glycine, which activates γGT.

In actual practice, while measuring γ-glutamyl-transpeptidase (γGT), problems arise as a result of the low solubility of glutamyl-p-nitroanilide. In manual assays it is even difficult to prepare a substrate solution in which the substrate concentration is optimal. Heating up to about 50°C can be required to prepare this substrate solution, but this is associated with autolysis. Automation of the determination, particularly with devices which measure according to the kinetic principle (which implies that the reaction is started by addition of a small volume of a concentrated substrate solution), is impossible without applying 'artifices'. A well-known artifice is the use of glycylglycine as starting reagent, but this has the important disadvantage that, in sera with increased γGT activity, the initial extinction during determination is fairly high(particularly if the sera in question are the last in a batch).

The γGT assay, therefore, can do with some improvement. We have studied a number of readily soluble alternative compounds as possible substrate for γGT. (see Table 1).

Table I. *Alternative substrates for γ-glutamyltransferase*

Group I	*Indicating reaction*
L-ν-glutamyl-adenosinediphosphate	ADP $\xrightarrow[\text{PEP}]{\text{PK}}$ pyruvate $\xrightarrow[\text{ATP}]{\text{LDH}}$ lactate NADH ⟶ NAD
L-ν-glutamyl-L-glutamate	glutamate $\xrightarrow{\text{GLDH}}$ ketoglutarate NAD ⟶ NADH
L-ν-glutamyl-D-alanine	D-alanine $\xrightarrow{\text{AO'ase}}$ pyruvate $\xrightarrow{\text{LDH}}$ lactate NADH ⟶ NAD
Group II	
L-ν-glutamyl-NO_2-anilide with substituents in the benzene ring	measurement of liberated substituted nitroaniline

The compounds of Table 1 can be divided into two groups: one group of compounds not derived from glutamyl-p-nitroanilide, for example: glutamyl-adenosine diphosphate, glutamyl-glutamate and glutamyl-D-alanine; and another group of compounds derived from glutamyl-p-nitroanilide by introducing substituents into the benzene ring to enhance solubility. Let us first consider the substrates of group 1.

Glutamyl-adenosine diphosphate

It should be borne in mind that various synthetic γ-glutamyl peptides have been found to act as donors of the glutamyl group and, therefore, as substrate for γGT. The idea was to convert the released adenosine diphosphate to lactate via coupled reactions. The amount of nicotinamide-adenine dinucleotide produced in the course of this process could then be used to measure γGT by monitoring the reaction at 340 nm. However, the compound glutamyl-adenosine diphosphate is not commercially available. Since according to experts the conjugation with the adenine-amino group would be unstable, no attempt at synthesis was made.

Glutamyl-glutamate

The second compound – γ-glutamyl-glutamate – was described as substrate for γGT as early as 1952 by Hanes[2]. The glutamic acid released could be converted to γ-ketoglutarate by glutamate dehydrogenase, and the reduced nicotinamide-adenine dinucleotide produced in the course of this process could be regarded as a measure of γGT activity. Determination of γGT could therefore be achieved by monitoring the increase of absorbance at 340 nm. Leucine was to be added for two reasons; it acts as acceptor and as such activates γGt and, secondly, it would stabilize the auxiliary enzyme glutamate dehydrogenase. Unfortunately the indicating reaction was ratelimiting and attempts to influence the balance favourably by adding a second auxiliary enzyme to withdraw the ketoglutarate formed from the reaction, were given up after a number of failures.

Glutamyl-D-alanine

With the compound γ-glutamyl-D-alanine as substrate it was expected that the alanine released could be deaminated with the aid of D-amino-acid oxidase, followed by reduction by lactate dehydrogenase in the presence of nicotinamide-adenine dinucleotide. The nicotinamide-adenine dinucleotide utilized in this reaction could be measured at 340 nm as a measure of γGT. D-alanine was chosen because, unlike L-alanine, it does not occur in human serum. In this case, too, the indicating reaction proved to be ratelimiting. Further experiments were given up because encouraging results had meanwhile been obtained in experiments with derivatives of glutamyl--p-nitroanilide; the compounds of group 2.

It is to be noted in this context that, given the same measuring arrangement (apart from the wavelength setting) in registration of the reaction the number of scale units per minute in the case of glutamyl-p-nitroanilide was almost twice that in the case of glutamyl-alanine. Moreover, it could not be excluded with certainty that the substrate glutamyl-alanine might inhibit the

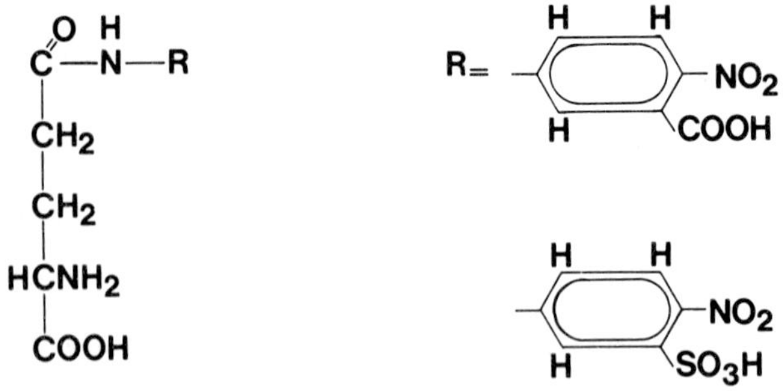

Fig. 1 – New substrates for γGT measurement.

auxiliary enzyme amino-acid oxidase. The substrates of group 2 were investigated. Fig. 1 shows the formulae of some substrates. The solubility of these compounds is more than 50 times that of glutamyl-p-nitroanilide. Favourable activities were measured only with the substituents sulphonate or carboxylate at site 3. The study was continued with these substrates γ-glutamyl-3-carboxy-4-nitranilide and γ-glutamyl-3-sulphonic-4-nitranilide (respective abbreviations: Glupac and Glupas).

Fig. 2 shows the molar absorptivities of the cleavage products between 300 and 500 nm. The new substrates are listed in the left-hand column of Table 2, and their cleavage products in the central column. The right-hand column shows that the absorptivity of the cleavage product of Glupac compares favourably with that of p-NO_2-aniline. These values are independent of pH in the range 7-9. Another favourable feature is that neither Glupac nor Glupas shows any measurable absorption at wavelengths exceeding 410 nm. As in the case of glutamyl-p-nitroanilide, therefore, γGT activity can be measured at, say 405 nm. Autolysis of the substrates is negligible.

Optimal conditions with Glupac and with Glupas were found to be identical (except for the pH).

Maximal activities (at 25°C) were measured at pH 8.2 with Glupac and at pH 8.3 with Glupas in 100 mmole/l Tris buffer. Higher Tris concentrations had an inhibitory effect. The optimal glycyl-glycine concentration was about 150 mmole/l. Maximal activities were measured at substrate concentrations between 4 and 8 mmole/l. These data are valid both at 25°C and at 37°C, albeit that pH optima are about 0.1 unit lower at 37°C. Substrate concentrations exceeding 8 mmole/l cause inhibition of the reaction.

γGT activities, therefore, cannot be measured at substrate concentrations which equal 40 x km.

Table 3 compares the activities and the calculated corresponding increase of absorbance per minute with the substrates Glupac and Glupas. We started with a serum which under optimal conditions has a γGT activity of 100 U/l

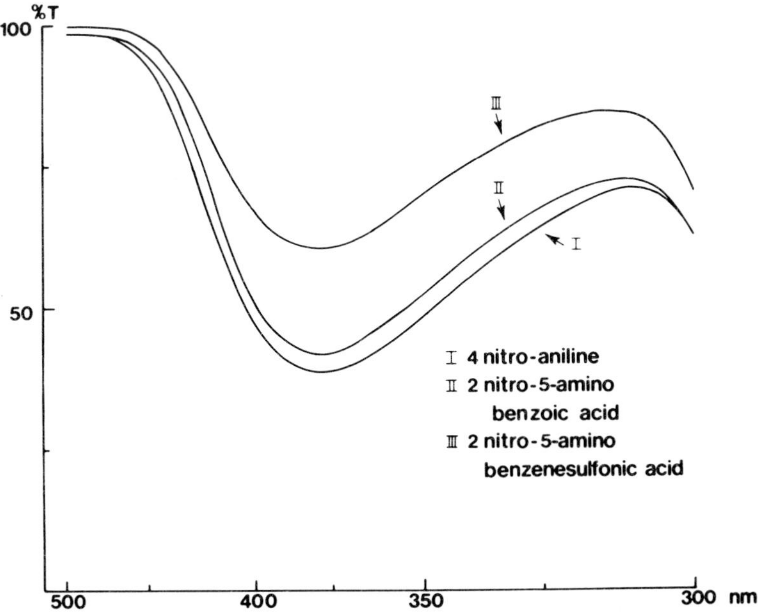

Fig. 2 – Absorption spectrum of cleavage products of γ-glutamyl pNO_2 anilide, γ-glutamyl-3-carboxy-4-nitanilide and γ-glutamyl-3-sulphonic-4-nitranilide.

Table II

Substrate	*Cleavage product*	*Molar absorptivity at 405 nm* $(cm^2 \cdot \mu Mol^{-1})$
L-ν-glutamyl-p-nitranilide	p-nitroaniline	9.92
L-ν-glutamyl-3-sulphonic-4-nitranilide	2-nitro-5-aminobenzen-sulfonic acid	5.45
L-ν-glutamyl-3-carboxy-4-nitranilide	2-nitro-5-aminobenzoic acid	9.49

with Glupac. Under optimal conditions such a serum sample gives a γGT activity of 155 U/l with Glupas. This table shows, however, that this higher activity is associated with a lower change of absorbance. In other words: when Glupac is used, the γGT activities in the borderline region between normal and pathological are measured with a stronger signal, that is with a higher degree of sensitivity, than when Glupas is used.

The substrate Glupac compares favourably with the Szasz' method, which uses glutamyl-p-nitroanilide as substrate. Under optimal conditions,

Table III

Activity	*Initial change of absorbance**
100 U/l with L-ν-glutamyl-3-carboxy-4-nitranilide	0.058
155 U/l with L-ν-glutamyl-3-sulphonic-4-nitralilide	0.048

*1 cm lightpath, 405 nm; 3 ml buffer substrate; 0.2 ml serum.

20% higher activities were measured with Glupac. The increase of absorbance was 14% higher with Glupac than with glutamyl-p-nitroanilide. Of course the higher activities with Glupac also imply that, with this new substrate, the upper limit of normal as generally valid with the substrate glutamyl-p-nitroanilide, must be increased by 20%. Some investigators may find this a disadvantage, but it can be neutralized by using suboptimal reaction conditions with Glupac; and in this case the sensitivity of the determination need not be less than that with glutamyl-p-nitroanilide. This is shown in Table 4.

Table IV. *Test-conditions*

		OLD		*NEW*
Tris		185 mM		100 mM
pH		8.25		8.25
Glycylglycine		40 mM		100 mM
Substrate	glupa	4 mM	glupa-3-COO	4 mM

The left-hand column indicates the concentrations of buffer, glycyl-glycine and substrate as used in some commercial test kits. The right-hand column indicates the concentrations at which the same activity is measured when Glupac is used. Proof is provided in Fig. 3.

The vertical axis shows the activities measured with glutamyl-p-nitroanilide, and the horizontal axis indicates the activities measured with Glupac under suboptimal conditions as indicated in Table 4.

To summarize:

The new substrate Glupac is characterized by high solubility and an acceptable molar absorptivity of the cleavage product. Consequently the use of Glupac provides us with a simpler method because various more or less

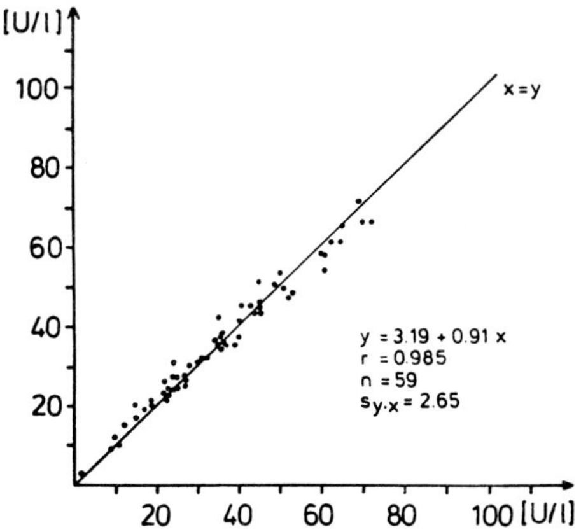

Fig. 3 – Comparison of γ-GT activities. (see Text.)

complicated attempts to prepare substrate solutions at optimal concentrations (as required when glutamyl-p-nitroanilide is used) are superfluous. We therefore propose as method of choice the procedure which utilizes Glupac under optimal conditions, as described. Glycine-free glycyl-glycine should be used, since glycine inhibits γGT activity. The above-mentioned advantages make it possible to adapt the assay to any kind of automation.

In our laboratory we have had over a year's experience with Glupac in determination of γGT activity in serum, making use of the LKB reaction rate analyser. The reaction is started by addition of 100 μl Glupac solution at a concentration of 72 mmole/l in Tris buffer. The cuvettes contain 1.1 ml buffered glycyl-glycine solution, which contains 0.1 ml serum. The measuring temperature is 37°C. The final concentrations are as described under optimal conditions.

On the basis of our experience we conclude that the assay is technically excellent.

Full details will be described elsewhere[3].

References

1. Szasz G.: Clin. Chem., *15,* (1969), 124-136.
2. Hanes C.S., Hird F.J.R. & Isherwood F.A.: Biochem. J., *51,* (1952), 25-35.
3. Persijn, J.P., van der Slik W.: Z. klin. Chemie u. Biochemie, (1976).

THE PROBLEM OF γ-GLUTAMYL-TRANSFERASE ISOENZYMES

J. Freise, P. Magersted and E. Schmidt

Summary

In recent years the determination of gamma-glutamyltransferase (gamma-GT, E.C. Nr. 2.3.2.2) activity in serum has become a widespread diagnostic method. Its value for the detection and differentiation of diseases of the hepatobiliary system and the pancreas is well established. The occurrence of the enzyme in parenchymal cells as well as in the brush border of littoral cells of several organs has raised the question whether specific isoenzymes of gamma-GT may exist. Many workers have succeeded in separating multiple forms of gamma-GT in serum, using different methods, e.g. electrophoresis on various supporting media, gel filtration, and column chromatography. Their results, however, are contradictory, and so are our own results, obtained by electrophoresis and column chromatography of numerous sera. Thus, at present, the utmost caution seems to be indicated before classifying these multiple forms of gamma-GT to be true isoenzymes.

Since the first description of γ-Glutamyl-Transferase (γGT), EC. Nr. 2.3. 2.2, by Hanes[1] in 1950 an ever-increasing interest has been shown in this enzyme. In the years between 1960-63, the behaviour of γ-GT in serum in hepato-biliary illnesses was reported by Goldbarg, Orlowski and Coll.[2, 3]. The methodic works of Orlowski, Meister, Szasz and others[4,5,6] led finally to the worldwide routine estimation of γ-GT activity in the serum, as the γ-GT is of considerable importance in the differential diagnosis of liver diseases[6-18].

As the γ-GT activity in serum is found to be elevated in different diseases of the liver and also is present in many tissue[19] such as kidney, pancreas, liver, prostata gland, small intestine, testes, spleen, lungs – listed in decreasing order of concentration – a lot of publications have dealt with the presence of γ-GT isoenzymes.

In 1965 Kokot and Kuska[20, 21] were the first to separate the γ-GT on paper electrophoresis into five bands. The results of publications between 1965 and 1974, concerning the multiple forms of γ-GT, differ vastly from one to

---

Div. of Gastroenterology and Hepatology, Dept. of Int. Medicine, Medical School, Hannover, G.F.R.

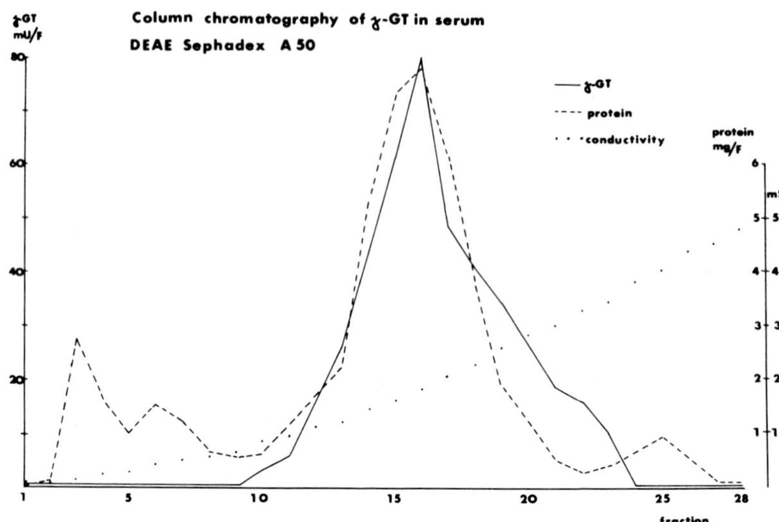

Fig. 1

eight so-called isoenzymes. The results cannot be compared at all. γ-GT is more sensitive than Alkaline Phosphatase in the diagnosis of cholestasis, but up to now it has not been possible to differentiate enzymatically between intra- and posthepatic cholestasis. By means of the ion exchange Sephadex DEAE A 50[22], we tried to separate the γ-GT into multiple forms and hoped to succeed in differentiating between δ-GT originating from hepatocytes and y-GT originating from the ductulus cells of the biliary system.

The separation of γ-GT by means of an ion exchange has not been published yet, and whatever we will find by our experiments cannot be wrong, if you compare our findings with the literature.

A few words on the method

Serum with increased γ-GT activity is dialysed for 24 hours at 4°C against 0.008 M Sodium Phosphate Buffer. Before and afterwards the γ-GT activity is measured. After equilibrating the ionexchange with 0.008 M SPB, 1 ml of the dialysed serum is placed upon the column, temperature 4°C. By means of a gradient mixer the elution had a conductivity change from 0.16 mS to 5.5 mS and a continuous gradient from 0.008 M to 0.4 M.

What our results show

A group of 20 patients with elevated γ-GT activity in serum and with the following diagnosis: liver metastases, bile duct obstruction by stones or tumors, tumor of the head of the pancreas, alcoholism with different hepatic

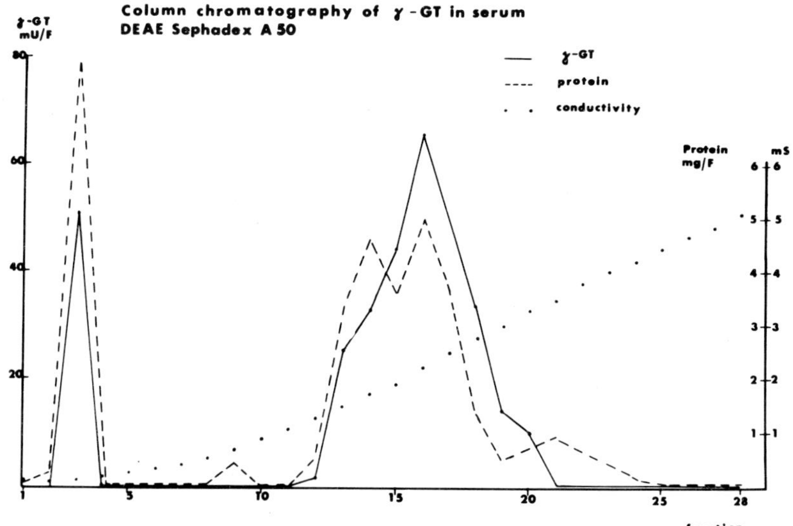

Fig. 2

		Sera with one y GT-Peak		Sera with two y GT-Peaks	
		$\bar{x}$	s_x	$\bar{x}$	s_x
y GT	U/l	620	411	851	527
Protein	g/l	67	8	64	10
Albumin	%	56	8	54	4,5
a_1 Globulin	%	5,4	1,8	6	1,8
a_2 Globulin	%	10,2	2,8	11	2
ß Globulin	%	11,3	1,0	13	1,4
y Globulin	%	18,4	6,8	16	4
Bilirubin	mmol/l	55	51	199	71
Cholesterol	mmol/l	5,5	2,2	12,5	3,8
Triglycerides	mmol/l	2,3	1,2	4,1	1,3
Unesterified Cholesterol	%	76	14,6	98,8	5,4
LP - X		7 of 14 positiv		4 of 6 positiv	
Chylomikrons		6 of 14 positiv		6 of 6 positiv	
HDL		13 of 14 positiv		6 of 6 negativ	

Tab. I

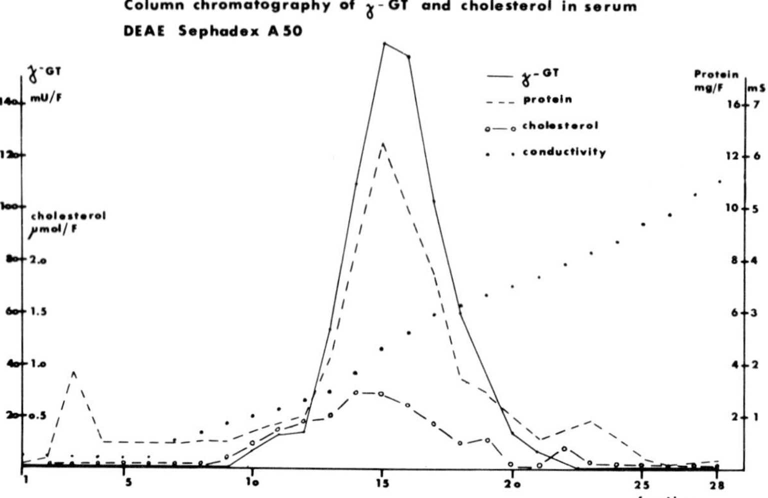

Fig. 3

involvement, liver damage through side-effect of drugs and primary biliary cirrhoses were examined. The column chromatography showed that in 11 sera one γ-GT peak could be found at a conductivity in the eluates from about 2.0-2.5 mS, and in 9 patients' sera two γ-GT peaks were found, the first at a conductivity around 0.2 mS, the second again from 2.0-2.5 mS. Figure 1 shows an individual example for the first group of patients with one γ-GT peak, Figure 2 the group with two γ-GT peaks. A correlation with the underlying diseases could not be seen, regarding the question, intra or posthepatic cholestasis. In both groups we examined additional parameters such as total protein, electrophoretic serum protein distribution, bilirubin, free and esterified cholesterol, triglycerides, electrophoretic lipoprotein distribution and lipoprotein X. Table I gives a summary of these results.

In both patient groups there was no significant difference in the γ-GT activity, the protein content and the protein distribution in the electrophoresis. In the second group, however, the serum cholesterol with 12.5 mmol/l was more than twice as high as in the first group (5.5 mmol/l). Triglycerides also showed this difference with 4.1 mmol/l and 2,28 mmol/l respectively.

We determined in both groups, in all chromatography fractions, cholesterol and triglycerides and detected cholesterol in all eluates with a conductivity between 1.0 and 4.5 mS (Fig. 3). In sera that showed a γ-GT peak of 0.2 mS, we also found a high cholesterol peak in this region (Fig. 4). In the case of the triglycerides there was no difference.

A separation of the serum cholesterol into free and esterified cholesterol showed that sera having two γ-GT peaks consisted predominantly of free

Column chromatography of γ-GT and cholesterol in serum
DEAE Sephadex A 50
γ-GT
— γ-GT
--- protein
•—• cholesterol
• • conductivity
Protein mg/F
mS
fraction

Fig. 4

Column chromatography of γ-GT in serum
DEAE Sephadex A 50
γ-GT mU/F
— γ-GT
--- protein
• • • conductivity
Protein mg/F
mS
fraction

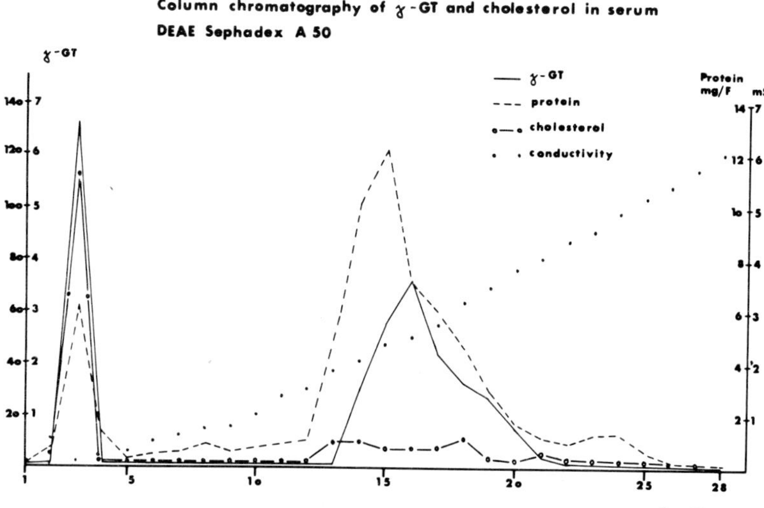

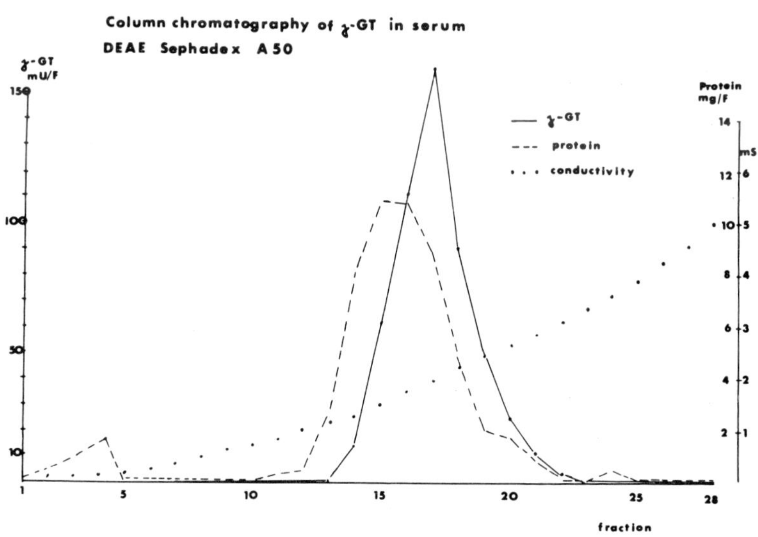

Fig. 5

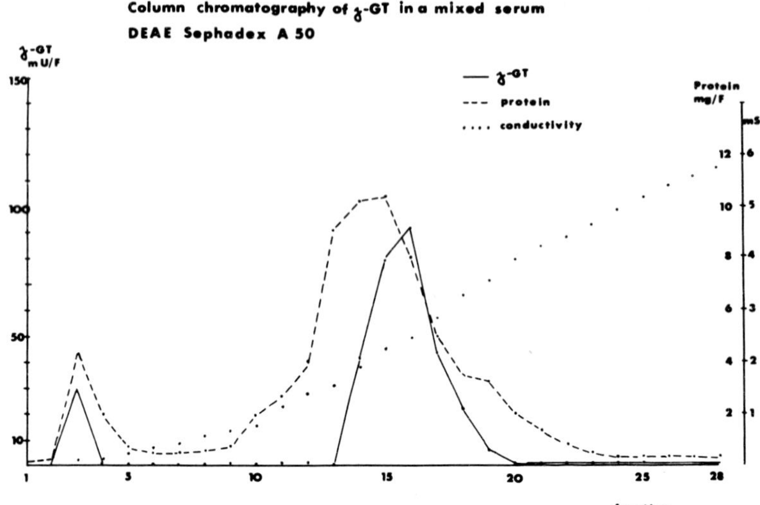

Fig. 6

cholesterol. The cholesterol peak at 0.2 mS conductivity showed only free cholesterol (Tab. I), as expected.

Pure cholesterol, which was added to a serum with only one γ-GT peak at 2.0-2.5 mS, failed to produce a second γ-GT peak at 0.2 mS. When, however, a serum with a highly elevated γ-GT (941 U/l) and cholesterol of 5.5 mmol/l and only one peak between 2.0-2,5 mS was mixed with a serum with a low γ-GT activity (25 U/l), but high cholesterol of 11.8 mmol/l the column chromatography of the mixed serum now showed two γ-GT peaks at 0.2 mS and 2.0-2.5 mS (Fig. 5) (Fig. 6).

On the other hand when Triton X 100 (5 μl Triton X 100 to 100 μl serum = 0.5%) was added to a serum with two γ-GT peaks, the peak with a conductivity at 0.2 mS disappeared completely (Fig. 7, Fig. 8). -

By immunological analysis of the lipoprotein X in the 20 sera, in 6 of the sera with two peaks, 4 times lipoprotein X could be detected and in 14 sera with only one peak, seven times lipoprotein X could be detected.

In lipoprotein electrophoresis, each serum with two γ-GT peaks contained chylomycrones, but α_1-lipoprotein, the high density lipoprotein, was completely absent in these cases. In the group with one γ-GT peak, lipid electrophoresis from 14 sera showed chylomycrones in 6 cases and α_1 - lipoproteins in 13 cases.

Many of the analyzed sera were turbid. It is common practice to clear a turbid serum with heparin and $MgCl_2$[23], which enables enzymes to be analyzed even if there is a low activity e.g. GLDH. After clarifying a serum with two γ-GT peaks, in this way only one peak at 2.0-2.5 mS in the clear serum

Column chromatography of γ-GT in serum
DEAE Sephadex A 50
γ-GT mU/F
γ-GT
protein
conductivity
Protein mg/F
mS
fraction

Fig. 7

Column chromatography of γ-GT in serum with 0.5 % Triton X100
DEAE Sephadex A50
γ-GT mU/F
Protein mg/F
mS
γ-GT
protein
conductivity
fraction

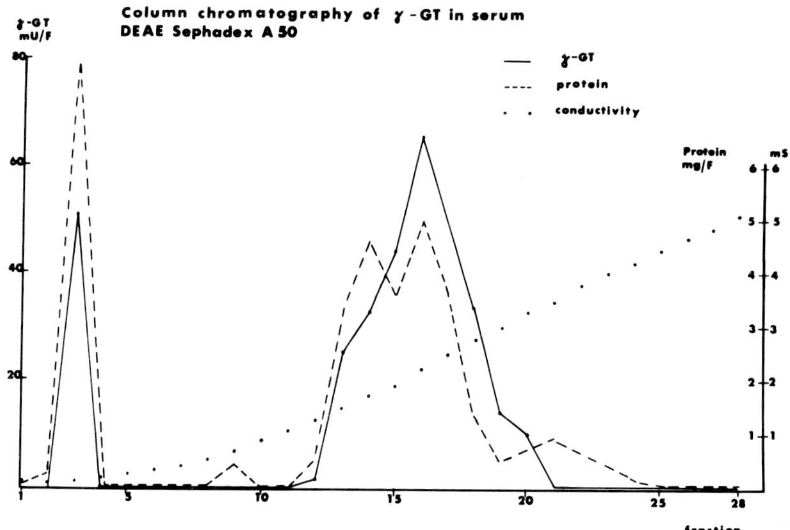

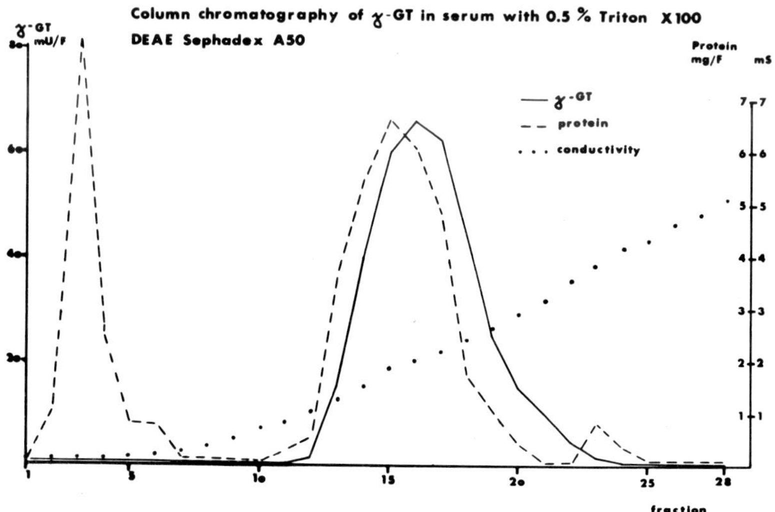

Fig. 8

pH – Optimum
%
100
80
60
40
20
6.0
6.5
7.0
7.5
8.0
8.5
pH
γ-GT of peak 1
γ-GT of peak 2

Fig. 9

Michaelis constant of γ-GT of Peak 1 and Peak 2 by Lineweaver-Burk
1/v
o,3
o,2
o,1
substrate:
L- Glutamyl-p-nitroanilide
peak 2
km = 1,33 × 10 -3 mol/l
peak 1
km = 1,25 × 10 -3 mol/l
-1,0
- o,5
- o,1
o,1
o,5
1,0
1/s

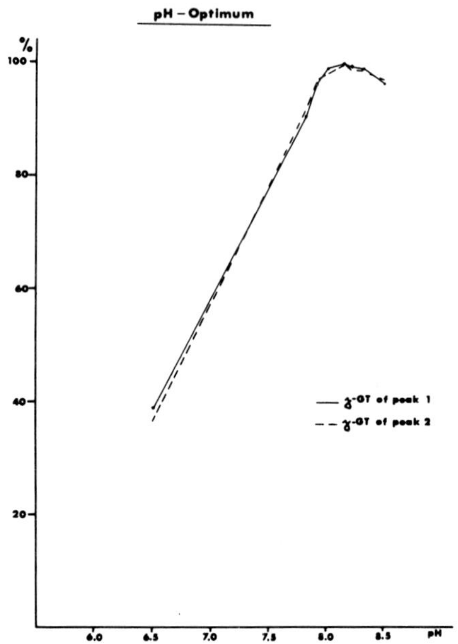

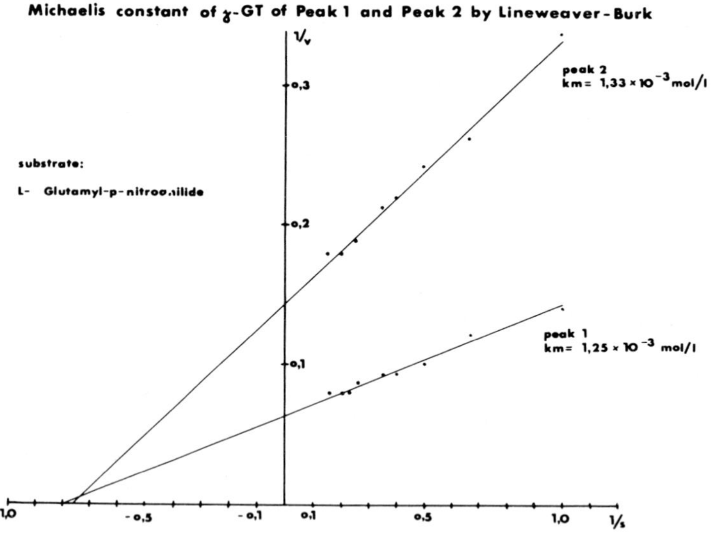

Fig. 10

was found. The other peak at 0.2 mS conductivity was found in the lipid rich supernatant.

In a further group of experiments we determined the pH-optimum of γ-GT and the respective Michaelis-constant for L-y-Glutamyl-p-nitroanilid, in the two eluates with the highest γ-GT activity at 0.1 mS and 2.0-2.5 mS. Figure 9 shows that the γ-GT in both fractions has a pH-optimum of 8.15. The Michaelis -constant we found to be 1.25 x 10^{-3} Mol/1. Figure 10 shows the results using the Lineweaver/Burk plot.

How can we explain our results? The column chromatographic separation of γ-GT from 20 sera, with a raised γ-GT activity, resulted in only 11 cases in one γ-GT peak between 2.0-2.5 mS. In 9 cases a second γ-GT peak at 0.2 mS conductivity was also seen. A correlation between these two groups with the underlying disease of the patients was not evident. Thus we were not able to distinguish, by this method, between different diseases or even between inter- and extrahepatic cholestasis. We achieved as little success in this point as previous publishers. A second aspect of our work was to suggest reasons as to why others succeeded in separating the γ-GT into multiple forms, but did not attempt to compare the results. γ-GT occurs 90% membrane-bound and 10% free in tissue. In serum it appears in a free form. Rosalki[24] has demonstrated that the free form and the membrane-bound form show such similar biochemical properties that one can assume the soluble and membrane-bound form to have a similar fundamental biochemical structure and so probably the soluble free form is part of the membrane-bound form. It may be possible that the free γ-GT aggregates with other serum components. Even by highspeed centrifugation (100.000 g/60 min) we have not yet succeded in reducing the γ-GT activity in serum as one would expect when the γ-GT is connected to a particle that can be sedimented by centrifugation with 100.000 g. If the γ-GT attaches itself to a plasma particle that cannot be centrifugated, e.g. molecules like albumins, globulins, lipids, lipoproteins, etc. then in serum electrophoresis the protein or the lipoprotein would be separated and by staining for γ-GT activity we would see bands corresponding to the distribution of the carrier. These stained bands can very easily be mistaken for isoenzymes. In different illnesses with raised γ-GT activity characteristic electrophoretic γ-GT bands are found; an explanation could be a concentration difference in the particular γ-GT carrier, e.g. a lipoprotein. This could also explain why by minimal changes in the separation technique, which alter the characteristic distribution of the carrier, vastly different results can be obtained.

In the sera which we examined no differences between the two groups regarding γ-GT activity, protein and protein distribution were found. In the group with the two γ-GT peaks bilirubin was found to be higher than in the group with the one γ-GT peak. However, bilirubin was ruled out as a carrier for the second peak at 0.2 mS, because there was no yellow colouring of the respective eluates. Both groups of sera however differed considerably in the amount of cholesterol and triglycerides. Cholesterol might be a theoretically possible carrier for the γ-GT in serum, as cholesterol is a basic constituent of the cell membranes. A constant exchange takes place between plasma cholesterol and membrane-cholesterol[25]. Cholesterol is an essential component of most lipoproteins in serum, and therefore it can be found in different concentrations in each lipid band separated by serum electrophoresis. In some patients sera cholesterol occurs in extremely high concentrations and in severe liver damage the relationship between free and esterified cholesterol can be upset.

A change in the cholesterol level would naturally affect the γ-GT part that is connected to the lipoprotein. A change in the multiple forms of the γ-GT because of an illness would then be only the expression of a change in the lipoprotein distribution.

As the second γ-GT peak at 0.2 mS may be produced by an abnormal lipoprotein, we tried to correlate our results regarding lipoprotein X, an atypical lipoprotein, consisting of up to 60% phospholipids, not found in healthy people. We have found no correlation, however.

There was no evidence of γ-GT isoenzymes in the ion exchange chromatographic separation. Naturally we cannot prove by our experiments that γ-GT isoenzymes will not exist. It is possible however that eventually the presence of isoenzymes will not be recognized by their electric conductivity, but for example by their molecular weights or by their antigenic behaviour. But we believe that the results of our experiments support the theory that lipoproteins containing cholesterol are responsible for the occurrence of the multiple forms of the γ-GT in serum.

References

1. Hanes G.S., F.J.R. Hird, F.A. Isherwood: Nature *166,* (1950), 288.
2. Goldbarg J.A., Friedman O.M., Pineda E.P., Smith E.E., Chatterji R., Stein E.H., Rutenberg A.M.: Arch. Biochem. Biophys., *91,* (1960), 61-70.
3. Szewczuk A., Orlowski M.: Clin. Chim. Acta,, (1960), 5680-688.
4. Orlowski M., Meister A.: Biochem. Biophys. Acta, *73,* (1963), 679.
5. Szasz G.: Clin. Chem., *15,* (1969), 124-136.
6. Szasz G., Rosenthal P., Writsche W.: Schweiz. Med. Wschr., *99,* (1969), 606-608.
7. Rutenberg A.M., Goldbarg J.A., Pineda E.P.: Gastroenterology, *45,* (1963), 43-48.
8. Szezeklik E., Orlowski M., Szewczuk A.: Gastroenterology, *41,* (1961), 353-359.
9. Rosalki S.B., Rau D.: Clin. Chem. Acta, *39,* (1972), 41-47.
10. Rollasen J.G., Pincherle D., Robinson D.: Clin. Chim. Acta, *39,* (1972), 75-80.
11. Zein M., Discombe G.: Lancet, *2,* (1970), 748-750.
12. Schmidt E.: Deutsche. Med. Wschr. *9,* (1975), 443-444.
13. Villa L., Dioguardi N., Agostoni A., Idéo G., Stabilinski R.: Enzymol. Biol. Clin. *7,* (1966), 109-114.
14. Idéo G., Morganti A., Dioguardi N.: Digestion, *5,* 326-336 (1972).
15. Konttinen A., Hupli V., Sulmenkivi K.: Acta Med. Scand., *189,* (1971), 529-535.

16. Szasz G., Rosenthal P., Writshce W.: Dtsch. Med. Wschr., *94,* (1969), 1911-1917.
17. Sczeklik E., Lukasik S., Orlowski M., Swiderski T.: Arch. Immunol. Ther. Exp., *13,* (1965), 573-579.
18. Jacobs W.L.W.: Clin. Chem. Acta, *38,* (1972), 419-434.
19. Goldbarg J.A., Friedman O.M., Pineda E.P., Smith E.E., Chatterji R., Stein E.H., Rutenberg A.M.: Arch. Biochem. Biophys., *91,* 61-70 (1960).
20. Kokot F., Kuska J.: Clin. Chem. Acta, *11,* (1965), 118-121.
21. Kokot F. et al.: Arch. Immunol. Ther. Exp., *13,* (1965), 549.
22. Sephadex Ionenaustauscher Leitfaden zur Ionenaustausch-Chromatographie der Firma Pharmacia Fine Chemicals, Uppsala, Schweden.
23. Undeutsch D.: in Fettstoffwechselstörungen, Editiones Roche by Hoffmann – La roche AG. Grenzach, page 113 (1974).
24. Rosalki S.B.: Gamma-Glutamyl Transpeptidase in "Advances in Clinical Chemistry". Academic Press New York – San Francisco – London 1975, volume 17, page 53.
25. Glomset J.A., Wright J.L.: Biochem. Biophys. Acta, *89,* (1964), 266.

ANGIOTENSIN CONVERTING ENZYME AND α_1-ANTITRYPSIN IN BLOOD PLASMA AFTER PULMONARY EMBOLISM

E. de Brito-Paiva, M. Roth, B. Vaudaux and J.P. Junod

Summary

Angiotensin converting enzyme, α_1-antitrypsin, LDH isoenzymes and a few other blood plasma constituents were assayed in a group of patients with a diagnosis of pulmonary embolism and another group of patients who were initially suspected of the same disease but left the clinic with a different final diagnosis. The angiotensin converting enzyme and α_1-antitrypsin were also determined in a control group without any pulmonary disease.

The plasma converting enzyme activity in the group with pulmonary embolism was not significantly different from that of the control group. On the other hand, a highly significant difference was observed for α_1-antitrypsin. 48 hrs after pulmonary embolism, the mean concentration of α_1-antitrypsin was 412 U/l, whereas the mean of controls was 246 U/l. An increase in α_1-antitrypsin was also observed in the group of those patients who, after having been first suspected of pulmonary embolism, finally left the clinic with another diagnosis, in general that of bronco-pulmonary disease.

Introduction

In Geneva's Geriatric Hospital, pulmonary embolism is the immediate cause of 27 % of all deaths. This diagnosis is recorded as the cause of the death of one third of the autopsied patients, and of one fifth of the others (62 % of the deceased patients are autopsied)[1]. It is believed that without authopsy, many lethal pulmonary embolism are actually not recognized by the clinicians. The morbidity is also still high in this hospital and the anticoagulant therapy represents an additional risk factor to older patients.

The diagnosis of pulmonary embolism is a constant challenge even to the most experienced physician. At present, this diagnosis is still based on clinical signs, additional information being sometimes provided by thoracic radiography and by the electrocardiogram. The latter, however, is often unreliable[2]. Right catheterism and gazometry may provide useful information but do not represent by themselves decisive diagnostic criteria. Lung scintigraphy, which may be falsely positive or negative, is considered by many au-

Hôpital de Gériatrie, Thônex, Genève (Suisse), Laboratoire central, Hôpital cantonal, Genève (Suisse).

thors as a good investigation[3,4]. However, angiography still remains the most reliable diagnostic method[2].

What about enzyme determinations? In 1960, Wacker and co-workers[5] stressed the usefulness of the triad LDH, bilirubin and aspartate aminotransferase. One year later, the same group claimed that the use of this triad permitted one to detect pulmonary embolism and infarction and to distinguish them from acute myocardial infarction and pneumonia, the two diseases which are mostly misdiagnosed for them[6]. In 1964, Stevens and co-workers[7] believed that the diagnosis of pulmonary infarction should still be based on clinical criteria and that the determination of LDH can provide some help with the differential diagnosis. Schonell and co-workers[8] investigated two groups of patients, one with pulmonary embolism and the other with pneumonia and were unable to observe any significant difference with respect to LDH, aspartate and alanine aminotransferases, alkaline phosphatase, bilirubin and thymol. Most authors, indeed, do not consider enzyme determinations very useful. In 1974, Konttinen et al.[9] determined aspartate aminotransferase, LDH isoenzymes and CPK in pulmonary embolism and concluded that these determinations are of little diagnostic value in this disease. Witchitz[2], who wrote last year that enzyme determinations have no practical interest, may also be quoted.

At present, therefore, despite the use of newer diagnostic methods, pulmonary embolic disease is often unrecognized.

This paper reports a study of the value of some traditional determinations (LDH, iso-LDH, aspartate aminotransferase, bilirubin and CPK) in the diagnosis of pulmonary embolism. In addition, the angiotensin converting enzyme, which is known to occur in particularly high amounts in lung[10,11], was determined, together with α_1-antitrypsin, as some indices suggest that the latter might increase after pulmonary embolism[12,13].

Material and methods

For the determination of the converting enzyme and of α_1-antitrypsin, a control group was established. Blood was drawn at 7 a.m. on the day following admission. The group consisted of 8 male and 21 female patients who, upon admission, presented no recent respiratory history, no febrile disease, no respiratory sign or symptom and who, after examination by two physicians, were considered as having a normal respiratory status. In particular, all cases whose auscultation was doubtful were eliminated. All selected controls had a satisfactory chest X-Ray picture; cases showing pulmonary foci, suspicious infiltrates, pleural effusions or a pattern of stasis were eliminated. However, subjects showing pictures of residual calcifications, pleural sequelae or slight emphysema were included in the normals.

Another group was made up of all patients who were suspected of pulmonary embolism during the period of study. In this group the converting enzyme, α_1-antitrypsin, LDH and its isoenzymes, CPK, aspartate aminotransferase and bilirubin were determined. Blood was drawn at 7 o'clock in the morning of the first and the second day following the day when the symptoms suggesting an embolism occurred.

After the analyses, the group with suspected pulmonary embolism was subdivided into two sub-groups: the first consisted of 18 patients for which the diagnosis had been confirmed, either by autopsy or by obvious clinical and paraclinical signs, and the second was formed by the other 20 patients who left the hospital with another final diagnosis.

The angiotensin converting enzyme was determined as reported previously with the artificial substrate Z-peh-his-leu[14]. For the assav of antitrypsin, 50 μl of trypsin solution (25 mg/l containing 1,5 g/l $CaCl_2 \cdot 2 H_2O$) were incubated with 3,5 ml benzoylarginine-β-naphthylamide solution (100 mg/l in Tris buffer of pH 7,8) in the presence of either 20 μl bovine albumin (1 g/l in NaCl 9 g/l) or from 10 to 80 μl of serum (5 ml/l in Tris buffer pH 7,8) for 1 h at 37° and the fluorescence was measured[15, 16]. The initial, straight part of the inhibition curve was extrapolated to 100% inhibition; from this, the trypsin amount equivalent to the serum was calculated, as U trypsin inhibited per 1 of serum.

Results

As shown in fig. 1, there is a highly significant increase of α_1-antitrypsin after pulmonary embolism. The increase apparently lasts for a long period,

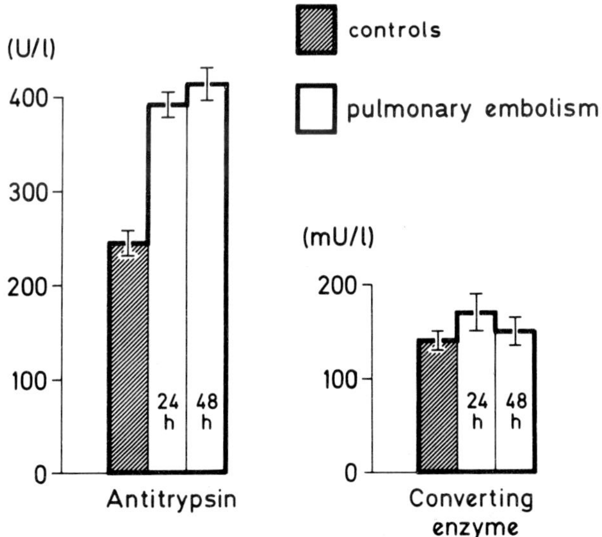

Fig. 1

Table I. *Chemical analyses in pulmonary embolism*

	α_1-*anti-trypsin U/l*	*Converting enzyme mU/l*	*Bilirubin mg/l*	*Aspartate amino transferase U/l*	*CPK U/l*
Pulmonary embolism					
mean	391	154	6,0	14,9	34,6
(s.d.)	(81)	(70)	(2,1)	(12,2)	(34,6)
(nb. of patients)	(18)	(17)	(18)	(18)	(17)
Reference range ($\overline{x}$ ±2s)	120-372	0-300	3-10	1-19	1-105

	Total LDH U/l	*LDH-1 %*	*LDH-2 %*	*LDH-3 %*	*LDH-4 %*	*LDH-5 %*
Pulmonary embolism						
mean	179	26,0	28,0	21,7	10,2	13,9
(s.d.)	(46)	(4,8)	(3,7)	(3,4)	(3,1)	(5,1)
(nb. of patients)	(17)	(12)	(12)	(12)	(12)	(12)
Reference range ($\overline{x}$ ±2s)	96-240	14-29	25-34	22-32	4-14	15-17

as the values do not decrease between 24 and 48 hrs. On the other hand, the mean values of the converting enzyme are only slightly higher after 24 hrs. than in normals, and the difference is not statistically significant. A survey of all constituents measured is shown in Table 1. The relative amounts of the 5 LDH isoenzymes lie practically within the reference range. Aspartate aminotransferase, total lactic dehydrogenase, creatine phosphokinase and bilirubin also show no significant difference from normals.

The increase in α_1-antitrypsin looks interesting. However, it is not specific for pulmonary embolism. In the group of patients who were first suspected of pulmonary embolism, but who subsequently turned out to have another disease, an increase of α_1-antitrypsin was also observed. This was again the only measured constituent that was markedly increased above the reference range of normals (Table 2).

Discussion

Although the angiotensin converting enzyme is a typical lung enzyme, it was not observed to be released into the circulation after pulmonary embolism. This is probably because, unlike the most sensitive enzymes used in clinical diagnosis, it is not a cytosol enzyme. It has been recently shown to be located on the luminal surface of pulmonary endothelial cells[17], and its bind-

Table II.

Final diagnosis		*Anti-trypsin 24 h*	*Converting Enzyme 24 h*	*Bilirubin mg/l*	*Aspartate Aminotransferase U/l*	*CPK U/l*	*Total LDH U/l*
Pneumonia and Bronchopneumonia	mean	378	172	6,2	14,5	24,2	141
	(Nb. of patients)	(10)	(10)	(10)	(10)	(10)	(10)
Others	mean	361	151	8,3	23,6	17,5	182
	(Nb. of patients)	(9)	(10)	(9)	(8)	(10)	(10)

ing to the solid structure of the cell may be sufficiently strong so as to prevent its release after embolism.

On the other hand, this study shows that the trypsin inhibiting capacity of blood plasma significantly increases after pulmonary embolism. This was already suggested by the work of Ritchie[13] using an immunologic method. The trypsin inhibitor that increases is obviously α_1-antitrypsin. The increase is probably due to a secondary effect, since α_1-antitrypsin is known to be synthesized in the liver.

The remainder of our results confirms the views of others who found that determination of aspartate aminotransferase and bilirubin is of no help in the diagnosis of pulmonary embolism[7-9, 18]. It is of interest that the CPK did not increase, which indicates that this test is a valuable help in differentiating myocardial infarction from pulmonary embolism. Whether the determination of antitrypsin is also useful in this differentiation is still an open question. Although Lundh[19] has reported increases in α_1-antitrypsin after myocardial infarction, we did not find such increases to occur consistently. A systematic study of antitrypsin values after myocardial infarction is presently being carried out in order to obtain more information regarding this question.

References

1. De Brito-Paiva E., Reymond J. Raymond L.: Les causes de mort en gériatrie. Psychol. med. (Paris), 7, 1771-1786 (1975).
2. Witchitz S.: Diagnostic et traitement d'une embolie pulmonaire. Rev. Prat., 25, 3249-3252 (1975).

3. Kistner R.L., Ball J.J., Nordyke R.A., Freeman G.C.: Incidence of Pulmonary Embolism in the Course of Thrombophlebitis of the Lower Extremities. Amer. J. Surg., *124,* 169-174 (1972).
4. Linton D.S., Bellon E.M., Bodie J.F., Rejali A.M.: Comparison of results of pulmonary arteriography and radioisotope lung scanning in the diagnosis of pulmonary emboli. Am. J. Roentgenol, *112,* 745-748 (1971).
5. Wacker W.E.C., Snodgrass P.J.: Serum LDH activity in Pulmonary Embolism Diagnosis. J.a.m.a., *174,* 2142-2145 (1960).
6. Wacker W.E.C., Rosenthal M., Snodgrass P.J. Amador E.: A Triad for the Diagnosis of Pulmonary Embolism and Infarction. J.a.m.a., *178,* 8-13 (1961).
7. Stevens L.E., Burdette W.J.: Enzymatic Levels in Pulmonary Infarction. Arch. Surg., *88,* 705-710 (1964).
8. Schonell M.E., Crompton G.K., Forshall J.M., Whitby L.G.: Failure to differentiate Pulmonary Infarction from Pneumonia by Biochemical Tests. Brit. Med. J., *1,* 1146-1148 (1966).
9. Konttinen A., Somer H., Auvinen S.: Serum Enzymes and Isozymes; Extrapulmonary Sources in Acute Pulmonary Embolism. Arch. Intern. Med., *133,* 243 (1974).
10. Ng K.K.F., Vane J.R.: Fate of angiotensin I in the circulation. Nature, *218,* 144-150 (1968).
11. Roth M., Weitzman A.F., Piquilloud Y.: Converting enzyme content of different tissues of the rat. Experientia (Basel), *25,* 1247 (1969).
12. Vairel E.G., Forlot P.: Evolution du pouvoir inhibiteur du sang vis-à-vis de certains enzymes au cours de syndromes d'hypercoagulation. 7th Int. Congr. Clin. Chem. Geneva/Evian 1969; vol. 2: Clinical Enzymology, pp. 40-45. S. Karger, Basel.
13. Ritchie R.F.: What abnormal serum proteins may indicate. Patient Care, May 15, 1974, pp. 2-6.
14. Depierre D., Roth M.: Fluorimetric determination of dipeptidyl carboxypeptidase. Enzyme, *19,* 65-70 (1975).
15. Roth M.: Dosage fluorimétrique de la trypsine. Clin. Chim. Acta, *8,* 574-578 (1963).
16. Roth M.: Methods biochem. anal., *17,* p. 262. J. Wiley, New York 1969.
17. Ryan J.W., Ryan U.S., Schultz D.R., Whitaker C., Chung A.: Subcellular localization of pulmonary angiotensin converting enzyme (kininase II). Biochem. J., *146,* 497 (1975).
18. Sasahara A.A., et al.: Clinical and physiologic studies in pulmonary thromboembolism. Am. J. Cardiol.,, *20,* 10 (1967).
19. Lundh B.: βIC-Globulin, trypsin inhibitors and haptoglobin in myocardial infarction and acute infectious diseases. Acta Univ. Lundensis, Sectio II, No. 24 (1965).

THE CLINICAL SIGNIFICANCE OF SERUM MB AND BB CREATINE KINASE ISOENZYMES

D.A. Nealon and A.R. Henderson

Summary

Three creatine kinase EC 2.7.3.2; CK isoenzymes have been isolated from human tissues. Brain contains predominantly the BB (CK_1) isoenzyme while healthy skeletal muscle contains the MM (CK_3) isoenzyme almost exclusively. Cardiac muscle appears to be the only tissue which contains, in health, appreciable quantities of the MB (CK_2) isoenzyme, and this knowledge has prompted many investigators to use the presence, or absence, of this isoenzyme as a diagnostic indicator of acute myocardial infarction. It is now apparent, however, that this diagnostic test depends on the insensitivity of the method used for detecting MB activity. Our experience, while using a DEAE-cellulose column technique, with the activity of the eluted fractions being determined by the Rosalki-Oliver assay at 340 nm, showed that the CK_2 isoenzyme can be readily detected in the serum of healthy persons.

Following an acute myocardial infarction the serum CK_2 isoenzyme activity increases but the differentiation between healthy persons and those with *minimal* myocardial damage may be impossible to make.

Not only does minimal myocardial damage present a problem but it is now becoming clear that there are many other conditions that will give rise to abnormal serum CK_2 activity such as malignant hyperpyrexia, Duchenne muscular dystrophy and also in some cases of polymyositis and dermatomyositosis. We have also observed that elevation of the serum CK_2 activity in *skeletal* muscle pathology is often accompanied by the appearance of appreciable quantities of serum CK_1 activity. Recently we have also observed elevated serum CK_1 activity following brain surgery.

The detection of serum CK_1 appears to be more readily accomplished with chromatographic techniques than by electrophoresis. We have observed that serum albumin inactivates CK_1 at 37°C and as albumin and CK_1 co-migrate in most electrophoretic systems, and as the isoenzymes are developed by incubation with substrate at 37°C, a likely explanation for the discrepancy between the two separate procedures may therefore be due to this protein-protein interaction.

Three creatine kinase [ATP: creatine N-phosphotransferase; EC 2.7.3.2.; CK] isoenzymes have been isolated from human tissues. Brain contains pre-

Department of Clinical Biochemistry, University Hospital (University of Western Ontario), London, Ontario, Canada.

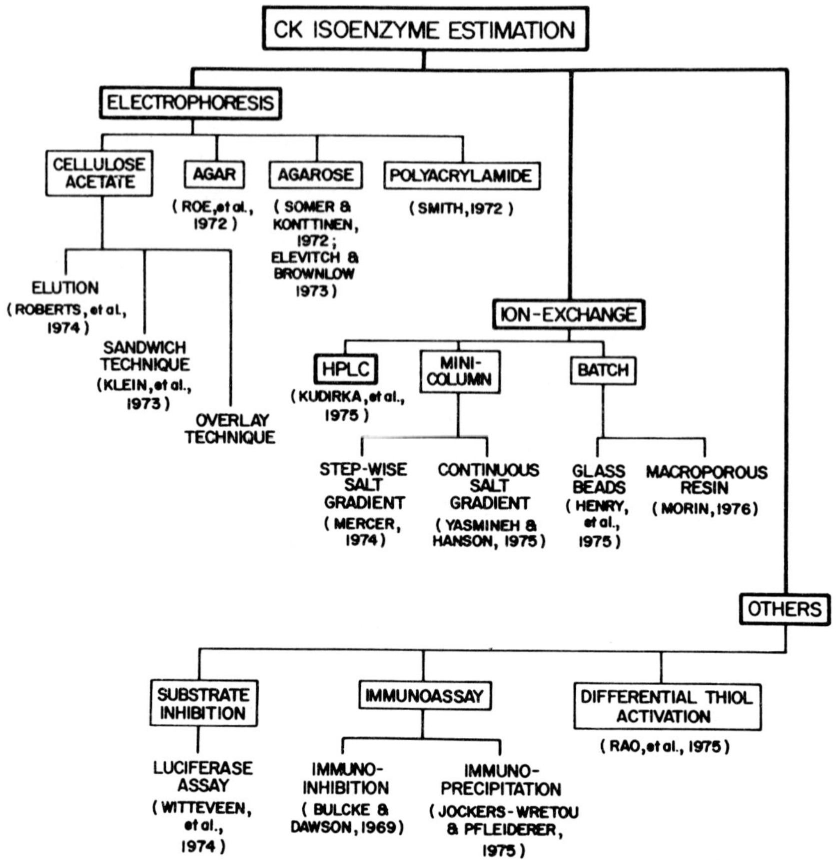

Fig. 1 – A Survey of Methods for Detecting the MB Isoenzyme of Creatine kinase in Serum.

dominantly the BB (CK_1) isoenzyme while healthy skeletal muscle contains the MM (CK_3) isoenzyme almost exclusively. Cardiac muscle appears to be the only tissue to contain, in health, appreciable quantities of the MB (CK_2) isoenzyme, and this knowledge has prompted many investigators to use the presence, or absence, of this isoenzyme in the serum as a diagnostic indicator of acute myocardial infarction.

In Figure 1 I have listed many of the techniques that have been used to determine the quantity of MB isoenzyme in the serum. I draw your attention particularly to the cellulose acetate electrophoresis technique which when combined with elution of the isoenzymes from the electrophoretic media for subsequent assay gives a technique of considerable sensitivity (Roberts et al., 1974). Notice also the mini-column procedure of Mercer (1974) using an ion-exchange separative method. I wish to spend some time discussing the

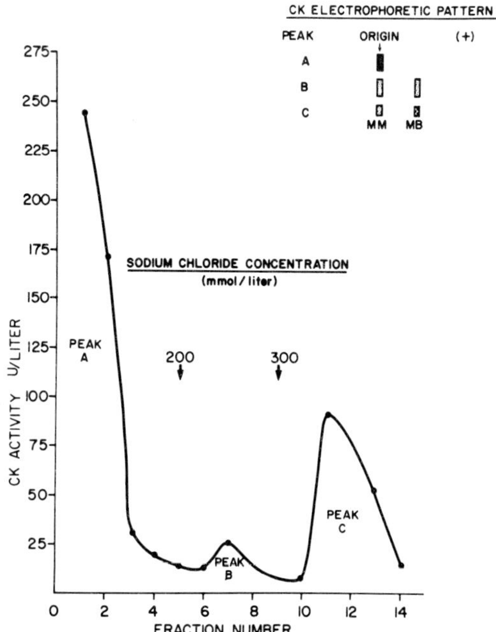

Fig. 2 – Elution Pattern of Creatine Kinase Activity from a proven case of myocardial infarction using Mercer's (1974) technique. For further experimental details refer to Nealon and Henderson (1975a).

latter technique and to compare the results obtained with it with those obtained by the cellulose acetate electrophoretic system.

We originally tried the method of Mercer (1974) but were unable to obtain separations of the isoenzymes (Fig. 2). No adjustments of buffer pH (Fig. 3) or column drainage (Fig. 4) could produce improvements in the separations. Finally we had to resort to step-wise gradient elution (Fig. 5) to obtain acceptable results (Nealon and Henderson, 1975a). However on using a DEAE-cellulose column (Keutel et al., 1972) we were able to obtain good separations as well as stable baselines between peaks (Fig. 6). Using this technique we were able to detect appreciable quantities of MB isoenzyme in serum from healthy persons and from non-cardiac patients (Nealon and Henderson, 1975a). These results were similar to those obtained previously by Roberts et al. (1974) using an electrophoretic technique. It therefore became clear that the technique used to detect MB isoenzyme could lead to diagnostic difficulties because some methods do not detect MB isoenzyme in the sera of normal persons whereas other techniques readily detect appreciable quantities and this could then lead to problems about establishing the diagnostic "cout-off" line between normal and myocardial infarction.

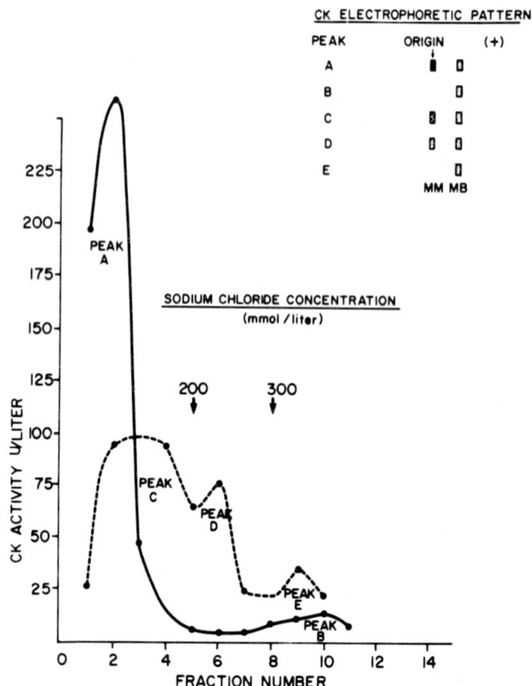

Fig. 3 – Elution Pattern of Creatine Kinase Activity from a proven case of myocardial infarction showing the effect of adjusting the pH of the Buffer before (●--●) and After (●——●) the Addition of NaCl. Refer to Fig. 2 for further experimental details.

Using the Cellulose-DEAE column technique we were able to separate the three isoenzymes (Nealon and Henderson, 1975b) in serum although we had originally, and erroneously, supposed that the BB isoenzyme was unimportant. We demonstrated the existence of the brain specific isoenzyme following handling of the brain (Fig. 7) and by use of this technique we were able to demonstrate the presence of BB and an increased MB in the sera of persons with skeletal muscle diseases.

For example, in a case of polymyositis in a young man of 17 13-26% of the serum CK activity was contained in the MB fraction and on admission 13% resided in the BB fraction. Total CK activity was elevated above one hundred times the upper limit of normal and in a case of dermatomyositosis (where heart involvement was suspended) we found MB of 3% and BB of 2% of the total serum activity. In absolute units these activities were each over 100 U/L with the total CK activity about eighty times the upper limit of normal. After exhaustive investigation of cardiac function it was concluded that, apart from the biochemical findings, there was no evidence of cardiac involvement.

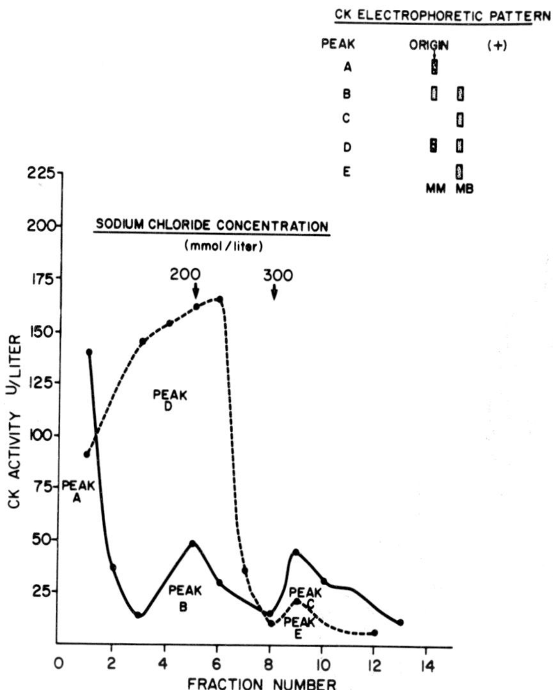

Fig. 4 – Elution Pattern of Creatine Kinase Activity from a proven case of myocardial infarction showing the effect of keeping the ion-exchange matrix immersed in buffer (●——●) or Draining After Application of each buffer (●------●). Refer to Fig. 2 for further experimental details.

In each of these cases not only were the MB values abnormal but there was a significant BB activity. This leads one to wonder if the muscle disease caused the normally inactive muscle B sub-unit gene to "switch on".

We have been able to obtain material from Dr. Beverley Britt's malignant hyperpyrexia cases and we have demonstrated the presence of significant levels of activity of BB, but in any one patient the activities found vary from the trivial to the definitely abnormal. These abnormal elevations do not always coincide with elevations of the total serum activity. Again, and not unexpectedly, the MB activities vary between those found in normal healthy individuals and the definitely abnormal. The clear conclusion is that these cases must be studied regularly – a single specimen is insufficient.

In passing I want to outline a possible source of discordance between electrophoretic and column chromatographic methods of CK isoenzyme estimations. We have been able to show that serum albumin interacts with CK_1

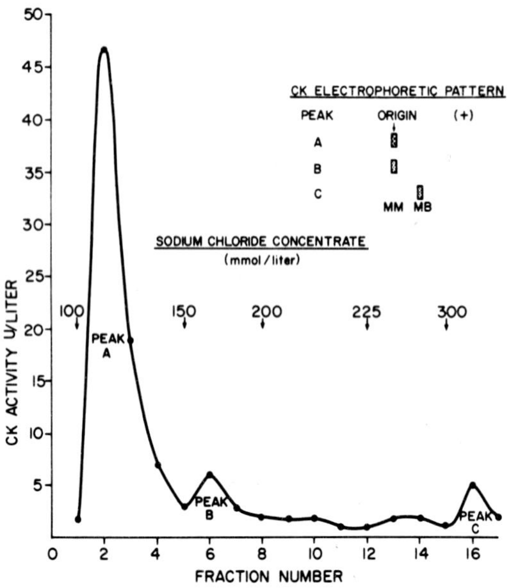

Fig. 5 – Elution pattern of creatine kinase activity obtained from a proven case of myocardial infarction obtained by applying a small step-wise increase in the NaCl Gradient. Refer to Fig. 2 for further experimental details.

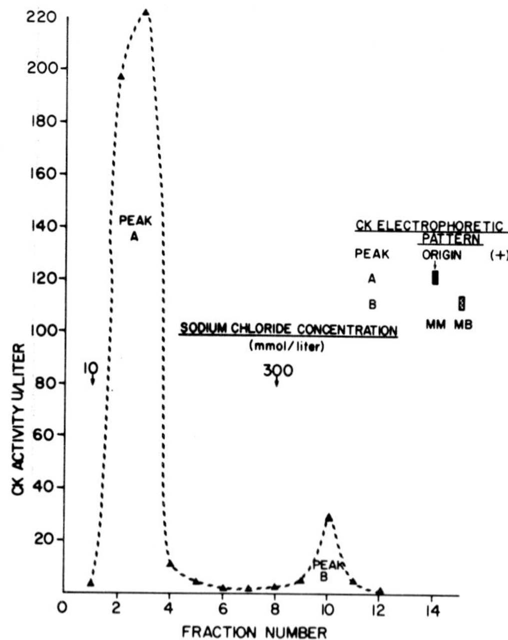

Fig. 6 – Elution patterns of creatine kinase activity using the technique of Nealon and Henderson (1975a).

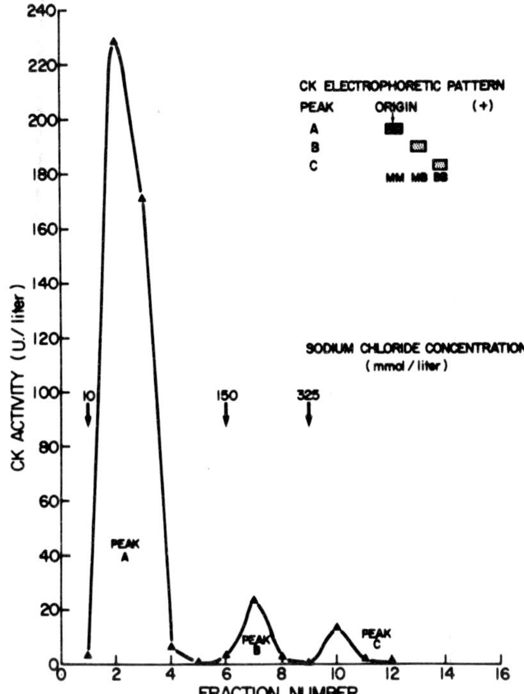

Fig. 7 – Elution Pattern of Creatine Kinase Activity from the serum of a patient after craniotomy and neurosurgical intervention. For further experimental details see Nealon and Henderson (1975b).

in vitro at 37°C (Nealon and Henderson, 1975c) and that enzyme activity is very quickly lost. This effect has not been observed with the other CK isoenzymes and as CK_1 and albumin co-migrate in many electrophoretic systems we wonder if CK_1 is may be being underestimated because of this effect.

Finally, I cannot resist raising the question of the temperature at which to assay CK isoenzymes. We have shown that 37°C would seem to be the only temperature at which to use the Rosalki-Oliver assay as there are inflexions in the Arrhenius plot at 30° or 25°C for the individual CK isoenzymes (Nealon and Henderson, 1976).

D.A.N. acknowledges the support of a Medical Research Council (Canada) Studentship, and the support of a grant from the Ontario Heart Foundation is gratefully acknowledged.

References

1. Bulcke J.A., Sherwin A.L.: Immunochem., *6,* 681-687 (1969).
2. Elevitch F.R., Brownlow K.: Amer. J. Clin. Path., *59,* 133-134 (1973).
3. Henry P.D., Roberts R., Sobel B.E.: Clin. Chem., *21,* 844-849 (1975).
4. Jockers-Wretou E., Pfleiderer G.: Clin. Chim. Acta, *58,* 223-232 (1975).
5. Keutel H.J., Okabe K., Jacobs H.K., Ziter F., Maland L., Kuby S.A.: Arch. Biochem. Biophys., *150,* 648-678. (1972).
6. Klein M.S., Shell W.E., Sobel B.E. : Cardiovasc. Res., *7,* 412-418 (1973).
7. Kudirka P.J., Busby M.G., Carey R.N., Toren E.C.: Clin. Chem., *21,* 450-452 (1975).
8. Mercer D.W.: Clin. Chem., *20,* 36-40 (1974).
9. Morin L.G.: Clin. Chem., *22,* (1976), 92-97.
10. Nealon D.A.,Henderson A.R.: Clin. Chem., *21,* 392-397 (1975a).
11. Nealon D.A., Henderson A.R.: Clin. Chem., *21,* 1663-1666 (1975b).
12. Nealon D.A., Henderson A.R.: J. Clin. Path., *28,* 834-836 (1975c).
13. Nealon D.A., Henderson A.R.: Clin. Chim. Acta, *66,* 131-136 (1976).
14. Rao P.S., Lukes J.J., Ayres S.M., Mueller H.: Clin. Chem., *21,* 1612-1618 (1975).
15. Roberts R., Henry P.D., Witteveen S.A.G.J., Sobel B.E.: Amer. J. Cardiol., *33,* 650-654 (1974).
16. Roe C.R., Limbird L.E., Wagner G.S., Nerenberg S.T.: J. Lab. Clin. Med., *80,* 577-590 (1972).
17. Smith A.R.: Clin. Chim. Acta, *39,* 351-359 (1972).
18. Somer H., Konttinen A.: Clin. Chim. Acta, *40,* 133-138 (1972).
19. Witteveen S.A.J.G., Sobel B.E., DeLuca M.: Proc. Nat. Acad. Sci., *71,* 1384-1387 (1974).
20. Yasmineh, W.G., Hanson N.Q.: Clin. Chem., *21,* 381-386 (1975).

CREATINE KINASE ISOENZYMES IN SERUM AFTER MYOCARDIAL INFARCTION AND IN VARIOUS EXTRACARDIAL CONDITIONS

G. Chemnitz, J. Lobers, L. Nevermann, P. Rathsack,
E. Schmidt and F.W. Schmidt

Summary

1. CK-MB in quite stable in vitro. Repeated thawing and freezing is tolerated. During storage at >4°C for one week no remarkable decrease of activity occurs.
2. CK-MB can be used for differential diagnosis not earlier than two hours and not later than 70 hours after myocardial infarction.
3. CK-MB activity is not correlated with the prognosis of an infarction. Our results support the hypothesis, that CK-MB activity in serum after myocardial infarction results from cell necrosis as well as from cardiac reparative processes.
4. CK-MB activity with normal total CK in serum can also be detected in valvular defects with cardiac failure.
5. CK-isoenzyme-BB could be demonstrated in serum by immuno-precipitation technique for the first time after delivery.

Introduction

Sometimes it is difficult to elucidate the source of high activity of creatine kinase (CK, EC 2.7.3.2) in serum using clinical and other biochemical parameters. According to Jockers and Pfleiderer (1975a) and Würzburg (1976) CK-isoenzymes can be differentiated quantitatively by two immunological methods: immunoprecipitation and immunoinhibition.

Using the immunoprecipitation method after Jockers and Pfleiderer one is able to differentiate between the three CK-isoenzymes CK-MM, CK-MB and CK-BB. The disadvantage is the incubation time of 24 hours.

With the immunoinhibition technique after Würzburg actual CK-B-subunits are measured, whether they originate from the hybrid dimer CK-MB or from the uniform dimer CK-BB. The advantage of this method is the incubation time of only 5 minutes.

Div. of Gastroenterology and Hepatology, Div. of Cardiology, Dept. Intern. Medicine and Dept. of Obstetrics and Gynaecology. Medizinische Hochschule Hannover, Germany.

We report our results on CK-isoenzyme determination:

1. Quality control and effect of storage on CK-MB-activity in serum.
2. A systematic study on CK-MB in myocardial infarction and other cardiac diseases.
3. Results of investigating CK-BB and CK-MB in extracardial affections.

Materials and methods

Blood was allowed to clot, centrifuged for 15 minutes at 7500 g at +4°C, serum removed and frozen at −25°C to be stored for a maximum of 72 hours. Prior to assay, serum samples were thawed at room temperature. CK was assayed with the optimized standard assay according to the recommendation of the German Society of Clinical Chemistry[2], with Boehringer-Monotest-CK or Merck-1-test-CK. Photometer Eppendorf, 334 nm, +25°C.

Immunoprecipitation[6, 7, 12]

For immunoprecipitation, total CK-activity in serum is determined. Aliquots of the serum are precipitated with anti-CK-BB and anti-CK-MM respectively. Anti-CK-BB precipitates quantitatively CK-MB and CK-BB. Anti-CK-MM precipitates quantitatively CK-MM and CK-MB. After precipitation and centrifugation, the remaining CK-activity is measured in the supernatants. The activities of CK-MM, CK-MB and CK-BB can be calculated from the difference between the first and the second value, determined from total CK.

Immunoinhibition[19]

With the immunoinhibition method, the reagents contain inhibiting CK-M-antibodies, blocking the activity of the CK-M-subunits completely. Incubation time is 5 minutes at +25°C. Sera with total CK-activity above 1000 U/l must be diluted.

Results

Quality control and effect of storage on CK-MB activity in serum

The precision of the immunoprecipitation method within the series was investigated in 10 sera by 6 fold determination of total CK and CK-MB activity. The coefficient of variation (CV) was 1.7% for total CK and 6.4% for CK-MB activity (Fig. 1).

The precision of the immunoinhibition method within the series was 1.7 for total CK and 1.9% for CK-MB respectively (Fig. 1).

Fig. 1 – Reproducibility of the method.

Immunoprecipitation	CV %	=	1.7 (CK-total)
	CV %	=	6.4 (CK-MB)
Immunoinhibition	CV %	=	1.1 (CK-total)
	CV %	=	1.9 (CK-MB)

Sera of 10 patients. 6 measurements at 25°C; Photometer Eppendorf, 334 nm, Merckotest.

Fig. 2 – Influence of storage (10 days) on CK and CK-MB activity in serum.

		1		*2*	
		CK-tot.	*CK-MB*	*CK-tot.*	*CK-MB*
Immunoprecipitation	CV	1.7	8.5	3.3	9.8
Immunoinhibition	CV %	1.2	1.5	1.2	1.7
	N	10		10	
		3		*4*	
		CK-tot.	*CK-MB*	*CK-tot.*	*CK-MB*
Immunoprecipitation	CV %	2.8	12.1	27	98
Immunoinhibition	CV %	2.1	7.7	29	63
	N	10		10	

Sera of 10 patients. Measurements at 25°C; Photometer Eppendorf, 334 nm, Merckotest.
1. Aliquots separately frozen at −25°C; 2. Serum repeatedly frozen and thawed;
3. Stored at + 4°C; 4. Stored at room temperature.

The influence of storing over a ten day period under various conditions is demonstrated by Fig. 2.

With freezing and thawing aliquots of the same sera consecutively, the CV was 8.5% for CK-MB, when immunoprecipitation was performed, and 1.5% using immunoinhibition.

When sera were repeatedly frozen and thawed ten times over a ten day period, the CV was 9.8% for immunoprecipitation and 1.7% for immunoinhibition.

In sera stored at +4°C, CK-MB exhibited a decrease of activity on the 8th day in both techniques. The CV for CK-MB were 12.1% and 7.7% respectively.

In samples kept at room temperature, a definite decrease of total CK and CK-MB activity was already observed by the second day.

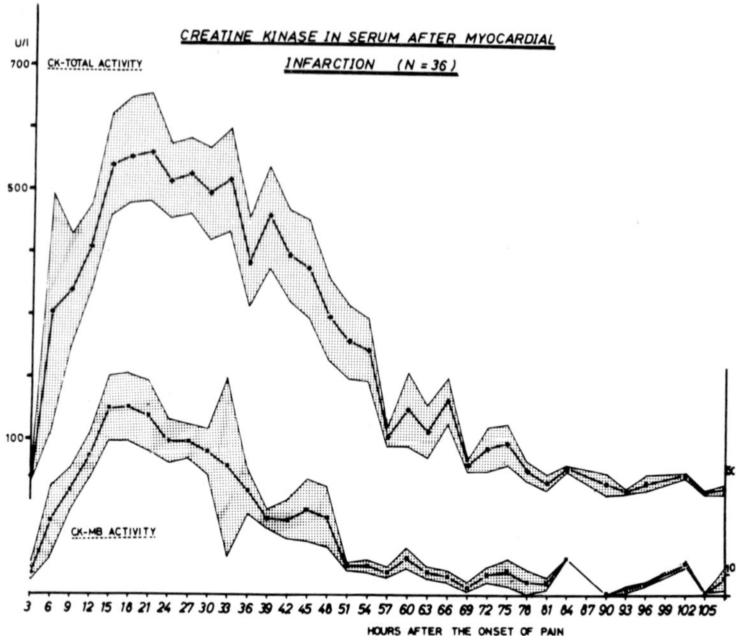

Fig. 3

CK-MB in myocardial infarction

In 36 patients with myocardial infarction blood samples were taken every three hours during the first 24 hours, and every 6 hours in the next three days. The commencement of blood sampling varied from 2 to 38 hours after the onset of pain, depending on the admission of the patients.

In the diagram (Fig. 3) total CK and CK-MB activity are shown. Two hours after the onset of pain the mean value of total CK activity was still in the accepted normal range (40 U/1). The average CK-MB was elevated at that time to 8.7 U/1. The relative amount of CK-MB was already as high as 22%.

Total CK showed highest activity 22 hours after the onset of pain with a mean value of 560 U/l, then decreased to normal values within 80 hours after infarction. In 19 patients maxima of total CK and CK-MB were observed at the same time. In 12 patients maximal CK and CK-MB activities were found at different times: in 5 patients maximal CK-MB occurred two hours, in 1 patient eight hours and in 2 patients even 17 hours before total CK peak value. 4 patients exhibited maximal CK-MB two hours later. Average CK-MB peak value 18 hours after the onset of pain was 60 U/l. In contrast to total CK normal values were reached within 70 hours after the onset of pain.

The course of CK and CK-MB activity classified to the prognostic index

CK AND CK MB ACTIVITY IN SERUM IN MYOCARDIAL INFARCTION REFERRED TO A PROGNOSTIC INDEX

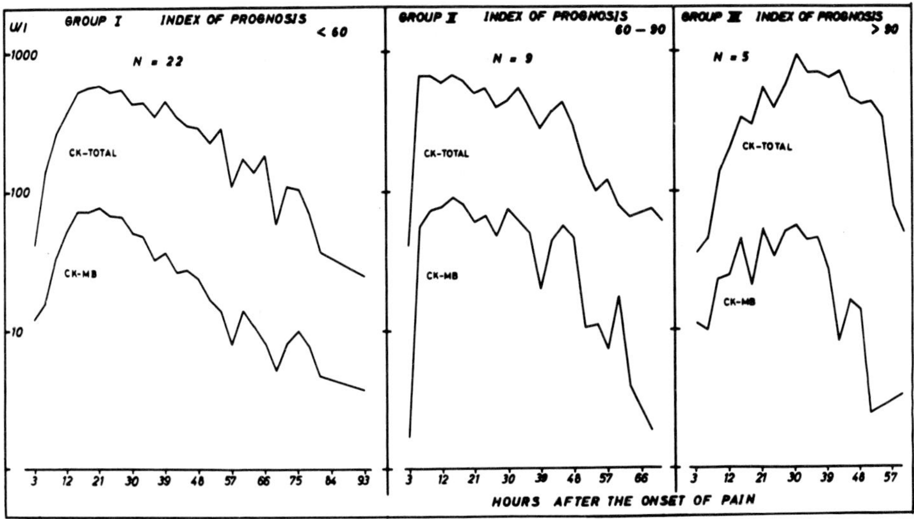

Fig. 4

according to Gallitz (1975) is shown in Fig. 4. The prognostic index used was calculated by the following clinical parameters: age, pulmonary congestion, leucocytosis, shock, blood pressure, infarct localisation and high pressure in anamnesis. The higher the prognostic index, the worse the prognosis. The prognostic index of group 1 is less than 60, in group 2 between 60 and 90 and in group 3 above 90. The CK-MB activity of group 3 – corresponding to a high prognostic index with bad prognosis – shows an abrupt decrease to almost normal values already after 48 hours. The curve of CK-MB activity of group 1 – corresponding to a low prognostic index and better prognosis – is enlarged over 93 hours.

CK-MB in surgery

Brain surgery: (Fig. 5)

Sera obtained throughout brain surgery were investigated by immunoprecipitation technique. CK-BB could not be detected in any case but we determined a small but significant activity of CK-MB in serum. The blood sample of patient M., obtained preoperatively but after the onset of anaesthesia, exhibited a CK-MB activity of 67 % (6.1 U/l).

Creatine Kinase in Serum in Brain Surgery
CK-TOTALACTIVITY
CK-MB-ACTIVITY
pr = praeoperative
in = intraoperative
po= postoperative
U/l
PONTINE ANGLE TUMOR
MENIGEAL TUMOR
GLIOMA
GLIOMA
PONTINE ANGLE TUMOR
P. Pl.
P. Ph.
P. O.
P. H.
P. K.

Fig. 5

CREATINE KINASE IN HEART SURGERY
1 - BEFORE ANAESTHESIA
2 - AFTER ANAESTHESIA
3 - AFTER STERNOTOMY
4 - AFTER CARDIOTOMY
5 - AFTER COMPLETION
U/l

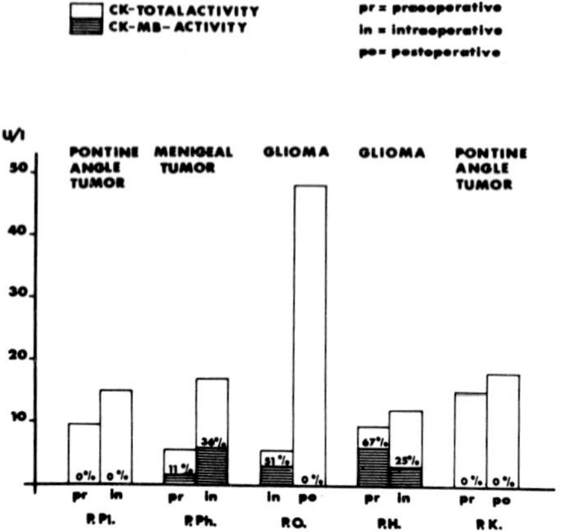

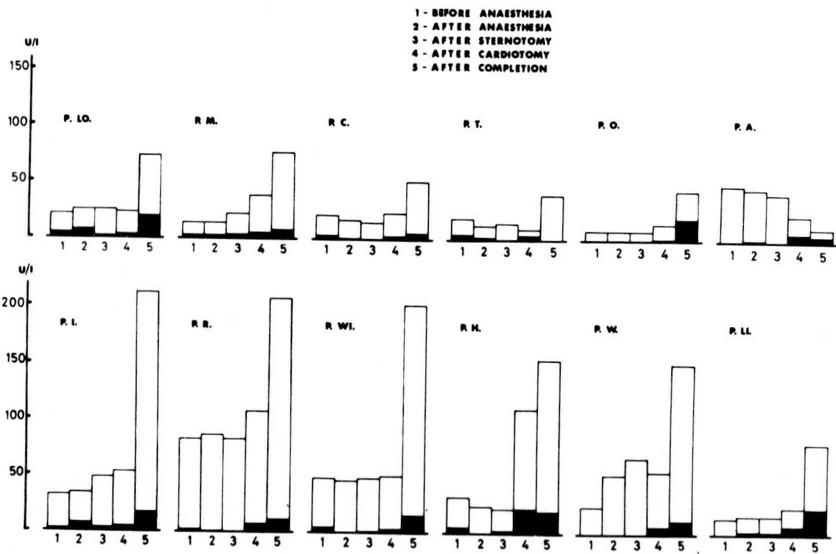

Fig. 6

Heart surgery: (Fig. 6)

We investigated sera of 12 patients who were operated by cardiotomy. Blood sampling was done at five exactly standardized time points. Only three patients exhibited a rise of CK-MB after the onset of anaesthesia. 7 patients, however, – all suffering from severe valvular defects and cardiac failure – had CK-MB up to 10 U/l already before anaesthesia. The increase of CK-MB found after the fourth blood sampling could be explained by the cardiotomy, performed just before and by hemolysis resulting from the heartlung-machine (Prellwitz, 1976).

In contrast blood samples of 11 patients with abdominal surgery exhibited only a slight rise of CK-MB up to 2 U/l after completion of surgery (Fig. 7).

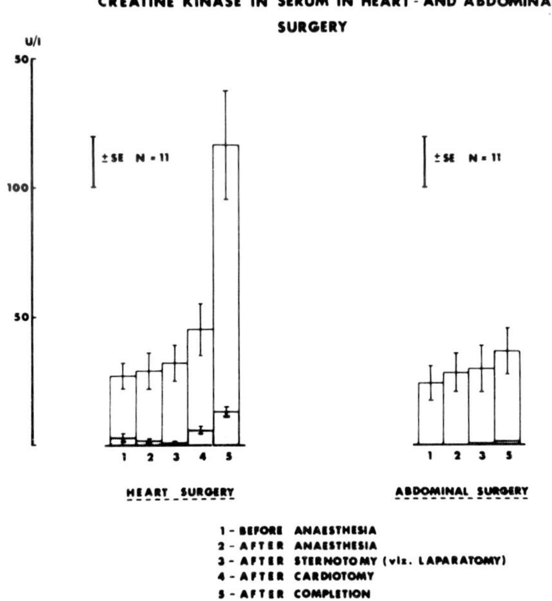

Fig. 7

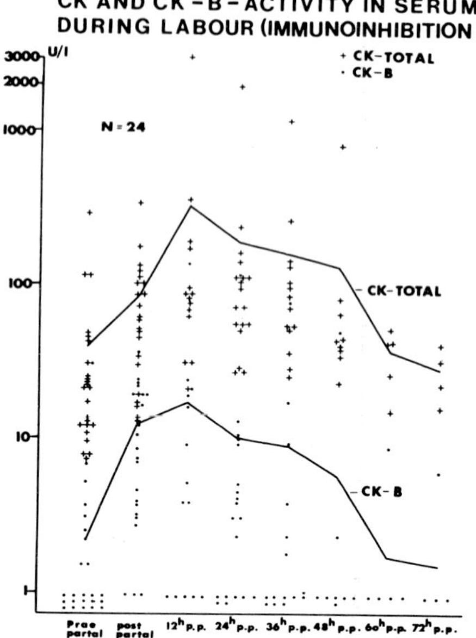

Fig. 8

CK-B during labour and after delivery: (Fig. 8)

In 6 out of 24 sera sampled during labour, CK-B-subunits could be detected with immunoinhibition technique already before delivery but after the onset of labour pains. After delivery we found high CK-B-subunit activity in 21 women. The mean values corresponded roughly to the curve of total CK.

In addition we examined 20 sera with immunoprecipitation. In the sera of eight women we found CK-isoenzyme-BB up to 136 U/l (Fig. 9).

Discussion

CK-MB is quite stable in vitro. Repeated thawing and freezing is tolerated. During storage at +4°C for one week no remarkable decrease of activity occurs.

In myocardial infarction the isoenzyme CK-MB – originating from cardiac muscle – can regularly be detected[1, 4, 10, 12, 14, 15, 16, 17, 19]. Increased CK-MB activity can therefore support the assumption of a myocardial infarction. The determination of CK-MB becomes important mainly when signs and

Fig. 9 – CK isoenzymes after delivery (immunoprecipitation)

	CK-total (U/l)	*CK-MB (U/l)*	*CK-BB (U/l)*	
1. praepartal	25	0	0	
12 h post	3032	0	136	(caesarian
20 h post	1946	0	121	section)
36 h post	1144	0	77	
45 h post	803	0	49	
2. praepartal	44	0	0	(caesarian
4 h post	151	10.5	5.1	section)
24 h post	217	9.7	7.6	
3. praepartal	12.1	0	0	
72 h post	46	8.6	6.0	
4. praepartal	115	30	0	
4 h post	197	9.1	6.0	
5. praepartal	9.2	0	0	
2 h post	88	36	14.5	
6. praepartal	15.0	0	0	
24 h post	14.6	0	3.0	
48 h post	23.6	0	0	
7. praepartal	11.1	0	0	
24 h post	60	2.0	0	
48 h post	50	0	3.0	
8; 1 h post	85	0	10.6	
24 h post	86	13.8	0	

symptoms, biochemical or ECG – parameters can neither exclude nor assure the assumption of an acute myocardial infarction. In such cases it is important to know how long CK-MB can be used for differential diagnosis. Up to 3 hours after the onset of pain 2 patients out of 7 showed no CK-MB activity. In 5 patients a clearly measurable CK-MB activity was observed with a mean value of 12.2 U/l although total CK activity of 40 U/l was still within the accepted range of normal values. 4 hours after the onset of pain CK-MB could be measured in the sera of each patient. The mean value was 29 U/l. CK-MB remained detectable even 70 hours after acute myocardial infarction.

The course of CK-MB based on the different prognostic index was rather surprising: considering the efflux of CK-MB as a result of muscle necrosis during cardiac infarction the severe courses with fatal prognosis should be found in group 1 with enlarged CK-MB activity over 93 hours. In contrast

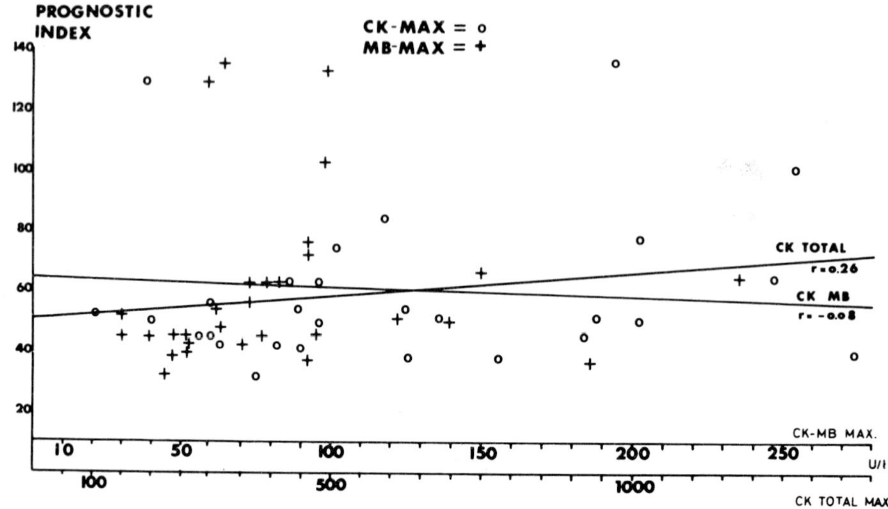

Fig. 10

group 3 – corresponding to a high prognostic index – shows an abrupt decrease of CK-MB already after 48 hours.

To comment on these rather surprising results, the following hypothesis can be advanced: the enzyme efflux during myocardial infarction is based on two overlapping processes:

1. The efflux of CK-MB from the damaged heart muscle cells – relevant for groups 1, 2 and 3.
2. Another enzyme efflux during delayed reparation processes.

In 1969 Wiesmann could demonstrate a higher amount of CK-MB in regenerating muscle cells.

These results also explain that it is impossible to correlate the prognostic index with total CK or CK-MB peak values (Fig. 10).

CK-MB activity is not only evident in serum in myocardial infarction. In accordance with Mercer (1975) we found CK-MB together with normal total CK in different heart diseases with cardiac failure.

Up to now CK-BB could not be detected in serum by immunoprecipitation techniques. According to the results of Jockers (1975 b) and Freise (1975) the CK of uterus smooth muscle consists of almost 100 % CK-BB. King (1972) reports a decrease of total CK in early pregnancy. In late pregnancy, during labour and after delivery the total CK activity in serum is increased[9]. This has been assigned to enzyme efflux from skeletal muscle during labour and from uterus in involution. We found CK-B-subunit activity in serum before and mainly after delivery. In 8 out of 20 sera we succeded in demonstrating CK-isoenzyme-BB after delivery by immunoprecipitation. At the present time we are investigating the clinical significance of these results.

References

1. Blomberg D.J., Kimber W.D., Burke M.D.: Creatine kinase isoenzymes. Predictive value in the early diagnosis of acute myocardial infarction. Americ. J. of. Med. *59,* 464-469 (1975).
2. Deutsche Gesellschaft für Klinische Chemie: Empfehlungen der Deutschen Gesellschaft für Klinische Chemie. Standardisierung von Methoden von Enzymaktivitäten in biologischen Flüssigkeiten. Z. Klin. Chemie *8:* 658-659 (1970); *10:* 182-192 (1972).
3. Freise J., Trautschold I.: Personal communication, 1975.
4. Galen R.S.: The enzyme diagnosis of myocardial infarction. Human Pathology, *6,* 141-155 (1975).
5. Gallitz T., Sander P. Jahrmärker H., Rackwitz R., Haider M.: Ein prognostischer Index bei akutem Myokardinfarkt. Dtsch. Med. Wschr, *49,* 2517-2523 (1975).
6. Jockers-Wretou E., Grabert K., Pfleiderer G.: Quantitative immunologische Bestimmung der Isoenzyme der Creatinkinase im Serum. Z. Klin. Chemie Klin. Biochemie, *13,* 85-88 (1975a).
7. Jockers-Wretou E., Pfleiderer G.: Quantitation of creatine kinase isoenzymes in human tissues and sera by an immunological method. Clin. Chim. Acta, *58:* 223-232 (1975b).
8. King B., Spikesman A., Emery A.E.: The effect of pregnancy on serum levels of creatine kinase. Clin. Chim. Acta *36,* 262-269 (1972).
9. Konttinen A., Pyörärä T.: Serum enzyme activity in late pregnancy, at delivery and puerperium. Scand. J. Clin. Lab. Invest. *15*: 429-435 (1963).
10. Konttinen A., Somer H.: Specificity of serum creatine kinase isoenzymes in diagnosis of acute myocardial infarction. Brit. Med. J., *1,* 386-389 (1973).
11. Mercer D.W.: Detection of cardiac specific creatine kinase isoenzyme in sera with normal or slightly increased total creatine kinase activity. Clin. Chem., *21:* 1088-1092 (1975).
12. Neumeier D., Knedel M., Würzburg U., Hennrich N., Lang H.: Immunologischer Nachweis von Creatinkinase MB in Serum beim akuten Myokardinfarkt. Klin. Wschr. *53,* 329-333 (1975).
13. Prellwitz, W. Neumeier D.: Such- und Schnellteste bei Herzerkrankungen und Kreislaufschock. Der Internist *17,* 436-444 (1976).
14. Roberts R., Gowda K.S., Ludbrock P.A., Sobel B.E.: Specificity of elevated serum MB creatine phosphokinase activity in the diagnosis of acute myocardial infarction. Americ. J. Cardiol., *36,* 433-437 (1975).
15. Sobel B.E., Roberts R., Larson K.B.: Estimation of infarct size from serum MB creatine phosphokinase activity: applications and limitations. The Americ. J. of Cardiology, *37,* 474-485 (1976).

16. Varat M.A., Mercer D.W.: Cardial specific creatine phosphokinase isoenzyme in the diagnosis of acute myocardial infarction. Circulation, *51*, 855-859 (1975).
17. Wagner G.S., Charles R.R., Limbird L.E., Rosati R.A., Wallace A.G.: The importance of identification of the myocardial-specific isoenzyme of creatine phosphokinase (MB-form) in the diagnosis of acute myocardial infarction. Circulation, *47*, 263-269 (1973).
18. Wiesmann U., Kaspar U., Mummenthaler M.: Necrosis and regeneration of tibialis anterior muscle in rabbit. Arch. Neurol., *21*, 373-380 (1969).
19. Würzburg, U. Hennrich, N. Lang H.: Bestimmung der Aktivität von Creatinkinase MB in Serum unter Verwendung inhibierender Antikörper. Klin. Wschr. *54*: 357-360 (1976).

CHANGES IN GLYCOLYTIC ENZYME LEVELS AND ISOZYME PATTERNS DURING BLAST TRANSFORMATION OF HUMAN LYMPHOCYTES*

R.W. Gracy**, T.L. Phillips and M.V. Kester

Summary

Glycolytic enzyme levels and their isoelectric focusing profiles were compared in extracts of human peripheral lymphocytes and mitogen-induced lymphoblasts. The levels of glyceraldehyde-3-phosphate dehydrogenase and lactate dehydrogenase increased almost ten-fold during blast transformation. Enolase, aldolase and phosphoglycerate kinase increased from four to five-fold, while hexokinase, pyruvate kinase, glucosephosphate isomerase, triosephosphate isomerase and phosphoglycerate mutase increased approximately two to threefold.

Three of the enzymes (hexokinase, glucosephosphate isomerase and enolase) yielded simple electrofocusing patterns of single peaks with no electrophoretic alteration resulting from blast transformation. Phosphoglycerate mutase exhibited two components which were unchanged after blast formation. Four enzymes (aldolase, glyceraldehyde-3-phosphate dehydrogenase, lactate dehydrogenase and glucose-6-phosphate dehydrogenase) were resolved into multiple isozymes in both lymphocytes and lymphoblasts and changes were observed in the relative contents of each isozyme as a result of blast transformation in the case of aldolase and lactate dehydrogenase. Phosphoglycerate kinase and triosephosphate isomerase from lymphocytes electrofocused as single species while pyruvate kinase exhibited two components. New isozymes were observed in lymphoblasts for all three enzymes. This new isozyme formation was studied in detail for the case of triosephosphate isomerase and shown to be the result of the activation of a structural gene for the isomerase which is not normally expressed in peripheral lymphocytes.

---

*This work was supported by research grants from the U.S. National Institutes of Health (AM-14638), The Robert A. Welch Foundation (B-502), and North Texas State University Faculty Research Funds.

**Recipient of National Institutes of Health Research Career Development Award (KO4 AM-70198) and Fellow of the Alexander von Humboldt-Stiftung. To whom correspondence should be addressed.

Department of Biochemistry North Texas State University, Denton, Texas 76203 U.S.A.

Introduction

Human peripheral lymphocytes, when treated with mitogenic plant glycoproteins such as phytohemagglutinin, concanavalin A or pokeweed mitogen, undergo blast transformation and mitosis between 48 and 72 hours[17]. This ability to induce large numbers of dividing cells has thus provided a readily available, convenient diagnostic system for the investigation of chromosomal morphology, as well as enzyme deficiency diseases[1, 21]. Although the phenomenon of mitogen-induced lymphocyte activation has been recognized for over 15 years[9, 17], the complex series of biochemical events which transform the small, quiescent lymphocyte into a metabolically active, morphologically distinct lymphoblast are the current topic of active investigations in many laboratories. The basic similarity to antigen-stimulated lymphocyte activation has resulted in the greatest activity being conducted in immunological laboratories, but an understanding of these processes may also help to explain the effects of enzyme deficiencies, tissue and cellular differentiation, and perhaps the basic errors associated with cancer cell proliferation.

A variety of studies have shown that among the earliest metabolic changes of blastogenesis are increased rates of glycolysis[8, 21, 23] phosphatidylinositol metabolism[6], acetylation of histones[14], and RNA synthesis[16]. Somewhat later, protein synthesis is stimulated[9], which is followed by DNA replication[15].Several studies[8, 18, 29, 22, 23] have demonstrated *in vivo* changes in the levels of glycolytic intermediates and have revealed drastic increases in the rate of lactate and pyruvate production during blast transformation. Increased carbohydrate metabolism is undoubtedly needed to meet the energy requirements of this highly active cell. The present study was designed to quantitatively compare the levels and isozyme expression of glycolytic enzymes in human peripheral lymphocytes and in mitogen-induced lymphoblasts.

Materials and Methods

Lymphocyte Isolation: Blood was obtained by standard venipuncture and collected into acid/citrate/dextrose, then diluted with 0.5 volumes of heparinized (20 units/ml) physiological saline. Lymphocytes were immediately isolated by centrifugation in Ficoll-Hypaque at 400 x g for 15 min[3]. The lymphocytes were carefully collected from the interface and washed by resuspending in saline and recentrifuging. Cells were counted in a hemacytometer. All glassware was siliconized by treatment with 1% Siliclad (Clay Adams).

Cell Culturing: Freshly isolated lymphocytes were transferred to Joklik's Modified Minimum Essential Media containing 15 % fetal calf serum, 20 units/ml heparin, and mitogens (10 ml each of phytohemagglutinin, PHA, and Pokeweed mitogen, PWM, per liter of media) and incubated at 37° in a

5 % CO_2/95 % air mixture. Cells were harvested by centrifugation at 400 x g for 15 min, followed by resuspension of the cell pellet in 10 ml of triethanolamine buffer (10 mM, pH 7.6) containing 0.5 M KCl, 1.0 mM EDTA and 2 % 2-mercaptoethanol. Cells were disrupted with a Potter-Elvejhem homogenizer, and centrifuged at 400 x g for 15 min.

Isoelectric Focusing: Isoelectric focusing was conducted in 110-ml columns at 2°, 350 volts for 96 hours in pH 3.5 – 10 ampholines (LKB) with a sucrose density gradient. Following electrofocusing, fractions of 0.6 ml were collected by pumping water onto the top of the focusing column. The pH and the enzymatic activities of the fractions were determined immediately following elution of the column.

Enzyme Assays: All enzymes were assayed by quantitative spectrophotometric assays by coupling with the appropriate purified dehydrogenases and measuring the absorbance change at 340 nm. The assays were carried out in thermostated (30°) recording spectrophotometers under standard assay conditions as described in the Boehringer/Mannheim catalogue.

Results

Figure 1 demonstrates the markedly increased levels of essentially all glycolytic enzymes in 72-hour mitogen-induced lymphoblasts when compared with peripheral lymphocytes. These data are expressed as international enzyme units per 10^{10} cells and are essentially identical to the relative specific activities of the enzyme in cell homogenates. The enzymes exhibiting the most marked elevations were glyceraldehyde 3-phosphate dehydrogenase (G3PD) and lactate dehydrogenase (LDH), each increasing almost ten-fold. Levels of enolase (Enol), fructose 1,6-diphosphate aldolase (Aldo) and phosphoglycerate kinase (PGK) showed increases of four to fivefold. Two kinases (hexokinase, HK, and pyruvate kinase, PK) and the three isomerases [glucose 6-phosphate isomerase (GPI), triosephosphate isomerase (TPI) and phosphoglycerate mutase (PGM)] showed smaller but significant increases of approximately two to three-fold. Although a number of attempts were made to evaluate levels of phosphofructokinase, the low levels of the enzyme and its instability precluded accurate measurements of the enzyme on small numbers of cells. In contrast to the elevated levels of these glycolytic enzymes, the apparent concentration of glucose-6-phosphate dehydrogenase was only marginally increased during blast transformation.

While the above data clearly revealed elevated levels of these enzymes in lymphoblasts, they did not distinguish between enzyme activation, increased *de novo* synthesis of the enzymes, or the synthesis of new isozymic forms of the enzymes. Thus, additional methods were necessitated in order to evaluate the bases for the increased levels of the glycolytic enzymes. Since isoelectric focusing provided a method to resolve each of the enzymes, as

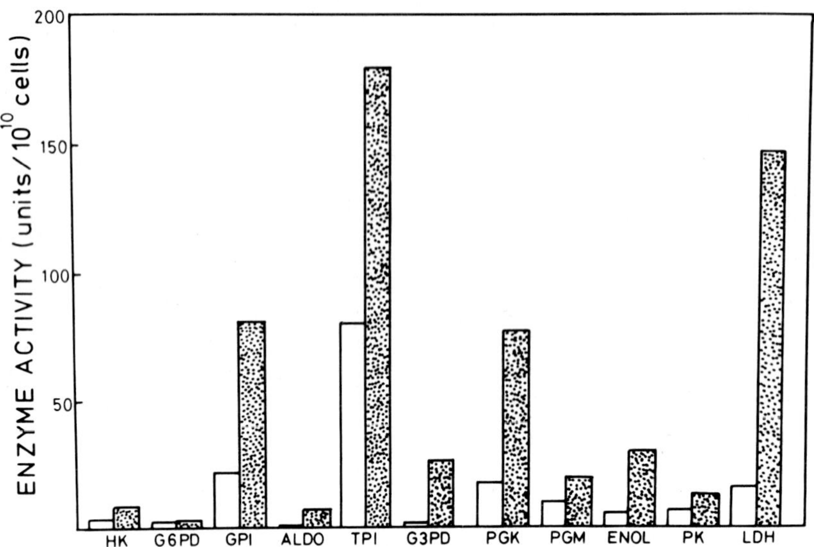

Fig. 1 – Levels of Glycolytic Enzymes in Human Lymphocytes and Lymphoblasts. Human peripheral lymphocytes (open bars) and lymphoblasts (stippled bars) were extracted and quantitatively assayed for the content of glycolytic enzymes as described in "Methods". Activity is expressed as international enzyme units per 10^{10} cells, and the abbreviations of each enzyme are as indicated in the text.

well as the various possible isozymic forms and still permit the quantitation of the relative amounts of each component, it was chosen for further studies. Table I summarizes the results of such isoelectric focusing experiments. Three of the enzymes, hexokinase, glucosephosphate isomerase, and enolase showed simple electrofocusing patterns and appeared as single species with no apparent changes in the electrophoretic profiles of lymphocyte and lymphoblast extracts.

Phosphoglycerate mutase exhibited an isoelectric pattern of two components which were not altered during blastogenesis.

Four of the enzymes, aldolase, glyceraldehyde 3-phosphate dehydrogenase, lactate dehydrogenase, and glucose-6-phosphate dehydrogenase, showed complex electrofocusing patterns with multiple peaks of enzyme activity. Aldolase from both lymphocytes and lymphoblasts focused as two components. Lymphocyte extracts routinely yielded approximately equal amounts of aldolase with isoelectric points of 9.2 and 8.8. Occasionally, a small amount of activity was observed at lower pH ranges. Isoelectric focusing of aldolase from lymphoblasts yielded predominantly a component of pH 9.2 with a shoulder (approximately 20-25 % of the total activity) at pH 8.8. Since detailed electrofocusing studies of human aldolases from various

Table I. *Isoelectric focusing of enzymes from human lymphocytes and lymphoblasts*

Enzyme	*Isoelectric pH*		*Remarks*
	Lymphocytes	*Lymphoblasts*	
Hexokinase	7.40	7.40	Single component, no change
Glucose-6-P-Dehydrogenase	7.60 7.10 5.80	7.60 7.10 5.80	Multiple components, no change
Glucose-6-P-Isomerase	9.35	9.35	Single component, no change
Aldolase	9.25 8.75	9.25 8.75	Concentration of the more basic isozyme may increase during blastogenesis
Triosephosphate Isomerase	5.61	5.61 5.20	New isozyme formed during blast transformation
Glyceraldehyde-3-P Dehydrogenase	4.96 5.20-7.50	4.96 5.20- 7.61	Multiple components, no change
Phosphoglycerate Kinase	9.00	9.00 8.74	New isozyme formed during blast transformation
Phosphoglycerate Mutase	6.63 5.80	6.63 5.80	Two components, no change
Enolase	8.30	8.30	Single component, no change
Pyruvate Kinase	8.73 8.50	8.73 8.50 7.60 5.48	Two new isozymes observed in lymphoblasts
Lactate Dehydrogenase	9.50 7.61	9.50 7.30 7.05 5.85	Most basic isozyme predominates in both cells but comprises larger percentage in lymphoblasts

tissues have not been completed, it is not certain which of the aldolase isozymes these represent. However, Harris[7] has observed that human aldolase crystallized from cardiac muscle exhibits an isoelectric value of approximately 9.3. Moreover, in rabbit, where the aldolase isozymes have been studied in detail, the liver isozyme (type-B aldolase) is the most basic with the muscle isozyme (type-A) having a lower isoelectric pH and the brain isozyme (type-C) being the most acidic[19]. Thus, based on these indirect observations, it would seem that the two forms of aldolase found in lymphocytes and lymphoblasts may represent the muscle and the brain isozymes. It would also appear that lymphoblasts synthesize primarily the muscle isozyme. It should, however, be pointed out that these interpretations must be verified by additional studies. Multiple forms of crystalline rabbit muscle aldolase have also been observed upon isoelectric focusing[19].

Isoelectric focusing of glyceraldehyde 3-phosphate dehydrogenase yielded rather complex patterns with activity being spread over a wide pH range. Lymphocyte and lymphoblast extracts revealed a major component at pH 5.0 and a series of smaller peaks in the range 5.2-6.0. Glyceraldehyde-3-phosphate dehydrogenase from human erythrocytes[7] and other mammalian tissues[24, 25] has exhibited pI values around 8.5. The discrepancy between these values and those found in the lymphocyte and lymphoblast extracts warrants further investigation.

Lactate dehydrogenase in peripheral lymphocytes electrofocused as two isozymes (pI 9.5 and 7.6) with the more basic isozyme comprising approximately 60% of the total LDH activity. The more acidic component accounted for most of the remaining activity, but small amounts of LDH activity were detectable at pH 6.3-6.5 and at 5.4 and 4.6. Lymphoblasts showed a higher relative content of the pI = 9.5 component (approximately 80%). Lesser amounts of activity were observed at pH 7.3, 7.1 and at 5.8. Thus, it appears that the primary gene product of LDH synthesized in blast cells is that from the M-cistron (i.e., M_4 or LDH 1 or A_4B_0).

Glucose-6-phosphate dehydrogenase from both lymphocytes and lymphoblasts electrofocused into multiple peaks over the range of pH 7.7 and 5.8. Major components were found at pH 7.6, 6.9, and 5.8.

Three of the enzymes showed relatively simple electrofocusing patterns which were clearly changed during lymphoblast transformation. Phosphoglycerate kinase from lymphocytes electrofocused as a single peak of activity with an apparent isoelectric point of 9.0. Extracts from lymphoblasts, on the other hand, contained PGK activity with isoelectric points of 9.0 and 8.7. Yoshida[27] has shown that purified human phosphoglycerate kinase is a nondissociable monomer of molecular weight 49,600. Furthermore, Chen et al.[5] have postulated a single structural gene for this enzyme in all human tissues. Thus, the molecular basis for this apparent change in the electrophoretic properties of phosphoglycerate kinase during blast transformation is not understood. It seems possible that the new form of the enzyme found in lymphoblasts may represent some form of post transcriptional modification.

Pyruvate kinase in peripheral lymphocytes consists of double components with isoelectric points of 8.8 and 8.6. Traces of more acidic forms of pyruvate kinase were noted on occasion, but the levels were very low. Lymphoblasts contained the isozymes with isoelectric pH's of 8.8 and 8.6, but also exhibited two other electrophoretic forms of the enzyme with isoelectric pH values of 7.6 and 5.5. Human pyruvate kinase has been shown to exist as tissue-specific isozymes[10], and the new isozymes observed in lymphoblasts may be the results of altered gene expression. However, the enzyme has recently been reported to exist in oxidized and reduced forms, both with catalytic activity[2], and the multiplicity may reflect this type of post translational modification.

Triosephosphate isomerase from human lymphocytes exists exclusively as a single component with an isoelectric pH of 5.6 (designated here as TPI B). On the other hand, 72-hour lymphoblasts contain essentially equal amounts of this isozyme and a new, more acidic, form (TPI A, pI = 5.2). Triosephosphate isomerase, from most human tissues is found in multiple electrophoretic forms, the molecular basis of which has been somewhat puzzling[26]. The clear differences in the triosephosphate isomerase isozyme expression in lymphocytes and lymphoblasts offered an ideal system in which to study the molecular basis for this multiplicity. Kester and Gracy[12] showed that when human lymphocytes were cultured in the absence of mitogens, TPI-A was not formed. Although the formation of TPI-A was coincident with DNA synthesis, if replication was postponed for approximately 16 hours by treatment with hydroxyurea, TIPI-A was still formed as usual. Thus, the formation of the new isozyme of triosephosphate isomerase does not appear to require DNA replication. The formation of TPI-A does, however, require protein and RNA synthesis[13]. When cells were treated with puromycin (10 μg/ml), both protein synthesis (as measured by the ability to incorporate ^{3}H-leucine into acid insoluble material) and the formation of TPI-A were blocked. Similar studies utilizing actinomycin D to inhibit the synthesis of RNA also resulted in blocking formation of TPI-A. Although these observations strongly suggested that the formation of TPI-A was the result of activation of a new structural gene during blastogenesis, they were not unequivocal. It was possible, for example, that during blastogenesis a modifying enzyme (e.g., protein kinase, protease, etc.) was synthesized which modified the TPI-B to TPI-A. However, extracts from lymphoblasts were not capable of converting TPI-B to TPI-A. The possibility of direct enzyme interconversion was also tested in pulse-chase radiolabeling studies (Fig. 2). Lymphocytes were supplied with ^{3}H-leucine immediately prior to the normal formation of TPI-A (i.e., isotope added at 22 hours post mitogen) and allowed to undergo blastogenesis by culturing for 50 more hours. Subsequently, the two isozymes of triosephosphate isomerase were isolated and their specific radioactivity determined. Under these conditions only TPI-A was radiolabeled. When the cells were grown with the radioactive amino acid prior to blastogenesis, then washed free of the isotope before blast transformation, essentially no label was found in TPI-A. Similarly, when cells were radiolabeled during blast transformation, then subsequently grown in isotope free media for an additional 72 hours, no transfer of radiolabel from TPI-A to TPI-B could be demonstrated. These data indicate that TPI-A is

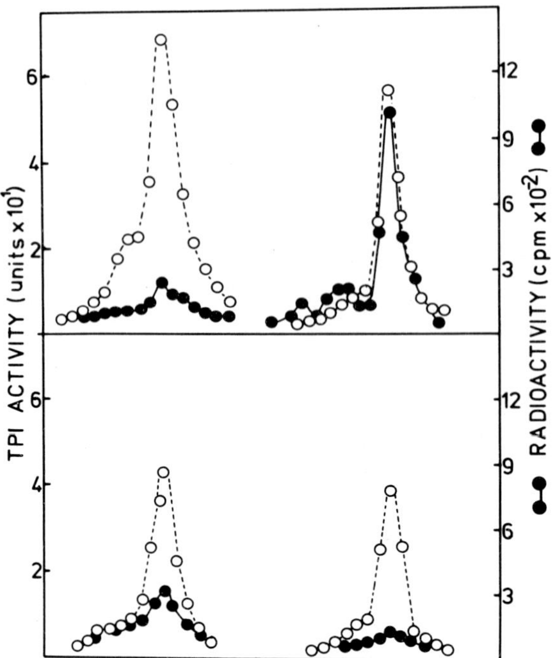

Fig. 2 – Isolation of Triosephosphate Isomerase Isozymes After Culturing in the Presence of ^{3}H-Leucine. Human peripheral lymphocytes were transformed with PHA and PWN in the usual way, but in the top figure ^{3}H-leucine was added to the culture media at 22 hours and the cells were allowed to grow until 72 hours. Cells were then harvested and the extracts were electrofocused. Each of the triosephosphate isomerase isozymes was collected and refocused twice and the enzyme activity and radioactivity of each fraction measured. In the lower figure, lymphocytes were incubated with ^{3}H-leucine prior to blastogenesis, then the isotope washed from the media and the cells allowed to undergo blast transformation in the absence of the free radioactive amino acid. The open symbols represent triosephosphate isomerase activity, and the closed symbols radioactivity.

synthesized *de novo,* that during blastogenesis TPI-A is the isozyme preferentially synthesized, and that the two isozymes are not interconvertible. Thus, it appears that the altered isozyme expression of triosephosphate isomerase during blast transformation does, indeed, represent the expression of an isomerase structural gene which is not normally active in peripheral lymphocytes.

Discussion

Cells which have undergone blast transformation were shown to contain markedly increased levels of all glycolytic enzymes. This increase appears to result, in some cases, from the increased synthesis of the same forms of the enzymes found in lymphocytes. In other cases, different isozymes are produced in the blast-transformed cells. This situation is not unlike the altered enzyme expression observed in highly dedifferentiated tumor cells.

Additional studies are required before the diagnostic significance or the clinical applications of these observations are fully appreciated.

The unequivocal identification of the specific isozymes of aldolase, lactate dehydrogenase and glyceraldehyde-3-phosphate dehydrogenase require additional studies, and the molecular basis for the new isozymic forms of phosphoglycerate kinase and pyruvate kinase need further investigation. It is possible that, like triosephosphate isomerase, they represent the activation of structural genes not normally expressed in the lymphocyte. However, it is also possible that they may be the result of post transcriptional modifications. The methodology required to resolve this question has been developed in the course of the studies on the triosephosphate isomerase.

References

1. Allison A.C., Hoyi T.: "The role of *de novo* purine biosynthesis in the responses of lymphocytes to mitogenic and antigenic stimulation", Trends in Biochemical Sciences, *1,* 36-37 (1976).
2. Badwey J.A., Westhead E.W.: "Sulfhydryl oxidation and multiple forms of human erythrocyte pyruvate kinase", Isozymes I. Molecular Structure (C.L. Markert, Editor), Academic Press, New York, 1975, pp. 509-521.
3. Boyum A., "A one-stage procedure for isolation of granulocytes and lymphocytes from human blood", Scand. J. Clin. and Lab. Invest., *21,* Suppl. 97, 77-90 (1968).
4. Cardenas J.M., Dyson R.D., Standholm J.J.: "Bovine and chicken pyruvate kinase isozymes intraspecies and interspecies hybrids", Isozymes I. Molecular Structure (C. L. Markert, editor) Academic Press, Inc., New York, 1975, pp. 523-541.
5. Chen S.H., Malcolm L.A., Yoshida A., Giblett E.R.: "Phosphoglycerate kinase: an X-linked polymorphism in man", Amer. J. Human Genet., *23,* 87-91 (1971).
6. Fisher D.B., Mueller G.C.: "An early alteration in the phospholipid metabolism of lymphocytes by PHA", Proc. Natl. Acad. Sci. (U.S.), *60,* 1396-1402 (1968).
7. Harris B.G., personal communication.
8. Hedeskov C.J.: "Early effects of phytohemagglutinin on glucose metabolism of normal human lymphocytes", Biochem. J., *110,* 373-380 (1968).
9. Hungerford D.A., Donnelly A.J., Nowell P.C., Beck S.: "The chromosome constitution of a human phenotype intersex", Amer. J. Human Gen., *11,* 215-236 (1959).
10. Ibsen K.H., Trippet P., Basabe J.: "Properties of rat pyruvate kinase isozymes", Isozymes I. Molecular Structure (C.L. Markert, editor), Academic Press, Inc., New York, 1975, pp. 543-559.
11. Kay J.E., Ahern T., Atkins M.: "Control of protein synthesis during the activation of lymphocytes by PHA", Biochim. Biophys. Acta, *247,* 322-334 (1971).
12. Kester M.V., Gracy R.W.: "Alteration of human lymphocyte triosephosphate isomerase isozymes during blastogenesis", Biochem. Biophys. Res. Commun., *65,* 1270-1277 (1975).

13. Kester M.V., Jacobson E.L., Gracy R.W., Manuscript in progress.
14. Kleinsmith L.J., Allfrey V.G., Mirsky A.E.: "Phosphorylation of nuclear protein early in the course of gene activation in lymphocytes", Science *154*, 780-781 (1966).
15. McIntyre O.R., Ebaugh F.G., "The effect of PHA on leukocyte cultures as measured by ^{32}P-incorporation in the DNA, RNA, and acid-soluble fractions", Blood *19*, 443-453 (1962).
16. Mueller G.C., Le Mahieu M.: "Induction of ribonucleic acid synthesis in human leucocytes by phytohemagglutinin", Biochim. Biophys. Acta, *114*, 100-107 (1966).
17. Nowell P.C., "Phytohemagglutinin: An initiator of mitosis in cultures of normal human leukocytes", Cancer Res. *20*, 462-466 (1960).
18. Parenti F., Franceschini P., Forti G., Cepellini R.: "The effect of phytohemagglutinin on the metabolism and γ-globulin synthesis of human lymphocytes", Biochim. Biophys. Acta, *123*, 181-187 (1966).
19. Penhoet E.E., Kochman M., Rutter W.J.: "Molecular forms of fructose diphosphate aldolase in mammalian tissues", Biochem., *8*, 4396-4402 (1969).
20. Polgar P.R., Foster J.M., Cooperband S.R.: "Glycolysis as an energy source for stimulation of lymphocytes by phytohemagglutinin", Exptl. Cell. Res., *49*, 231-237 (1968).
21. Quaglino D., Hayhoe F.G.J., Flemans R.J.: "Crytochemical observations on the effect of PHA in short term tissue cultures", Nature, *196*, 338-340 (1962).
22. Rabinowitz Y., Schimo I., Wilhite A.: "Metabolic responses of separated leucocytes to PHA: effects of anaerobiasis, actinomycin D and puromycin", J. Haematol., *15*, 455 (1968).
23. Roos D., Loos J.A.: "Changes in the carbohydrate metabolism of mitogenically stimulated human peripheral lymphocytes I. Stimulation by phytohemagglutinin", Biochim. Biophys. Acta, *222*, 565-582 (1970).
24. Smith C.M., Velick S.F.: "The glyceraldehyde 3-phosphate dehydrogenases of liver and muscle", J. Biol. Chem., *247*, 273-284 (1972).
25. Susor W.A., Kochman M., Rutter W.J.: "Structure determinations of FDP aldolase and the fine resolution of some glycolytic enzymes by isoelectric focusing", Ann. N.Y. Acad. Sci., *209*, 328-344 (1973).
26. Snapka R.M., Sawyer T.H., Barton R.A., Gracy R.W.: "Comparison of the electrophoretic properties of triosephosphate isomerases of various tissues and species", Comp. Biochem., Physiol., *49B*, 733-741 (1974).
27. Yoshida A., Watanabe S., "Human phosphoglycerate kinase. I. Crystallization and characterization of normal enzyme", J. Biol. Chem., *247*, 427-432 (1971).

ENZYMES AND HORMONES IN PLASMA OF PATIENTS WITH HYPERTHYROIDISM

A.G. Hilvers, L.J.M. Eilermann, J. Zondervan, and H.J. Piso

Summary

In a previous communication we suggested that the (elevated) alkaline phosphatase in hyper- and hypothyroid patients is not of hepatic origin. Studies of the levels of marker-enzymes in serum of hyper- and hypothyroids and of normal controls further supports this hypothesis. It has been demonstrated that the elevated alkaline phosphatase is electrophoretically similar to the bone fraction. Together with the observation of Rubegni et al., Kivirikko et al., and Lee and Lloyd, who reported elevated hydroxyproline secretion in hyperthyroids, our findings suggest that bone metabolism may be affected either by physical pressure on the parathyroids or *via* a disturbed calcitonin synthesis and/or secretion.

Enzymes and hormones in plasma of patients with hyperthyroidism

Hyperthyroidism is an extensively investigated diesease characterised by an increased secretion of thyroxine and/or triiodothyronin, usually caused by a hyperfunction of the thyroid itself and in rare cases by a tumor of the pituitary. The prominent symptoms, loss of weight and the relatively high body temperature, have been explained in the past as an effect of the thyroid hormones on mitochondrial oxidative phosphorylation. In vitro experiments, however, have shown that this effect occurs only at extremely high hormone concentrations which do not occur physiologically. It is now generally assumed that the action of thyroid hormones must be explained on the basis of the second messenger principle developed by Sutherland[1].

According to this general theory, hormones interact with a specific receptor-molecule located in the cellular membrane and this interaction causes an increase of the synthesis of intra-cellular cyclic AMP. The latter compound acts as a trigger of intra-cellular metabolism.

It is well known that patients with hyperthyroidism show normal or even lower fasting glucose-levels whereas their glucose-tolerance-tests resemble those of diabetics. Marks and co-workers[2] have explained this effect as

Central Hospital, Medical Centre Alkmaar Metiusgracht 30, Alkmaar, The Netherlands.

due to an increased intestinal absorption of glucose probably associated with the increased intestinal turnover rate. During the 5th Congress on Clinical Enzymology in Venice, Hilvers and co-workers[3] presented evidence that patients with hyperthyroidism show an increased oxidation of fat and proteins whereas carbo-hydrates are predominantly converted into lactic acid via anaerobic pathways. Low respiratory quotients and increased levels of intermediates of fatty-acid-oxidation indicated an increased fat metabolism[4, 2]. The increased quotient of urea: creatinin as well as the presence of increased levels of glutamate and tyrosine indicated an increased protein catabolism as has been reported earlier by Melmon et al.[5] and Sierbaej/Nielsen[6]. The increased tyrosine levels, however, may be explained also by different processes. Furthermore we reported increased lactate and pyruvate concentrations in blood plasma and a low cytoplasmatic redox state as reflected by the $NAD^+/NADH$ quotient. These findings could be explained by a decreased anabolic activity and it has been suggested that the thyroid hormones might have an effect on the secretion of insulin by the pancreas resulting in the observed decreased glucose tolerance. Secondly the observed shift in the metabolism of carbohydrates, fats and proteins might be caused by effects of the thyroid hormones on the pituitary or the adrenal cortex resulting in altered secretions of growth-hormone, adrenal-corticotrophormone (ACTH) and cortisol.

Table 1 shows the plasma-concentrations of thyroxine triiodothyronine, growth-hormone, cortisol, insulin as well as the fasting glucose-levels of hyperthyroid patients and of a group of euthyreotic controls. The differences in the hormone levels are not significant according to the chi-square-test.

From these data it can be concluded that there are no changes in those hormones which might give an indirect shift in the catabolism of carbo-hydrate, proteins and fats. The most probable explanation for our previous findings therefore is a shortage of glucose due to a decreased carbohydrate reserve which may be expected in these patients. In this respect the situation is comparable to that of starvation and it should be noted that histo-chemical investigations in liver byopts by Pipher and Poulson support this conclu-

Table I. *Fasting hormone levels in patients with hyperthyroidism*

		Hyperthyreotics (n = 41)	*Controls (n = 63)*
T_4	(nM)	204	126
T_3	(nM)	4.56	2.66
Glucose	(mM)	4.8	4.9
Insulin	(ng/l)	28.7	31.5
H.G.H.	(ng/l)	10.8	13.9
Cortisol	(μM)	0.44	0.51

Table II. *Comparison of patient - and control groups*

		Hyperthyreotics	*Controls*
Females/males		17/3	17/3
Mean age		43.4	42.7
T_4	(nmoles/l)	292	125
T_3	(nmoles/l)	8.0	3.1
T_3	(as index)	0.86	0.99

sion. Since patients with hyperthyroidism usually show a normal intravenous glucose-tolerance-test and sometimes even an increased half-value-time for glucose[8-9] a direct effect on the shift of metabolism is not likely.

At the previous symposium we also suggested that the increased levels of alkaline-phosphatase in plasma of hyperthyroid patients could originate from the thyroid itself. In coōperation with dr. M.W. Bosch and drs. M. Bezey from the Joannes de Deo Hospital in Haarlem the iso-enzymes of alkaline-phosphatase were separated by means of poli-acryl amide gel electrophoresis; furthermore we measured a number of enzymes with a certain organ specificity in particular with regard to the liver. The second table shows the composition of the patient and the control groups. The patient group was composed of 17 females and 3 men with severe hyperthyroidism.

Table 3 shows that both aminotransferases are within the normal levels although slightly elevated in the patient group. There are no significant

Table III. *Plasma enzyme levels (U/l, 25°C) - (mean values)*

	Hyperthyreotics	*Controls*
Asp. aminotransf.	15	10
Ala. aminotransf.	18	14
Pyruvate kin.	11	14
Aldolase	3.4	2.3
Sorbitol deh.	0.4	0.4
Glutamate deh.	1.0	1.0
Malate deh.	51	51
Lactate deh.	180	181

differences in the liver-specific enzymes sorbitol and glutamate dehydrogenase. The same applies to the other enzymes shown in this table. Table 4 shows the activity of alkaline-phosphatase, gamma GT, 5'NT and LAP.

Table IV. *Plasma enzyme levels (U/l, 25°C) - (mean values)*

	Hyperthyreotics	*Controls*
Choline esterase	2800	2860
Leuc. aminopept.	16	15
5'-nucleotidase	9	9
γ-glutamyl tr. p.	19	13
Alk. phosphatase	183	109

In the first place it should be noted that all mean values are within the normal limits but both gamma GT and alkaline-phosphatase levels are significantly higher in the patient group. The fact that alkaline-phosphatase levels are increased in hyperthyroidism has been known for some time[10-7] whereas an increased gamma GT-value has been reported[3]. Most authors have suggested that the increased alkaline-phosphatase should originate in the liver.

Table 5 shows the individual values for the activity of bone - and liver alkaline-phosphatase for the patients and the control groups in comparison with

Table V. *Alkaline phosphatase (U/l, 25°C)*

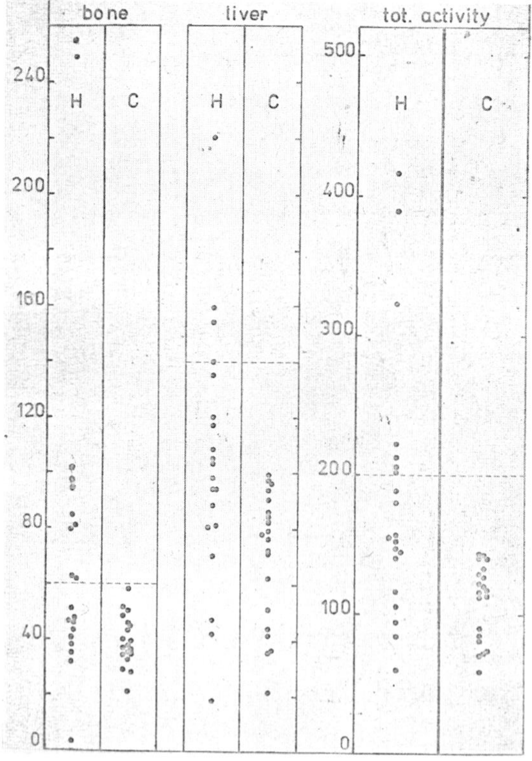

Table VI. *Bone and liver fractions of plasma alkaline phosphatase in patients with hyperthyroidism*

	N.	*per cent*	*Activity (U/l)*		
			bone	*liver*	*total*
Controls	20		39	70	109
Patients	20	(100)	78	104	183
Bone and liver normal	9	(45)	37	77	114
Bone incr.	8	(40)	136	103	235
Bone and liver incr.	2	(10)	82	190	272
Liver incr.	1	(5)	51	154	205

the totale activity. It is clear that in the patient group the bone-fraction is more frequently increased than with the controls, whereas the liver-fraction is usually normal. These data are summarized in the next table. About 50% of the patients show an increased bone-fraction even in those cases where the total activity is normal. In two cases the liver fraction was also increased whereas only 1 patient showed only an increased liver iso-enzyme. The latter patient was known as an alcoholic, however, as was supported by the increased value of the gamma GT-level. One of the patients with the increased liver and bone fraction had a demonstrated primary liver disease.

This patient had a decreased cholinesterase concentration whereas the LDH 5, gamma GT, 5'NT, malate dehydrogenase and pyruvate kinase were found to be increased. The second patient in this group suffered from an ulcus duodeni and after treatment of this disorder the liver fraction of the alkaline-phosphatase became normal, whereas the bone-fraction remained increased. The data presented here indicate that there is no direct association between hyperthyroidism and liver disease. The increased alkaline-phosphatase levels in patients with hyperthyroidism must be explained by an effect on bone turnover. Increased osteoporotic activity can occur in hand, skull and pelvis. This has been confirmed by scintigrafic investigation in one of our patients and we shall try to verify this aspect in further research.

In view of the results presented here, we considered it necessary to examine a possible influence of parathormone and calcitonine.

It has been reported that patients with hyperthyroidism may have an elevated serum calcium[11] as well as an increased urinary secretion of hydroxyproline[12-14].

There are four possible explanations for these observations:

a. Thyroxine and triiodo and thyronine may have a direct effect on bone tissue.

b. (Physical) pressure on the parathyroids might induce an altered secretion of parathormone.
c. The synthesis and/or secretion of calcitonine by the thyroid might be impaired.
d. Finally, these factors might occur simultaneously.

Observations of the data of three patients indicated that after a therapeutic improvement of the hyperthyroidism, the levels of alkaline phosphatase did not alter after a period of approx. 2 months. The thyroxine levels decreased from 460-140 nM whereas the alkaline phosphatase values were 143 and 172 U/l respectively.

A direct effect of T4/T3 on bone seems therefore less probable, unless one has to assume that the induced bone effect is irreversible.

The observed hypercalcemia could be due to an increased secretion of parathormone. Although measurements of the levels of this hormone have not been reported yet, there are some indirect indications that its concentrations are even decreased. The phosphate clearance is often decreased in hyperthyroidism, whereas the effect of parathyroid entracts in these patients on the phosphate tubular resorption is more pronounced than with normal individuals.

Parathormone induces an increase of the phosphate clearance via an inhibition of the back-resorption of phosphate resulting in a lower plasma phosphate concentration. This is rather in contrast with our observations and those of other authors summarized above with respect to hyperthyroidism.

On the other hand, calcitonine promotes calcium deposition in bone and therefore has a depressing effect on plasma calcium. If we assume that the calcitonine secretion is decreased in patients with a thyroid disease, which is in fact not unlikely at all, the equilibrium between calcitonine and parathormone will be disturbed and this could result in the observed osteoporosis and hypercalcemia.

Consequently an increased plasma phosphate concentration could be expected if we assume that the absolute tubular resorption of phosphate, which is a function of the parathormone level, is unaltered.

In view of our limited knowledge of the plasma levels of parathormone and calcitonine at this time, we can only propose that the latter hypothesis, i.e. an alteration of the ratio of these hormones, induces the secondary bone effects reported here.

We intend to obtain more direct evidence for this hypothesis in the future.

References

1. E.W. Sutherland, JAMA, *214*, 1281 (1970).
2. B.H. Marks, 1 Kiem and G. Hills: Metabolism, *9*, 1133 (1960).
3. A.G. Hilvers, H.J. Piso, J.H.H. Bonnier, L.J.M. Eilermann: in A. Burlina, Proc. Vth International Symposium on Clinical Enzymology, 1974.

4. E.F. Dubois, Arch. Intern. Med., *17*, 915, (1916).
5. K.L. Melmon, R. Rivlin, J.A. Oates, A.S. Sjoerdsma: J. Clin. Endocrin. *24,* 691 (1964).
6. K. Sierbaek-Nielsen,: Acta Med. Scand., *179*, 171 (1966).
7. J. Pipher, E. Poulsen: Acta Med. Scand., *127*, 439 (1947).
8. B.A. Lambery: Acta Med. Scand., *178*, 351 (1965).
9. D.S. Amatuzio, A.L. Schultz, M.J. Vanderbilt, E.D. Rames, S. Nesbitt: J. Clin. Invest., *33,* 97 (1954).
10. H.P. Dooner, J. Parada, C. Alliagra, C. Hoyl: Arch. Intern. Med., *120,* 25 (1967).
11. S.M. Krane, in The Thyroid, p. 598, Harper and Row, Publishers, New York, Editors S.C. Werner, S.H. Ingbar, 1971.
12. M. Rubegni, G. Ravenni, L. Del Giovane: Boll. Soc. Ital. Biol. Sper., *38*, (1962) 879.
13. K.I. Kivirikko, O. Laitinen, B.A. Lamberg: J. Clin. Endocrinol., Metab., *25* (1965), 1347.
14. C.A. Lee, H.M. Lloyd: Med. J. Australia, *1* 992, (1964).

SERUM ENZYMES IN THE DIAGNOSIS AND EVOLUTION OF PRIMITIVE NEOPLASIAS OF THE LIVER AND/OR IN HEPATIC METASTASES

E. De la Morena

Summary

In view of the enzymatic richness and the plurality of the metabolic functions of the liver, this constitutes an extensive and important field for various enzymatic tests in battery form.

In this context we have been studying in our Laboratories for several years a series of enzymes, such as GGTP, a-HBDH, GLDH, MDH, SDH, ICDH, LAP, 5'ND, Ph. Alk, ALT, and APT, in patients who are affected by primitive tumours of the liver and/or patients with hepatic metastases.

The average values found, as well as the use of the coefficients (those of de Ritis, Schmidt, etc., etc.) have made it possible for us to establish a table of values indicative of the progress of the malignant process. We have reached the conclusion that in these types of patients, who are already in a critical state, enzymatic diagnosis is efficacious, and it is thus possible to avoid the use of surgical methods.

Likewise, the study of enzymes makes it possible for us to determine the functional state of the liver in those patients who, without having hepatic disease, are being treated with hepatotoxic drugs.

Introduction

Early diagnosis of malignant diseases of the liver generally constitutes a serious problem for physicians. We are of the opinion that enzymatic diagnosis in battery form of the serum of patients affected by primitive tumours of the liver, hepatic metastases, or hepatic diseases produced by drugs, is of great use. In the present work we analyse the study of various enzymes (γ-Glutamyl-transpeptidase, α-Hydroxy-butyrate-dehydrogenase, Glutamate-dehydrogenase, Malic-dehydrogenase, Sorbitol-dehydrogenase, Isocitric-dehydrogenase, Alkaline Phosphatase, Aspartate-amino-transferase, Alanine-amino-transferase, Lactic-dehydrogenase, and Leucine-aminopeptidase) and the anatomic-pathological study by means of biopsy of the liver, checking the aid provided by the enzymatic diagnosis by means of histological results.

Biochemistry Department (Clinical Enzymology), Fundación Jiménez Diaz, School of Medicine of the Universidad Autonoma, Madrid – 3, Spain.

Table I. *Hepatic affections caused by Hodgkin's disease. Enzymes = I/U*

Name	*Sex*	*γ-GT*	*α-HBDH*	*GLDH*	*MDH*	*SDH*	*ICDH*	*ALK-Ph.*	*AST*	*ALT*	*LAP*	*LDN*
A.D.O.	M	129	197	12.2	148	3	19	140	215	197	155	230
I.D.M.	M	162	197	7.5	138	1	9	105	35	15	45	185
I.D.M.	M	97	197	6.8	98	0.9	11	135	40	25	75	175
M.S.E.	M	162	207	8.5	118	1	11	60	75	45	77	190
J.C.J.	M	1980	1880	18.8	158	2	13	300	185	105	103	190
A.R.R.	M	129	197	9.8	98	1	9	105	225	197	75	185

Table II. *Hepatic affections (Steatosis)*

Name	*Sex*	*γ-GT*	*α-HBDH*	*GLDH*	*MDH*	*SDH*	*ICDH*	*ALK-Ph.*	*AST*	*ALT*	*LAP*	*LDN*
M.E.S.	M	68	282	11.3	118	0.9	11	50	55	32	50	200
D.S.O.	F	129	197	8.5	118	0.7	9	175	40	31	87	90
M.S.E.	M	162	207	8.5	118	0.8	11	60	75	45	77	190
E.M.S.	M	65	225	5.3	98	0.8	11	55	45	40	39	175
E.C.M.	M	178	197	8.5	118	1	11	105	35	30	45	165
M.M.R.	M	65	197	9.4	118	0.9	13	75	40	30	47	205

Material and methods

Patients from the Department of Oncology affected by Hodgkin's disease.

We grouped them into four tables according to the grade of the hepatic lesion: the first group with a liver affection caused by Hodgkin's disease; the second group with the liver affected by steatosis; the third group suffering from a disease of the liver caused by toxic hepatitis; and the fourth group without histological alteration of the liver.

We determined the γ-GT by the Szasz method.

α-HBDH was determined by the method of Wroblewski.

The GLDH was determined by the Schmidt method. The SDH by the Gerlach method, and AST and APT by the Wroblewski method; the ICDH by the Wolfson method, ALK-Phos by the Szasz method; the LDH by the Wroblewski method and the LAP by the Nagel method. The enzymatic kinetics was effected with a UNICAM-SP-825 with bath thermostatically controlled at 25°C.

Results

In the corresponding tables we found that all the patients with hepatic histopathology showed an increase in the number of enzymes in the serum, principally γ-GT, GLDH, SDH, and MDH and, in a smaller proportion, α-HBDH, ICDH, ALK-Phos. The rest were increased but not so significantly, although in general they showed figures higher than normal or within normal range.

In neoplastic patients without hepatic alteration, slightly increased figures were shown compared with the enzymatic standard, especially γ-GT; the rest did not show significant figures; the values of the GLDH and SDH were very high.

We have defined γ-GT (of the smooth reticulum) as the "WARNING enzyme" since we have proved that when it is high and the rest of the battery remains within normal limits, it may be an acute process or it may be that a disease is beginning which may affect the hepatic parenchyma.

In several of the patients the first datum was the elevation of the γ-GT and we found that after a certain time the other enzymes also showed an increase.

MDH is a mitochondrial and cytoplasmic enzyme, all that of the liver being cytoplasmic. We named this enzyme "Timid", since we found its increase to be very high at the beginning (figures higher than 158 I/U) and then there was a return to lower figures which never reached normal levels, and it was always high in a liver affected by Hodgkin's disease. GLDH and SDH are enzymes that we may consider to be organ-specific to the liver, GLDH being mitochondrial and SDH being principally hepatic; their elevation is persistent in chronic processes when we find ourselves with a decrease

Table III. *Hepatic affection caused by toxic hepatitis*

Name	*Sex*	*γ-GT*	*α-HBDH*	*GLDH*	*MDH*	*SDH*	*ICDH*	*ALK-Ph.*	*AST*	*ALT*	*LAP*	*LDN*
D.A.O.	M	580	197	12.2	148	3	17	140	215	197	66	230
D.R.P.	M	145	197	16.9	197	1	17	45	40	19	99	260
C.P.V.	M	65	188	7.5	98	0.9	9	70	35	19	60	215
F.R.P.	F	4.200	470	15.8	197	17	19	350	170	85	169	350
A.D.O.	M	5.000	310	20.7	197	2	19	115	205	120	155	320
I.L.M.	F	64	254	7.5	98	0.9	7	145	15	15	71	160
L.L.M.	F	170	197	7.5	98	0.8	9	350	45	35	106	200
M.G.G.	M	81	197	7.5	98	0.9	11	350	85	205	127	175

Table IV. *Patients suffering from Hodgkin's disease without hepatic affection*

Name	*Sex*	*γ-GT*	*α-HBDH*	*GLDH*	*MDH*	*SDH*	*ICDH*	*ALK-ph.*	*AST*	*ALT*	*LAP*	*LDH*
M.B.I.	M	16	188	3	98	0.2	4	135	70	35	66	275
M.M.L.	M	31	150	3.8	89	0.6	4	135	160	35	103	255
P.G.M.	F	16	134	2.8	69	0.3	6	175	75	35	71	225
L.M.B.	M	32	188	3.8	79	0.4	6	130	55	45	67	180
A.S.F.	M	16	134	3.8	89	0.5	6	85	35	30	57	190
F.S.S.	F	16	144	2.8	69	0.4	4	55	35	30	195	170
A.F.M.	M	22	188	3	79	0.5	8	150	25	20	71	175
J.M.P.	M	16	144	3	89	0.4	6	210	45	35	57	250
F.M.G.	M	16	144	4.8	79	0.5	6	110	42	16	61	175
M.G.G.	M	32	188	4.8	89	0.3	6	120	20	30	73	200
A.B.B.	M	37	207	6.8	89	0.6	6	75	45	50	47	215
M.D.T.	M	32	188	4.7	98	0.6	7	70	30	25	57	150

Table V. *Enzymatic standard in serum.*

	Mu/ml		
	Men		*Women*
γ-GT	6-32		6-28
α-HBDH		Up to 188	
GLDH		Up to 4.5	
MDH	48-96		48-96
SDH		Up to 0.5	
ICDH		Up to 6	
LAP	8-28		8-32
LDH		Up to 282	
ALK-Phosphat.		Up to 198	
AST		Up to 17	
ALT		Up to 17	

in AST and ALT, the ICDH is always high in processes with hepatic metastases; ALK-Phos, LDH and LAP were increased in almost all the cases studied although we were also confronted with figures that were very low in relation to the rest of the battery.

Discussion

If we compare the enzymatic values with the pathological anatomy we arrive at the conclusion that by carrying out this enzymatic study it is not necessary to use surgical means, since our results are highly significant, as were those of Kolavic, who studied the comparative effect between liver scanning and various enzymes in the serum of patients with malignant hepatic processes and found that γ-GT, and the rest of the enzymes LDH, '5 ND, AST and ALT were increased in all the patients studied, the scanning of the liver being positive.

Rutenborg studied 98 patients suffering from cancer and found an increase in the activity of γ-GT in 91% of the cases; Baden found in 45 patients with hepatic metastases that γ-GT is increased by 50% and ALK-Phos by 60%.

Conclusions

We are of the opinion that enzymatic diagnosis using a battery of 11 enzymes in serum indicates the functional state of the liver, thus it is not necessary, to our way of thinking, to use surgical means that are more traumatic for this type of patient who is already psychologically affected in himself.

Acknowledgements

We wish to thank Dr. J. Vicente, Head of the Department of Oncology for his continuous help and advice in carrying out this work.

Bibliografy

1. Baden H., Adersen B.: Surg. Gyn. Obst, *133,* 769 (1971).
2. Kolavic K., Rosulgic A.: Clinica Chemica Acta, *60,* 109 (1975).
3. Rutenborg A.M., Pinelda E.P.: Cancer, *17,* 781 (1964)

DESIGN AND PERFORMANCE OF AN ULTRAMICRO KINETIC ANALYZER

J.G. Atwood, J.L. Di Cesare, and A. Burlina*

Summary

The performance of a Model KA-150 Kinetic Analyser is evaluated by analyzing serum enzyme activities such as lactate dehydrogenase, aspartate aminotransferase, alanine aminotransferase, creatine phosphokinase and alkaline phosphatase. The design of the apparatus is illustrated together with a quality control program with control sera and pooled human sera.

The Model KA-150 Kinetic Analyzer is an automated single channel analyzer which performs all of its tests by kinetic methods using ultramicro volumes of samples and reagents. It has a number of novel design features which have been described elsewhere[1] in more detail. We now have more than a year's experience with performance of these designs in daily use in several clinical laboratories. In this paper we assess the performance that has been attained in routine clinical use.

For those who are not familiar with the KA-150, it prepares kinetic reaction mixtures with an end volume of 200 microliters for measurement at a rate of 150 samples per hour as shown schematically in Figure 1. The system automates classical two-reagent kinetic methods, with a preincubation step to permit interfering reactions between serum components and the first reagent to go to completion before the desired reaction is started by adding the second reagent. Samples are processed and measured at a rate of one every 24 sec. Through time for a single sample is 7 to 9 min.

For all serum enzyme tests, the overall dilution of the sample in the reaction mixture is 20-fold.

Reaction mixtures, which have been prepared in separate cups, are transferred into the photometer's flowcell and brought to thermal equilibrium at the temperature at which the reaction is measured, between 25°C and 37°C. About 25 sec after each reaction has been triggered, its absorbance is measured four times in an 8.8 sec interval centered about a 30 sec interval after triggering. The average rate of change of absorbance and the average curva-

The Perkin-Elmer Corpo, Norwak, Conn., –* Centro Enzimologia, Ospedale Civile, Conegliano, Italy.

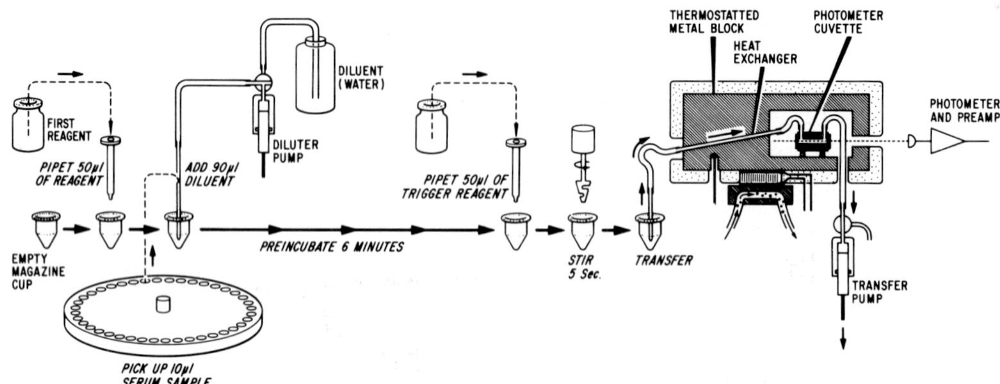

Fig. 1 – *Preparation of reaction mixtures.* Reaction mixtures are assembled in cups shown schematically here which travel through the instrument. Following one cup from left to right, the first step is addition of 50 ul of reagent. Then an automatic diluter picks up 10 ul of serum from the sample tray and washes it into the cup with 90 ul of water diluent. After at least 6 min of preincubation, 50 ul of the triggering reagent is added. The mixture is stirred, then transferred through a thermostatted heat exchanger into the photometer's flowcell and brought to rest and thermal equilibrium at the measurement temperature.

ture over the 8.8 sec interval are calculated digitally. Also, the highest and lowest absorbances are compared with thresholds set at 1.5 and 0.5 A, respectively.

For measurement of serum enzyme activities, the rate of change of absorbance is scaled by a fixed factor to print out activity directly in the desired units referred to the undiluted sample, usually International Units per liter (U/l).

The instrument is pre-programmed to measure lactate dehydrogenase (LD), aspartate aminotransferase (AST), alanine aminotransferase (ALT), creatine phosphokinase (CK), and alkaline phosphatase (AP). Other kinetic tests can be programmed by the user with an auxiliary control panel accessory. These include other enzymes, substrates by kinetic methods[1], and drugs by EMIT* methods.

Photometry

One of the unusual design aspects of the KA-150 is that its photometer uses hollow cathode lamps with interference filters as sources. These lamps emit spectra consisting of bright lines. The filters isolate groups of lines near the desired wavelengths. A cobalt lamp is used to generate wavelengths near 340 nm. A manganese lamp is used near 404 nm. Figure 2 shows a spectrum

*Trademark of SYVA Corp.

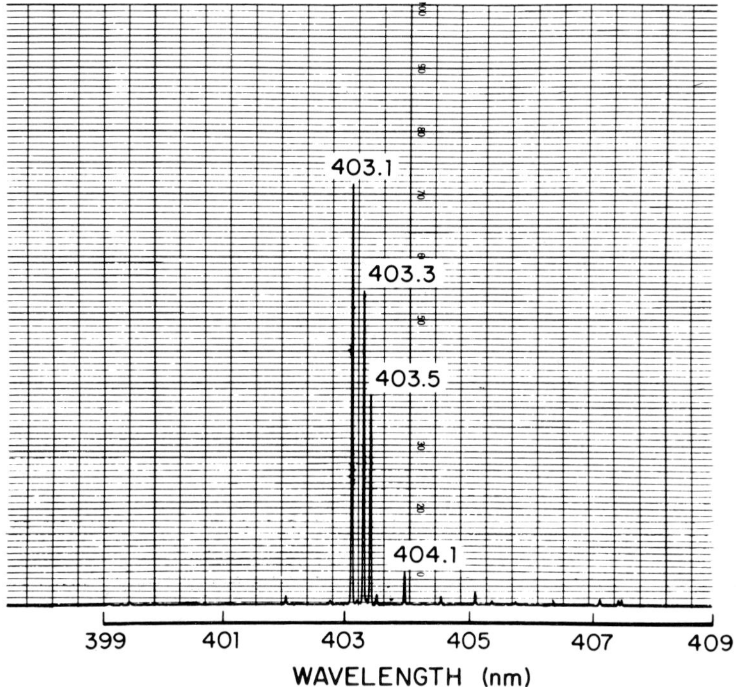

Fig. 2 – Spectrum of manganese hollow cathode lamp and interference filter used as source for AP and other tests run at 404 nm.

of the manganese lamp and filter combination. One design advantage of these sources is that they are very bright and of very great stability. This is necessary to get good signal to noise ratio with an ultramicro flowcell 10 mm long but of only 20 microliter volume. Another design advantage is good long term spectrophotometric repeatability because of freedom from wavelength drift. The double beam photometer used is shown schematically in Figure 3.

Results actually obtained have conformed to expectation. The combination of bright line lamps and filters has provided linearity of working curve for NADH to an absorbance of about 2.0, as shown in Figure 4. If this shape of working curve were produced by a conventional monochromator system, it would correspond to a level of scattered light of about 0.2%.

Typical linearity to high enzyme activity is shown in Figure 5, for ALT. It shows that over the range of NADH concentrations which occur in this reaction mixture, departure of the photometry from linearity is not measurable.

Similar results are obtained near 404 nm, where alkaline phosphatase and gamma glutamyl transpeptidase are measured.

The life of the lamps has proved longer than a year of typical field use.

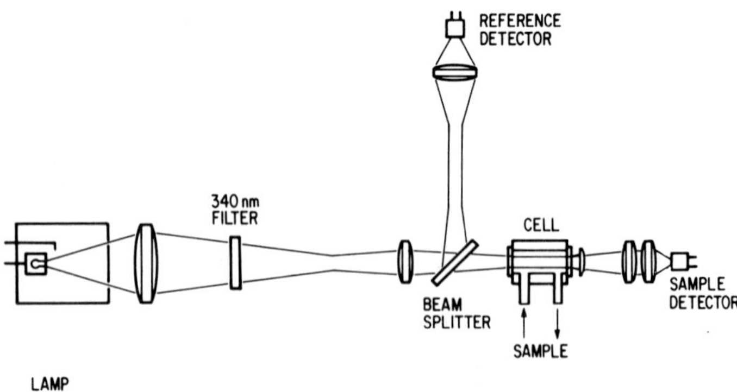

Fig. 3 – *Schematic of photometer.* The photometer is double-beam, with a 10 mm flow-cell in one beam. Detectors are silicon PIN diodes. There is no modulation fo thė beams. One source is a cobalt hollow-cathode lamp followed by an interference filtered at 340 nm. A manganese lamp is used for wavelengths near 403 nm.

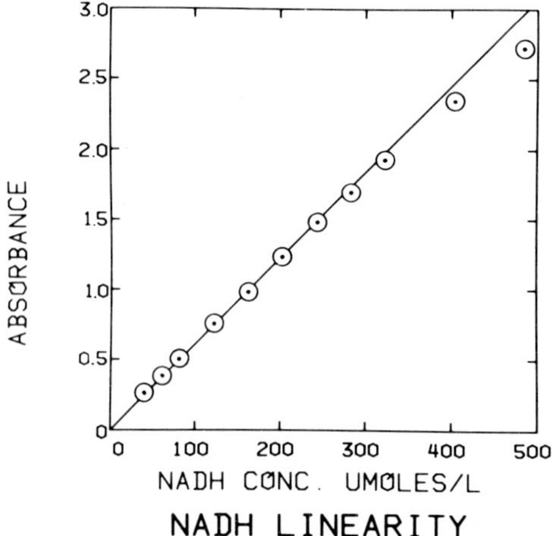

Fig. 4 – Photometric curve for NADH, using cobalt hollow-cathode lamp and 340 nm interference filter.

As for signal to noise ratio, the within-run repeatability, expressed as standard deviation, for low activity serum enzymes run at 340 nm is typically less than 2 U/liter, referred to the undilutes sample. This corresponds to

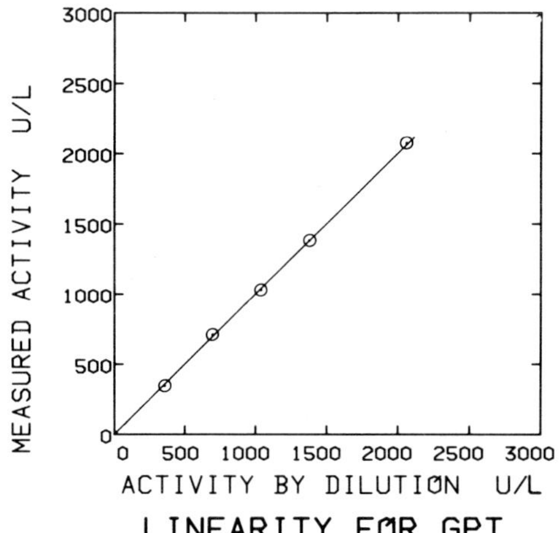

Fig. 5 – Measurements on dilutions of a serum with very elevated ALT show linearity to over 2000 U/L.

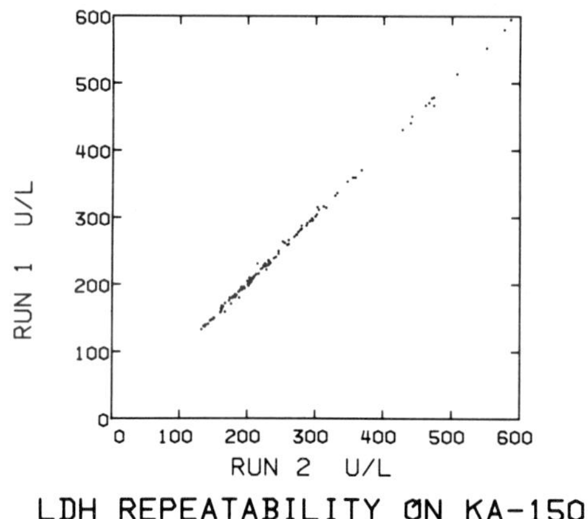

Fig. 6 – Two runs for LD activity about 1 hour apart on 150 routine clinical sera.

0.1 U/liter activity in the photometer flowcell, or, in terms of absorbance change, less than 1×10^{-4} A in the 8.8 sec measuring time.

An example of within-run repeatability is shown in Figure 6, two runs on 150 serum samples for LD.

Calibration

The KA-150 has a mode of operation designed to ckeck the accuracy of the enzyme activity calibration[2]. This includes checking the effects of small errors in overall dilution factor, absorbance accuracy, and electronic scale factor, in a single measurement. To make this check, a solution of accurately known molarity of a stable absorbing substance is introduced at the sample diluter input. It should produce a certain reading on the printout in calibration mode. This is achieved by printing one of the absorbance readings on the absorbing sample, multiplied by an appropriate factor to convert absorbance differences to enzyme activity units. In calibration mode, the printout is of one reading only, *without* subtracting successive absorbance readings to form an absorbance difference, as is done in routine kinetic measurement mode.

Thus, calibration mode makes it very easy for the user to perform what is otherwise a tedious task of separately checking the dilution and pipetting steps and the photometric scale by use of standard absorbing solutions.

Because the International Unit of enzyme activity is based on a rate of change of molarity, and not a rate of change of absorbance, this method is in fact more free of errors in knowledge of molar extinction coefficients, and spectrophotometric parameters such as wavelength accuracy and spectral purity, than are methods based on absorbance measurement of standards such as dichromate. This is because the method as implemented on the KA-150 uses a known molarity of the *same* absorbing substance as is measured in the kinetic reaction.

For example, to calibrate the alkaline phosphatase channel, we used a 434 micromolar solution of paranitrophenol, which is the absorbing product of the AP-catalyzed reaction in the Bowers-McComb method.

Field experience with the paranitrophenol method of calibration has been excellent. The material is easy to obtain with purity better than 99.8%. Purity of paranitrophenol was checked by thermal analysis of its melting curve[2]. Figure 7 shows daily results of calibration of the alkaline phosphatase channel during a typical day to day study for 25 days. Each day a new aliquot of the paranitrophenol solution was used. The result recorded is the percent of the correct value that was obtained. We obtained a day to day coefficient of variation of 0.8%. The calibration setting of the instrument was not disturbed during the entire testing period.

For the 340 nm enzymes, the desired absorbing substance is NADH, but it is too difficult to routinely produce standards having an exactly known molarity of this labile material. Therefore, we use as a standard, a material which can react with NADH in a special calibration reagent during the six-minute preincubation time. There is stoichiometric conversion of NADH to NAD, in the reaction with the calibration material in the sample. Thus, an accurately known change in molarity of NADH in the reaction mixture is produced, and should produce a certain change in the printout.

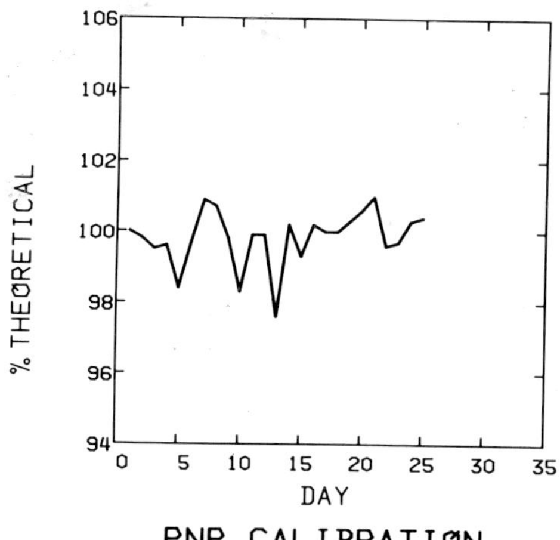

Fig. 7 – Day to day repeatability of calibration check of the KA-150's 404 nm channel, using a paranitrophenol solution as calibrator. Results are expressed as percent of the correct printout value.

For calibration of the 340 nm enzymes, we have gathered field experience using 400 micromolar sodium pyruvate solutions as the calibration sample, with a reagent containing LDH and NADH in buffer, and also 400 micromolar glucose as the calibration sample, with a regular commercial hexokinase glucose kit as the reagent[3].

Our experience has shown that pyruvate is not a fully satisfactory material for long term use as a calibrator. We found difficulty in maintaining stability and good bottle to bottle repeatability of a 400 micromolar solution over a long term. It also proved difficult to be certain of molarity and purity of this material.

The experience with glucose as a calibrator has been highly satisfactory. We used a 400 micromolar solution of glucose Standard Reference Material #917 from the U.S. National Bureau of Standards, prepared by weighing into 0.2% benzoic acid. This solution showed no detectable instability. In a 31 days study, using the glucose calibrator, we achieved a day to day coefficient of variation of 0.9%. Figure 8 shows results obtained.

Day-to-day repeatability

In the same day to day study of calibration stability, day to day repeatability on LD, ALT, AST, CK, and AP was measured[4].

Materials and Methods

A single lot of Perkin-Elmer reagents was used throughout the study for each enzyme. One vial pair of each reagent was reconstituted approximately one half hour before tests were initiated. Reconstitution was with KA-150 diluent for all tests except AP (AP-2 is reconstituted with AP-1 buffer), and all reagents were filtered through a 0.45 micron filter before use.

Single lots of Hyland Multi-Enzyme A, a synthetic clear control and of Hyland Control Serum I, a serum-based, turbid control were used as samples. The Hyland A samples were analyzed in various configurations such as a frozen pool, a fresh daily pool of 2 vials or, as a single vial reconstituted daily. The Hyland Control Serum I was used only as a frozen pool. The frozen pools were prepared by reconstituting several bottles of control, pooling them, aliquoting, and freezing. Each day, one aliquot was thawed by immersion in a beaker of water for 15 minutes, then mixed well. All enzyme assays were performed between 1 and 4 hours after reconstitution or thawing.

All assays were performed on the same production model KA-150 in Perkin-Elmer's laboratory on the same days as the calibration study. The instrument was operated and maintained according to the KA-150 User's Manual.

The reagent blank rate on each enzyme test was set to zero on the first day of testing, and not changed thereafter. The instrument's calibration setting was never changed during the study. The analysis temperature was 30°C.

Day-To-Day Enzyme Precision

Day-to-day precision on enzyme measurements was evaluated using Hyland Multi-Enzyme A and Hyland Control Serum I. Hyland Multi-Enzyme A is a synthetic, clear control material. This control was analyzed daily in several configurations: a frozen pool; a single bottle freshly reconstituted; a pool of two bottles freshly reconstituted; and reporting the average of 3 measurements compared to reporting a single measurement.

Table 1 shows the results obtained with Hyland A. The results tend to indicate that there is little vial-to-vial sample variations in the control and that averaging of 3 results added little, if any, improvement. Use of a frozen pool appeared to give similar results, except on CK, where the precision was poorer. The mean value of the frozen pool was the same as for the freshly reconstituted vials, except for CPK, where a decrease of about 7% was observed.

The results with Hyland CSI, a serum-based turbid control material, were about a factor of two poorer, on all tests except AP, where the precision was about the same. The reason for this behavior is that turbid samples generate results on the KA-150 which are not as precise as those obtained using clear samples[6].

Table I. *Day to day precision on Hyland multienzyme A: 31 days of measurements*

Sample	*LD Mean = 270 U/l*	*SGOT Mean = 50 U/l*	*SGPT Mean = 9 U/l*	*CPK Mean = 165 U/l*	*ALKP Mean = 110 U/l*
1	2.1 C.V.	2.0 S.D.	1.8 S.D.	4.0 C.V.	4.8 C.V.
2	1.8	2.6	3.3	2.8	4.5
3	2.2	2.4	1.5	2.4	4.9
4	2.1	2.5	1.7	2.4	4.4
5	2.2	1.9	1.4	2.5	4.5
6	2.1	2.7	1.8	1.9	5.6
7	2.1	2.0	1.5	2.4	4.9

Sample Key:

1. Frozen pool, 1st value
2. Freshly reconstituted vial #1, 1st value
3. Freshly reconstituted vial #2, 1st value
4. Freshly reconstituted vial #1, ave. of 3 values
5. Freshly reconstituted vial #2, ave. of 3 values
6. Freshly reconstituted pool of vials 1 & 2, 1st value
7. Freshly reconstituted pool of vials 1 & 2, ave. of 3 values.

Results with pooled human serum controls

It is interesting to compare the results described above with day to day repeatability obtained using a different control material on different KA-150 instruments in a hospital laboratory[5].

In Table 2, the means and coefficient of variation described earlier for a single reading on one vial of Hyland control are compared with the same quantities for 31 days of routine control data from the hospital laboratory. The control material used was frozen aliquots of a large pool of filtered human serum, prepared at the hospital. The reagents used were a different lot of Perkin-Elmer reagents, with the same formulation.

The data include routine control values obtained on two different KA-150 instruments used side by side, interchangeably in the hospital laboratory, so an instrument to instrument effect is included.

The results tend to show that with these very different conditions, there is no major difference in day to day repeatability performance.

The apparent difference for % C.V. for ALT is caused entirely by the very low, and different mean values for the control. In both cases, the *standard deviation* is about 1.8 U/liter.

Table II. *Typical quality control data*

Enzyme	*Hospital Lab 2 KA-150's Frozen Pool*		*P-E Lab 1 KA-150 Hyland Control*	
	Mean U/l	*C.V. per cent*	*Mean U/l*	*C.V. per cent*
LD	306	2.7	270	2.1
AST	54	7.1	50	4.0
ALT	18	9.1	9	20.0
CK	506	3.0	165	4.0
AP	121	4.8	110	4.8

Acknowledgement

We are grateful to Dr. Herbert Malkus of Jackson Memorial Hospital, Miami, Fl., for the control data based on a frozen human serum pool.

References

1. Atwood J.G., DiCesare J.L.: "Making enzymatic method optimum for measuring compounds with a kinetic analyzer". Clin. Chem., *21,* 1263 (1975).
2. Atwood J.G., Di Cesare J.L.: "Calibration of the Model KA-150 Kinetic Analyzer". Laboratory Medicine Application Study No. 67, The Perkin-Elmer Corp., Norwalk Ct., U.S.A. (1974).
3. Glucose Kit, Cat. No. 969404, Calbiochem., La Jolla, Ca., U.S.A.
4. DiCesare J.L.: "Day-To-Day Repeatability of the KA-150 Kinetic Analyzer". Clinical Products Application study No. 90, The Perkin-Elmer Corp., Norwalk, Ct., U.S.A. (1976).
5. Malkus H.: Jackson Memorial Hospital, Miami, Fl., U.S.A., Private communication.
6. Atwood J.G., Boyd L.L.: "Repeatability with 10 freeze-dried control materials on the KA-150". Laboratory Medicine Application study No. 75, the Perkin-Elmer corp., Norwalk, CT, U.S.A. (1975).

ALKALINE PHOSPHATASE, LDH ISOENZYMES AND CARCINOFETAL PROTEINS IN LIVER DISEASES

A. Malesci, M.A. Tommasini, G. Fiorelli and N. Dioguardi

Summary

Alkaline phosphatase isoenzymes with properties similar to placental ones have been found in sera of patients with various tumors and have been hypothesised to be carcinofetal proteins.

The incidence of placental-type alkaline phosphatase (PAP) in a group of patients with primary hepatic carcinoma (PHC) and in a group with liver cirrhosis was studied. The presence of PAP was correlated to isoenzymatic LDH patterns, to the occurrence of alfa$_1$-fetoprotein (AFP) and to carcinoembryonic antigen (CEA). PAP was identified by its heat-stability to 65°C thermoinactivation and characterized by gel-electrophoresis. PAP was found in 60% of PHC, whereas it was never detected in patients with cirrhosis. Heat stable alkaline phosphatase isoenzymes showed an electrophoretic mobility between placental and liver phosphatases. No correlation was found between concentration of PAP and AFP. AFP was elevated in 90% of PHC whereas CEA only in 20%; variable results were observed in patients with cirrhosis. In all the patients an increase of LDH_5 and LDH_4 fractions was found.

Introduction

Placental alkaline phosphatase (PAP) is normally synthetized in trophoblastic tissue during pregnancy; its own physico-chemical and immunological properties enable us to differentiate it from all other alkaline phosphatase isoenzymes detectable in the serum of not pregnant healthy individuals.

Recently there have been reports of phosphatases with heat-stability and antigenicity indistinguishable from those of PAP, being found in patients with various non trophoblastic tumors.[1, 2, 3, 4]

This work was undertaken to evaluate the incidence of PAP in the sera of patients with primary hepatic carcinoma (PHC) and liver metastases. A group of acute and chronic non malignant liver diseases was also studied.

Detection of PAP was then correlated with the occurrence of two well known fetal antigens: alfa$_1$-fetoprotein (AFP) and carcinoembryonic antigen (CEA).

Istituto di Patologia Speciale Medica e Metodologia Clinica, Università di Milano.

Lactate deydrogenase (LDH) isoenzymes, whose pattern can be significantly modified during cellular differentiation[5], were also evaluated.

Cases Investigated

112 patients were divided into the following groups on the basis of clinical, laparoscopic and histological findings:

	N. patients
PHC	13
Liver metastasis	10
Acute hepatitis	10
Chronic active hepatitis	31
Postnecrotic cirrhosis	15
Alcoholic cirrhosis	33

Methods

Alkaline phosphatases from human liver, bone, intestine, placenta and bile were partially purified by homogenization with tris-buffer at pH 8.6 followed by an extraction with butanol and ammonium-sulphate fractionation, after which a Sephadex G 200 gel filtration was performed[6, 7].

Gel-electrophoresis, heat inactivation and 5mM L-phenylalanine inhibition[8] were used to characterize tissue extracts and serum alkaline phosphatase isoenzymes.

Electrophoresis was carried out on poliacrylamide gel columns[9], modifying the gel concentration to 8% and using a discontinuous buffer system (tris-citrate pH 8,8 for gel tubes and borate buffer pH 9.2 for tanks); cellogel strips electrophoresis at 260 Volts in sodium veronal buffer for 1 hour, was also performed.

Heat inactivation at 56°C[10] was performed on tissue extracts and sera to identify bone and liver phosphatases: residual activity was measured in each serum at three different incubation times (5.15 and 25 min). PAP was determined by the 65°C heat inactivation test of Fishmann (II): at this temperature (5 min incubation) PAP is unaffected, whereas all other AP isoenzymes are denaturated; residual activity of PAP was then measured using 72 mM disodium-phenylphosphate as a substrate in 5 mM carbonate-bicarbonate buffer at pH 10.7.

AFP was measured in the sera of all the patients by the quantitative immunoelectrophoresis of Laurell[12] and also by double antibody radioimmunoassay (Abbott I^{125}AFP RIA).

CEA was evaluated by radioimmunoassay (RIA CEA-test Roche).

LDH isoenzymes were determined by cellogel electrophoresis in sodium veronal buffer at pH 8.6.

Results

Table I shows the electrophoretic characterization of PAP and other AP isoenzymes in the patients tested.

PAP was found in 23% of PHC and in 10% of the liver metastases. In the other liver diseases PAP was electrophoretically detected in only one out of 10 cases with acute hepatitis.

Table I. *Electrophoretic characterization of alkaline phosphatase isoenzymes*

	Cases *N.*	*Bile* *N.* *(%)*	*Liver* *N.* *(%)*	*Bone* *N.* *(%)*	*Intestine* *N.* *(%)*	*Placenta* *N.* *(%)*
Normal	20	0	20 (100)	19 (95)	4 (20)	0
Primary Hepatic Carcinoma	13	2 (15)	13 (100)	12 (93)	2 (15)	3 (23)
Liver metastases	10	3 (30)	9 (90)	10 (100)	2 (20)	1 (10)
Acute Hepatitis	10	3 (30)	10 (100)	9 (90)	2 (20)	1 (10)
Active Chronic Hepatitis	31	4 (13)	31 (100)	30 (96)	7 (23)	0
Postnecrotic Cirrhosis	15	1 (7)	10 (67)	14 (93)	6 (40)	0
Alcoholic Cirrhosis	33	2 (6)	31 (94)	33 (100)	11 (33)	0

Table II. *PAP, AFP and CEA in liver tumors*

	Cases *N.*	*PAP* *N.* *(%)*	*AFP* *N.* *(%)*	*CEA* *N.* *(%)*
Primary Hepatic Carcinoma	13	9 (70)	10 (77)	4 (30)
Liver metastases	10	4 (40)	1 (10)	8 (80)

Table 2 reports the incidence of PAP (measured by heat inactivation), AFP and CEA in the patients with PHC and liver metastases.

PAP (70%) showed a very high incidence which was similar to that of AFP (77%), whereas CEA was positive in only 30% of these patients. In metastatic liver patients CEA was found positive in 80%, PAP in 40% and AFP in only 10%.

Table 3 shows the incidence of the same three proteins in patients with acute and chronic liver diseases. In the group with acute hepatitis PAP was detected in 40% and AFP in 50%, whereas CEA did not show abnormal values. PAP and AFP showed the same incidence in chronic active hepatitis (7%); in patients with cirrhosis AFP was detected with an incidence of about 20%, whereas PAP and CEA were always negative.

Table 4 shows the LDH_4/LDH_5 ratio in PHC and liver metastases; this parameter is also evaluated in a group of patients with postnecrotic cirrhosis. A decrease of this ratio was found mainly in PHC, whereas in liver metastases the ratio was normal.

Table III. *PAP, AFP and CEA in non malignant liver diseases*

	Cases N.	*PAP N. (%)*	*AFP N. (%)*	*CEA N. (%)*
Acute Hepatitis	10	4 (40)	5 (50)	0
Chronic Active Hepatitis	31	2 (7)	2 (7)	0
Postnecrotic Cirrhosis	15	0	3 (20)	0
Alcoholic Cirrhosis	33	0	6 (18)	0

Table IV. *LDH_4/LDH_5 ratio in liver cancer and chronic liver diseases*

	Cases N.	*LDH_4/LDH_5 (mean)*
Normal Controls	15	1.16
Primary Hepatic Carcinoma	13	0.64
Liver Metastasis	10	0.96
Postnecrotic Cirrhosis	15	0.84

Discussion

A very high incidence of PAP was found in patients with PHC (70%), whereas only 40% of patients with metastatic liver cancer showed abnormal values of PAP by using the heat inactivation test of Fishman[11]; the lack of sensitivity in serum PAP identification by electrophoresis (23% in PHC and 10% in liver metastases) is probably due to the very small amounts of serum isoenzymes to be detected.

Recently electrophoresis has allowed us to characterize a few unusual variants of PAP in liver carcinoma extracts, having the same antigenicity but less heat-stability and mainly a slower electrophoretic mobility[14-15]. In our study on sera, electrophoresis did show a band corresponding to the Higashino isoenzyme (fast $alpha_2$ serum globulins) in only one case of PHC, whereas a band with identical mobility to PAP (beta globulins) was found in two other patients with PHC. It has been proposed that the polimorphism of all PAP variants reflects a gradual modification of normal hepatic phosphatase towards placental isoenzyme during neoplastic cellular differentiation; the mechanism probably involved in the occurrence of heat stable PAP in neoplastic patients could be a gene derepression of alleles which provide a normal synthesis of PAP in fetal life[13,1].

It is possible, however, that the occurrence of PAP is not only related to malignant degeneration but also to liver cell regeneration, as our findings in non cancerous liver diseases (40% in acute hepatitis, 7% in chronic active hepatitis) suggest.

In this report we must emphasize that PAP levels in these patients were always significantly lower than values obtained in neoplastic subjects. The comparative study of PAP and other carcinofetal markers in liver tumors and non malignant liver diseases did not show any significant correlation. AFP, which is known to be a highly specific marker of PHC, has shown an incidence similar to PAP in patients with PHC; however, no statistical correlation between PAP and AFP levels was found. The associated study of PAP and AFP showed that no patient with PHC was negative for both these tests. This finding suggests that the two proteins are synthetized in different neoplastic cells.

The high incidence of CEA in metastatic liver tumors (80%) can be related to the very frequent gastrointestinal origin of these cancers; in these patients no correlation was found between CEA and PAP values, whereas only in 10% of cases were the two tests negative.

As concerning LDH isoenzymes, the increase of LDH_5 in PHC seems to be related to hepatic massive necrosis rather than to cancer degeneration; in fact, in metastatic cancer, which showed the highest LDH total activity, the increase involved all LDH fractions.

These preliminary findings should induce further investigations into the role of PAP as a marker of malignancy, its usefulness in screening, and the therapeutic management of patients with primary or metastatic liver tumors.

References

1. Fishman W.H., Inglis N.R., Stolbach L.L.: A serum alkaline phosphatase isoenzyme of human neoplastic cell origin. Cancer Res. *28*:150-154; (1968).
2. Stolbach L.L., Krant M.S., Fishman W.H.: Ectopic production of an alkaline phosphatase isoenzyme in patients with cancer. N. Engl. J. Med.; *281*:757-762; (1969).
3. Nakayama T., Yoshida M.: L-leucine sensitive heat stable alkaline phosphatase isoenzyme detected in a patient with pleuritis carcinomatosa. Clin. Chim. Acta, *30*:546-548 (1970).
4. Inglis N.R., Kirley S., Stolbach L.L., Fishman W.H.: Phenotypes of the Regan isoenzyme and identity between the placental D-variant and the Nagao isoenzyme. Cancer Res. *33:* 1657-1661.
5. Zondag H.A., Kline F.: Clinical application of LDH isoenzymes: alteration in malignancy. Ann. NY Acad. Sci. *151*:578 (1968).
6. Eaton R.H., Moss D.W.: Partial purification and some properties of human alkaline phosphatase. Enzymologie, *35*:31-39 (1968).
7. Nimai K., Ghosh , Fishman W.H.: Purification and properties of molecular-weight variants of human placental alkaline phosphatase. Biochem. J., *108*:779 (1968).
8. Moss D.W.: The estimation of intestinal alkaline phosphatase in human blood serum. Clin. Chim. Acta, *25*: 117-125 (1969).
9. Smith I., Lightstone P.J.: Separation of human tissue alkaline phosphatases by electrophoresis on acrylamide disc gel. Clin. Chim. Acta, *19*: 499-505 (1968).
10. Moss D.W., Whitby L.G.: A simplified heat inactivation method for investigating alkaline phosphatase isoenzymes in serum. Clin. Chim. Acta, *61*:63-71 (1975)
11. Fishman W.H.: The placental isoenzyme of alkaline phosphatase in sera of normal pregnancy. A.J.C.P., *57*:65-74 (1971)
12. Laurell C.B.: Electroimmuno–assay. Scand. J. Clin. Lab. Invest., *29*:124 (1972)
13. Suzuki H., Iino S.: Tumor specific alkaline phosphatase in hepatoma. Ann. NY Acad. Sci., *259*:307 (1975)
14. Warnock M.L., Reisman R.: Variant alkaline phosphatase in human hepatocellular cancers. Clin. Chim. Acta, *24*:5-II (1969).
15. Higashino K., Ohtani R.: Hepatocellular carcinoma and a variant alkaline phosphatase. Ann. Intern. Med. *83*:74-78 (1975).

7th INTERNATIONAL SYMPOSIUM ON CLINICAL ENZYMOLOGY
Venezia, Italy, 5-7th April, 1976

LIST OF PARTICIPANTS

AGUZZI FRANCESCO
Via Montebello, 17
27100 PAVIA

ALBERTINI ALBERTO
3° Servizio di Analisi Chimico Cliniche-
Ormonologia-Tossicologia
Spedali Civili
25100 BRESCIA

ALIANELLO SERGIO
Via Costantino, 95
00154 ROMA

ALIBERTI UMBERTO
Via Stresa, 117
00135 ROMA

ALIMENTI GIULIO
Via Emilia, 31
53042 CHIANCIANO

AMENTA FRANCESCO
Via Lgt. Pietra Papa, 179
00146 ROMA

ANGELICO RAFFAELE
Viale di Villa Massimo, 33
00161 ROMA

ANICHINI MARIO
Via B. Croce, 2
63100 ASCOLI PICENO

ANTONUCCIO ORAZIO
Corso G. D'Agata, 69
96012 AVOLA

APOLITO EUGENIO
Via L. Cadorna, 40
87075 TREBISACCE

AQUILI CECILIA
Viale G. Cesare, 180
28100 NOVARA

ASTALDI GIOVANNI
The Blood Center
Ospedale Civile
15057 TORTONA

ATWOOD JOHN
Perkin-Elmer
NORWALK, Conn. U.S.A.

AUGUSTINSSON KLAS-BERTIL
Arrhenius Laboratory
Biochemical. Dept. University of
STOCKHOLM, Sweden

AURELI GIOVANNI
Istituto Rizzoli
40100 BOLOGNA

AUS H.M.
Max Planck Institute for
Biophysical Chemistry
GOTTINGEN

AVRAMIDIS HERCULES
Mavromateon 6
ATHEN-147

BARBARINO FEDORA
Piata Mihai-Viteazul 11
Bloc B1, Scava 1, apart. 6
3400 CLUJ-NAPOCA (Romania)

BADIELLO ROBERTO
Via Castagnoli, 1
40100 BOLOGNA

BALLESTA GIMENO A.M.
Ganduxer 18-2°-1a
6-BARCELONA

BANCHERI SALVATORE
Primario Laboratorio
Ospedale Civile
34170 GORIZIA

BARALDO LUIGI
Ospedale Civile
35044 MONTAGNANA

BARBIERI FRANCESCA
Via Ludovico il moro, 78
27100 PAVIA

BARBIERI RENZO
Via dei Boderi, 16
21100 VARESE

BARDSLEY WILLIAM G.
Dept. of Obstetrics and Gynaecology
University of Manchester
St. Mary's Hospital
MANCHESTER, M 13 OJH

BARTOLI FRANCESCO
Via Ramarini
00015 MONTEROTONDO (Roma)

BASSAN EMMA
C.so V. Emanuele, 116
35100 PADOVA

BASTIAANSE A.J.
Kruisbergseweg 25
DOETINCHEM – The Netherlands

BATTISTA LINA
Via Acaia, 76
00183 ROMA

BELLIA GIUSEPPE
Via Notarbartolo, 35
PALERMO

BELLINI LUCIANA
Via Novacella, 28
39100 BOLZANO

BELLOMI DANIELA
P.zza Bertazzolo, 4
46100 MANTOVA

BERGMEYER H.U.
Boehringer Mannheim GmbH
D-8132 TUTZING

BERNARDI SILVANO
Via Zenson di Piave, 8
31100 TREVISO

BERNO GIULIANO
Ospedale Civile
31049 VALDOBBIADENE

BERTI GIOVANNI
c/o Ames-Miles
Via F.L. Miles, 10
20040 CAVENAGO BRIANZA

BETTINSOLI GIOVANNI
Ospedale Gardone
Val Trompia
25100 BRESCIA

BIAGI RENATO
Ospedale Malpighi
40100 BOLOGNA

BIANCHI MARIA ADELAIDE
Via Masaccio, 219
50132 FIRENZE

BIGI LEONARDO
Via S. Giovanni, 15
41037 MIRANDOLA

BLUM JOSEFA
Stadtspital Triemli
ZURICH

BOCCATO PAOLO
Laboratorio Analisi
Ospedale Civile
30027 S. DONA' DI PIAVE

BOELAERT JOHAN
Veedhoek, 19
LOPPEM, Belgium

BOIVIN PIERRE
100, Bld du Général Leclerc
Service d'Hématologie Clinique
Hopital Beaujon
92110 CLICHY

BOLANDRINA ERNESTO
Via Martinoli, 5
24085 LOVERE

BOLCATO ANTONIO
Via Adige
36041 ALTE MONTECCHIO

BONGIOVANNI ANNA
Istituto Igiene e Profilassi
20100 MILANO

BONOMINI EUGENIO
Via Asse, 5
29100 PIACENZA

BORTOLOZZI MENENIO
Via Terraglio, 443
31022 PREGANZIOL

BOTTI LUIGI
Via T. d'Aquino, 9
57023 CECINA

BRAGADINI ANNA
Ospedale Civile
PIEVE DI CADORE

BRAMBILLA ANTONIA
Laboratorio Analisi
Ospedale Broni e Stradella
Spessa Po (PAVIA)

BRAMEZZA MARIA
Laboratorio Analisi
Ospedale Civile
33100 UDINE

BRETAUDIERE JEAN-PIERRE
149, rue de Sevres
75730 PARIS Cedex 15

BRUNO GIUSEPPE
Viale Fratelli Vivaldi, 1
95123 CATANIA

BUFFO DINO
Via Facchinetti, 6
MILANO

BUGIARDINI GIUSEPPE
Policlinico S. Orsola
40100 BOLOGNA

BURIGHEL ALDO
Via Teodorico, 22
20100 MILANO

BURLINA ANGELO
Laboratorio Chimica Clinica
Istituti Ospitalieri
37100 VERONA

CACCIATORE LUIGI
Via Merliani, 20
80127 NAPOLI

CAGNIN VALENTINO
Laboratorio Analisi
Ospedale Civile
35012 CAMPOSAMPIERO

CALDERA LUCIANO
Ospedale Civile
37012 BUSSOLENGO

CAM GILBERTE
B.P. N. 5
95.380 LOUVRES (France)

CAMBIAGHI STEFANO
c/o Carlo Erba
Via Imbonati, 24
20100 MILANO

CAMBIERI FERDINANDO
Via Vivai
27100 PAVIA

CAMBON M.me Maddey
4, rue Grimaldi
06000 NICE

CAMBON PIERRE
4, rue Grimaldi
06000 NICE

CAMPO GIOVANNI
P.zza Stefani
37100 VERONA

CANTABONI ANGELO
Via Boccaccio, 45
20100 MILANO

CAPPADONA GIUSEPPE
P.zza S. Pio X, 4
31100 TREVISO

CAPPUZZO GIAN MARIA
Via Chilesotti, 18
35100 PADOVA

CAPUTO ENZO
Viale Caldara, 16
20100 MILANO

CAPUZZO MARIA
Ospedale Civile
30016 JESOLO

CARLEVARO GIANCARLO
Via Moroni, 7/A
20099 SESTO S. GIOVANNI

CASANOVA MARCO
Via Lanzone, 36
20123 MILANO

CASSETTI PAOLA
Via E. Vendramini, 7
35100 PADOVA

CASTELLANI LUIGI
Via Malta, 2/7
16121 GENOVA

CASTELLI ADRIANO
Istituto Chimica Biologica
dell'Università Cattolica S. Cuore
00100 ROMA

CATTANEO SPESSA MARIA TERESA
Via Volturno, 29
36100 VICENZA

CAVALIERE TOBIA
Via S. Leonardo, 1
80078 POZZUOLI

CECCHI LUIGI
Ospedale S. Giovanni di Dio
50100 FIRENZE

CENNI GIULIANO
Via Tolosano, 34
48018 FAENZA

CERINI RAIMONDO
Policlinico Cappella dei Cangiani
80100 NAPOLI

CERIOTTI GIOVANNI
Ospedale Civile
35100 PADOVA

CHABAS LOPEZ JOSE
Bertran, 96-98
BARCELONA, Espana

CHEMNITZ GERHARD
Fuhrberger Strasse, 13
3, HANNOVER

CHIRILLO ROSARIO
Ospedale Civile
31100 TREVISO

CILIA MARIA TERESA
Via S. Paolo, 13
20100 MILANO

CIMINO FILIBERTO
Via S. Pansini, 5
80131 NAPOLI

COLAJANNI LUCIANO
Via Padova, 41
00161 ROMA

COLALONGO GAETANO
Via P. Vizzani, 78
40100 BOLOGNA

COLETTI PREVIERO M.A.
U-147, Inserm 34
MONTPELLIER, France

COLOMBO GIOVANNI
Via Rismondo, 8
21049 TRADATE

COLOSIO MARIA GRAZIA
Ospedale Gavardo
25100 BRESCIA

CORTESI SERGIO
Via Paolo Sarpi
36100 VICENZA

CRISTANTE GIANNI
Via Carducci
35100 PADOVA

CURRADO NATALINA
Via Manzoni, 7
14100 ASTI

D'ALESSANDRO LUCIANA
Via Benini, 7
00191 ROMA

DAMIANI SILVIO
Via Nomentana, 41
00161 ROMA

DANIELI ALDO
Via dei Mille, 57
TRENTO

DECKX RAPHAEL
Appelmansstraat 29
ANTWERPEN

DEFRANCESCHI ANTONIO
c/o Carlo Erba
Via Imbonati, 24
20100 MILANO

DE LA MORENA ENRIQUE
c/ Capitan Haya 20
MADRID-20 - Spain

DELLA CROCE ENZO
Boehringer-Biochemia
Viale Monza, 270
20126 MILANO

DELLAROLE FLAVIO
Laboratorio Analisi
Ospedale Civile
31015 CONEGLIANO

DE LUCA UGO
Piazza G. Cesare, 14
20145 MILANO

DEL PIANO MARILENA
Via Monselice
09100 CAGLIARI

DEL PIANO ELIO
Via Monselice
09100 CAGLIARI

DE MARCO FRANCESCO
Parco Comola, 122
80122 NAPOLI

DEMATTE' PAOLO
Via S. Antonino, 257/a
31100 TREVISO

DE NEGRI MARIO
Via Ivrea, 1/2
35100 PADOVA

DENT NIGEL JOHN
Inveresk Research Interdatt.
EDINBURGH

DE PAOLIS GIANFRANCO
Via Monte Madonna, 11A
00060 FORMELLO

DE PAULA FRANCISCO
Freixa Larga, s/n
BARCELONA, Espana

DE VECCHI MARIO
Via Sidoli, 1
20129 MILANO

DIBELLA LUIGI
Via G. Campi, 287
41100 MODENA

DI MARTINO GENNARO
Via Stendhal
80133 NAPOLI

DIOGUARDI NICOLA
Via Pace, 15
20122 MILANO

DOMPE' MARIANO
c/o Anic-ASS/CS
20097 S. DONATO MILANESE

DOTTORINI SILVIO
Via XX Settembre, 47/r
06100 PERUGIA

DRAGO EMILIA BRUNA
Ria Rodolfo Lanciani, 24
00162 ROMA

DUBLIC VLASTA
Pavleka Miskine 64
41000 ZAGREB

D'URSO GIORGIO
Via G. Lorenzoni
50134 FIRENZE

EID M.
Laboratory of General
Biochemistry University of
GHENT

ELIA MARCO
Ospedale Civile
14100 ASTI

ESTBORN B.D.
Tabuk Hospital
c/o Whittaker Corp. P.O. Box 3838
JEDDAH, Saudi Arabia

FALOMO ROBERTO
Via Unità D'Italia
MANIAGO

FANFANI MANFREDO
Piazza Indipendenza, 18
50100 FIRENZE

FARINA MARIA PINA
Via Isnardi
16016 COGOLETO

FAROLFI ANTONIO
L. Gonzaga, 4
00100 ROMA

FAVARO WALLY
Laboratorio Analisi
Ospedale Civile
31015 CONEGLIANO

FAVENTO RENATO
Sal. Promontorio, 11
34123 TRIESTE

FELLA COSIMO
Via Luzzati
31046 ODERZO

FERRAGUT CANALS ANTONIO
General Mitre 193
BARCELONA – Espana

FERRARA MARIA ANTONIETTA
Via A. De Gasperis
00143 ROMA

FINCO BRUNO
Via E. Vendramin, 7
35100 PADOVA

FIORENTINO MARIO
Via G. Verdi, 35
80133 NAPOLI

FIORI G. PAOLO
Via Venezia, 18
ALESSANDRIA

FISCHER ERNST J.
Haagsche Allee, 35
417 GELDERN/Ndm.

FONTANA ANGELO
Via Marzolo, 1
35100 PADOVA

FONTANIN MARISA
Viale N. Bixio, 23
31100 TREVISO

FORCINA BRUNO
Via Gallipoli, 9
23013 GALATINA

FORGHIERI MARIA ENRICA
Via Grazioli, 4
46100 MANTOVA

FORNASIER DE CARLI CECILIA
Via degli Alpini
31100 TREVISO

FRANCO GIUSEPPE
Viale G. Matteotti, 55
88018 VIBO VALENTIA

FRANK DE NADAI ANNA
Ospedale Civile
35100 PADOVA

FREDIANI M. GIOVANNA
Via S. Anna la Foce
LA SPEZIA

FREISE JURGEN
Nobelring 21
3-HANNOVER

GABRIELE ANTONIO
Via Daunia, 37
71016 SAN SEVERO

GABBRIELLI GABRIELE
Laboratorio Analisi
Ospedali Riuniti
89100 REGGIO CALABRIA

GALFANO MARIELLA
Laboratorio Analisi
Ospedale Civile
31015 CONEGLIANO

GALLINA GIORGIO
Via Maglio, 18
25034 ORZINUOVI

GALLO DONATO
Piazzale Brunelleschi, 30
80055 PORTICI

GALVANI MARCELLO
Via Parmeggiani, 3
40131 BOLOGNA

GALZIGNA LAURO
Istituto Chimica Biologica dell'Università
35100 PADOVA

GASPA UMBERTO
Via Roen, 59
39100 BOLZANO

GATTA LUIGI
Via della Mendola, 131
00100 ROMA

GAVELLA MIRJANA
Petrinjska 34
ZAGREB

GAZZOLA FAUSTO
Via Maglio del Lotto, 2
24100 BERGAMO

GENNARELLI EMILIO
Via Battistello Caracciolo, 38
80136 NAPOLI

GENNARI ADRIANO
Via C. Battisti
21036 GEMONIO

GERHARDT WILLIE
Klin. Kem. Lab. Lasarettet
251 87 HELSINGBORG

GEYER HELMUT
Unit. Frauenklinik
FREIBURG

GIAGNONI PIERO
Ospedale S. Bartolomeo
19038 SARZANA

GILHUUS-MOE CARL CHR.
Nycov. 2
Nyegaard Div. Diagn.
OSLO

GIORMANI VIRGILIO
Dorsoduro 3895
Fondamenta della Frescada
30123 VENEZIA

GIOVANOLA ANDREA
c/o General Diagnostics
Via Balzaretti, 9
20133 MILANO

GIRONI ANGELO
P.zza Trento Trieste
MONZA

GIULIANI DANTE
Corso Italia, 12
20100 MILANO

GIUNTA RICCARDO
P.zza Miraglia
80100 NAPOLI

GIUNTA SONETTI LUIGINA
Via Jacopo Nardi, 50
50132 FIRENZE

GOBBATO LUCIANO
Ospedale Civile
33053 LATISANA

GOLDBERG DAVID M.
Biochemist-in-Chief
Dept. of Biochemistry
Hospital for Sick Children
555 University Avenue
TORONTO ONTARIO M5G 1X8

GONANO FABIO
Via Bainsizza, 15
33100 UDINE

GOZZO MARIA LUISA
Via Guerrazzi
00100 ROMA

GRACY R.W.
Dept. of Biochemistry
North Texas State University
DENTON, TEXAS 76203 U.S.A.

GREGOLIN CARLO MARIA
Istituto Chimica Biologica dell'Università
35100 PADOVA

GREGORI PAOLO
Ospedali Riuniti
34100 TRIESTE

GRIFFITHS J.
Division of Clinical Chemistry
University of California
SAN DIEGO, LA JOLLA, CALIFORNIA

GRIS FURIO
Ospedali Riuniti
34100 TRIESTE

GROSSI GIUSEPPE
Lab. Cent. Policlinico
41100 MODENA

GRÜNDIG ELSE
Med. Chem. Institut
WIEN 9, Währingerstrasse 10

GUALA' LUIGI
P.zza Sannanano, 57
80122 NAPOLI

GUARGUAGLINI MAURO
Ospedale Neuropsichiatrico
56048 VOLTERRA

GUARINO AMEDEO
Ospedale Civile
83100 AVELLINO

GUARNIERI GIANCARLO
Ospedale Civile
30015 CHIOGGIA

GUERRIERI MATTEO
Via Giardino, 45
64021 GIULIANOVA

GUNZER ULRICH
Med. Univ. Klinic Hematology
87 WURZBURG, Germany

HAGEN ALEXANDER
Boehringer-Mannheim GmbH
Biochemica Werk Tutzing
Bahnhostrasse 5
8132 TUTZING

VAN DER HEIDEN C.
Nieuwe Gracht 137
Wilhelminakinder-Ziekenhuis
UTRECHT – Nederland

HENDERSON A.R.
Dept. of Clinical Biochemistry
University Hospital
LONDON, ONTARIO N6A 5A5

HENKEL EBERHARD
Institut fur Klinische Chemie
Med. Hochschule
HANNOVER – West Germany

HENKEL RENATE
Poclielstristrasse 380
Institut fur Klinische Chemie
Med. Hoschule
HANNOVER – West Germany

HOFFMANN HEINZ
Saronweg 4
D-48 BIELEFELD 13 – West Germany

HOMOLKA JIRI
Karlovo nam 32
12111 PRAHA 2

HORNBY W.E.
University of St. Andrews
Dept. of Biochemistry
ST. ANDREWS, FIFE KY16 9AL

HORNE THOMAS
Coulter Electronics
Coldharbour Lane
HARPENDEN HERTS. U.K.

HRISOHO RADMILA
Mediciniski Fakultet
OOZT Klinicka Biohemija
SKOPYE – Jugoslavia

IAVARONE ALBERTO
Via Fossati, 26
10141 TORINO

IMPERIALE I. FRANCO
Via Petrarca, 3
20026 NOVATE MILANESE

IZQUIERDO JOSE M.
Hospital General de Asturias
OVIEDO – Espana

KAISER ERICH
Währinger Strasse 10
A-1090 WIEN

KAMMERAAT CORNELIS
Langendyk 75
BREDA, Holland

KAPELI STEFANIA
Via Ferrarese, 4
40128 BOLOGNA

KATHER HORST
Bergheimerstrasse, 58
HEIDELBERG

KHAN P.M.
Wassenaarseweg 72
LEIDEN – The Netherlands

KING J.
Dept. of Biochemistry Royal Infirmary
GLASGOW

LAMEDICA GIUSEPPE
P.zza Alessi
16100 GENOVA

LAMPIASI GIANCARLO
Via Lambruschini, 7
54100 MASSA

LANCIOTTI GABRIELLA
Via Fiorenzuola, 235
47023 CESENA

LANCIOTTI LUCIANO
Via Colle San Michele
63012 CUPRAMARITTIMA

LA PAGLIA SALVATORE
Via Nuvolato, 5
35042 ESTE

LA TORRE MANZO ORAZIA
Via Croce Rossa, 34
90144 PALERMO

LATTUADA LUIGI
Via Gramsci, 24
25100 BRESCIA

LEMONNIER DANIEL
St. Martin, 292
75141 PARIS Cedex 03

LEONDEFF ILIA
Via Murge, Trav. 47/10E
70100 BARI

LESS GIULIO
Via Novacella, 28
39100 BOLZANO

LEVI DELLA VIDA BRUNO
Via Magalotti, 15
00197 ROMA

LIETTI ANDREA
Istituto Inverni della Beffa
Via Ripamonti, 99
20100 MILANO

LIGABUE AMEDEO
Via P. Palagi, 9
40100 BOLOGNA

LIPOVAC KSENIJA
Institut for Diabetes Petrinvska 34
ZAGREB

LIPPI UGO
Laboratorio Analisi
Ospedale Civile
35012 CAMPOSAMPIERO

LISI SALVATORE
Viale delle Regioni, 21
93100 CALTANISSETTA

LOLLIS BRUNO
Laboratorio Ricerche Cliniche
Ospedale Regionale
34100 TRIESTE

LOMBARDO ADRIANA
Istituto Chimica Biologica dell'Università
Viale Saldini, 50
20100 MILANO

LOOSER SIEGFRIED
Boehringer-Biochemica
8132 TUTZING

LORA ANTONIO
Primario Laboratorio Ospedale Civile
37044 COLOGNA VENETA

LORI AGOSTINO
Laboratorio Tecnologie Biomediche
C.N.R.
Via Morgagni, 30/E
00100 ROMA

LOSIO EMILIO
Via Frua, 12
20146 MILANO

LUCE GIUSTINO
Via G. Frua, 14
20146 MILANO

LUMBROSO CLAUDE
4, rue Thovin
PARIS – France

MACCIONI OTTAVIO
Via Barone Rossi
09100 CAGLIARI

MACHA HILDE
Elisabethstrasse 30
8-MÜNCHEN

MAGLIONE SALVATORE
Piazza Benedetto XV, 12
80026 CASORIA

MANGIAROTTI MARIA ANGELA
Via Salita Madonna di Gretta, 19
34136 TRIESTE

MANNINO SIGILLO ANTONIO
Via Norvegesi, 5
89100 REGGIO CALABRIA

MANNUCCI ENZO
Via Fiorentina, 1
53100 SIENA

MANZO FRANCESCO
Via Croce Rossa, 34
90144 PALERMO

MARABOLI SERGIO
c/o Mondadori
Via Zeviani, 2
37100 VERONA

MARCHESI GIOVANNI
Laboratorio Analisi
Ospedale Civile
25087 SALO'

MARCHETTI CESARE
Via Roma
61028 SASSOCORVARO

MARCHI SPARTACO
Ospedale Civile
Laboratorio Analisi
46049 VOLTA MANTOVANA

MARCOVINA SANTICA
Ospedale S. Marta
26027 RIVOLTA D'ADDA

MARINIELLO PASQUALE
Via Aquileia, 88
58100 GROSSETO

MAROTTA MARIANO
Viale Tirreno, 187
00141 ROMA

MARSILI GUIDO
Via Camerini
62032 CAMERINO

MARTORANA G.E.
Via Serafini
00100 ROMA

MASCHIO CAMILLO
Ospedale Civile S. Salvatore
67100 L'AQUILA

MASI INES
Istituto Superiore di Sanità
Viale Regina Elena, 299
00161 ROMA

MAURO RAFFAELE
Ospedale S. Gennaro
80100 NAPOLI

MAZZA LUIGI
P.tta Lega Lombarda, 14
15100 ALESSANDRIA

MEI ENRICO
Via Garibaldi, 15
53045 MONTEPULCIANO

MEI VITTORIO
Via Dagnini, 44
40137 BOLOGNA

MEIATTINI FRANCO
I.S.T.V. Sclavo
Via Fiorentina, 1
53100 SIENA

MELZI D'ERIL GIANVICO
Piazzale Golgi
27100 PAVIA

MEROLLA RAFFAELE
Via V. Veneto, 7
81031 AVERSA

MERUCCI PAOLO
Istituto Regina Elena
per lo studio e la cura dei tumori
Via Regina Margherita, 278
00198 ROMA

MESSERI GIANNI
Vicolo della Pergola, 2
50100 FIRENZE

MIGNANI ENRICO
Via Battisti
51017 PESCIA

MIGNI ALDO
Via S. Andrea Fratte, 2
06100 PERUGIA

MINETTI RODOLPHE
Institut Therasse
B-5180 GODINNE/Belgio

MISANTONE NINO
Ospedale Civile
64015 NERETO

MONTAGNA DANIELA
Largo Perosi, 1
27058 VOGHERA

MONTANI AUGUSTO
Istituto di Chimica Biologica
Via Taramelli, 1
27100 PAVIA

MORACE GIUSEPPE
Piazza Istria, 3
00198 ROMA

MORATTI REMIGIO
Piazzale Golgi
27100 PAVIA

MORENO-NOGUEIRA JOSE' ANDRES
Via Dr. Gonzales Gramaje, 1
SEVILLA – Espana

MORETTI CATERINA
Via F. Bronzetti, 26
25100 BRESCIA

MORO VALENTINO
Ospedale Civile
30026 PORTOGRUARO

MORRESI GIANNI
Via del Sodo, 10
50100 FIRENZE

MOSS D.W.
Royal Postgraduate Medical School
LONDON W12 OHS

MURADOR ENZO
Carlo Erba
Via Imbonati, 24
20100 MILANO

MUSMECI MICHELE
Via Romana, 20
34074 MONFALCONE

NADOR GIORGIO
Viale Duodo
33033 CODROIPO

NASSI LELIO
Via Luca Giordano, 13
50132 FIRENZE

NECCHI-GHIRI LUCIANO
Via Europa, 9
54027 PONTREMOLI

NEUMANN ULRICH
Boehringer-Biochemica
8132 TUTZING

NIGITO ALOISE
Via Zanrè, 2
43043 BORGOTARO

NOEBELS HENRY
rue 41 Marziani
Institute Technical Center
GENEVA – Switzerland

NUTI RANUCCIO
Piazza Postierla, 2
53100 SIENA

ODDO NINO
I.L. S.p.A.
Viale dell'Industria, 3
20037 PADERNO DUGNANO

O'GORMAN PATRICK
Dept. of Clinical Chemistry
Brook Hospital
LONDON

ORDELL RUNE
Dept. Clin. Chemistry Div. Enzymology
V. Klin., Central Hospital
S-55185 JÖNKÖPING – Sweden

ORIENTE PASQUALE
Via S. Pansini
80100 NAPOLI

ORLANDO ANTONIO
Via Monviso, 32
80144 NAPOLI

OTTONE LUISA
Via Santena, 5 bis
10100 TORINO

OYVIND HETLAND
Dept. of Clinical Chemistry
Ulleval Hospital
OSLO - Norway

PALEANI-VETTORI PIERGIUSEPPE
Laboratorio Tecnologie Biomediche
del C.N.R.
Via G.B. Morgagni, 30/E
00189 ROMA

PALLINI-MALLABY FRANCA
Via G.C. Vanini, 28
50100 FIRENZE

PANDOLFI CESARE
Via Carducci, 42
80100 NAPOLI

PANTANO EMANUELE
Laboratorio Analisi
Ospedale Civile
29100 PIACENZA

PAOLINI RICCARDO
Ames-Miles
Via F.L. Miles, 10
20040 CAVENAGO BRIANZA

PARADOSSI GIORGIO
Via Washington, 18
20146 MILANO

PARENTE ANTONIO
Largo S. Martino
80129 NAPOLI

PARMIGIANI PIETRO
Ospedale Psichiatrico
36030 MONTECCHIO PRECALCINO

PASCA ENEA
Via A. De Gasperi, 31
73024 MAGLIE

PASQUINELLI FILIPPO
Viale Mazzini, 18
50132 FIRENZE

PECCHIO FERNANDO
Strada ai Ronchi, 116
10100 TORINO

PELTONEN VEIKKO
Aalto G B 10
OLLITUOTE – Finland

PENTTILA ILKKA
Dept. Clinical Chemistry
University Central Hospital
KUOPID, Finland

PERACINO ANDREA
Ospedale Consorziale
25047 TREVIGLIO

PERARI ALBERTO
Via Gasbarrini, 17
64022 GIULIANOVA

PERERA SERAPIA
Via Nicolò de Stefani, 3
32100 BELLUNO

PERINI PIETRO
Via Bergamo, 3
25015 DESENZANO

PERSIJN J.P.
Antoni Van Leeuwenhoek Ziekenhuis
Plesmanlaan 121
AMSTERDAM

PERUZZINI ALDO
Via Bonelli, 13
38100 TRENTO

PETTENE ALESSANDRO
Corso Cavour, 44
27100 PAVIA

PIASERICO PIERLUIGI
Ospedale Civile
31044 MONTEBELLUNA

PICARDI ROBERTO
Via di Priscilla, 35
00199 ROMA

PICCOLI PAOLO
Via E. Altino, 128
80127 NAPOLI

PIERALISI MARCELLO
Via Boito, 10
20026 NOVATE MILANESE

PIGHINI ALBERTO
Via Ospizi Civili, 8
43100 PARMA

PILGERSTORFEK HANNS W.
Schubertstrasse 9
A-4020 LINZ – Austria

PINAMONTI FERRUCCIO
Via del Parco
36016 THIENE

PIZZOFERRATO ARTURO
Via Dante, 11
40125 BOLOGNA

PORTA FRANCO
Laboratorio Analisi
Ospedale Civile
23100 SONDRIO

PORTOLAN ANTONIO
Via delle Palme, 18
35100 PADOVA

POSTIGLIONE LOREDANA
Via P. Castellino, 77
80100 NAPOLI

PUCHE JUAN
Avda. Filipinas, 1
MADRID 3

PULIDO EDOARDO
Ospedale Civile
31010 ASOLO

QUAGLIARIELLO E.
Istituto Chimica Biologica dell'Università
Facoltà di Scienze
70100 BARI

RACUGNO GIUSEPPE
Via Orsini, 26
74100 TARANTO

RAGUSA GIUSEPPE
Via Rosso di S. Secondo, 3
95128 CATANIA

RAMACCIOTTI PIERGIORGIO
Via Valsalva, 12
47100 FORLI'

RAMPA PIERLUIGI
Via Gottardo
10100 TORINO

RANA SAVERIO
Via Nino Bixio, 68
70017 PUTIGNANO

RASERA BERNA FRANCESCA
Via Montalban, 1
32100 BELLUNO

RASO MARIO
Direttore Istituto Anatomia Patologica
dell'Università
80100 NAPOLI

REY-ROMERO CONCEPCION
Via Dr. Gonzales Gramaje, 1
SEVILLA – Espana

RIGHINI ROBERTO
Via Marostica, 35
20077 MELEGNANO

RIGOLIN FRANCO
Via Don Minzoni, 4
37045 LEGNAGO

RIVANO GIANCARLO
Via Marconi, 80
08045 LANUSEI

RIVANO RENATO
Via Mura Cappuccine, 14
16100 GENOVA

RIZZO BRUNO
Via S. Giovanni Bosco, 4/B
35043 MONSELICE

RIZZOTTI PAOLO
Via Largo Caldera, 11
37100 VERONA

ROMITO SANDRO
Via M. Ortigara, 166
37100 VERONA

ROSSI FILIPPO
Direttore Istituto Patologia Generale
dell'Università
34100 TRIESTE

ROSSO CAMILLO
Primario Laboratorio
Ospedale Civile S. Giovanni
10100 TORINO

ROTH MARC
Hôpital Cantonal Central Laboratory
1211 - GENEVA

ROTOLO GRAZIELLA
Ospedale Civile
27049 STRADELLA

ROVERE RENATO
Via Santena, 5 bis
10100 TORINO

ROVESCALLI AURELIANO
Via Ettore Ciccotti, 8
20100 MILANO

RUSSO GENNARO
Via Meridionale, 10
80021 AFRAGOLA

RUSSO NICOLINO
Via Guido Cortese
80128 NAPOLI

SABOTTO BENITO
Via L. Grossi, 5
21100 VARESE

SALERNO C.
Istituto di Chimica Biologica dell'Università
00100 ROMA

SALVATI ANNA MARIA
Istituto Superiore di Sanità
Viale Regina Elena, 299
00161 ROMA

SALVATORE FRANCESCO
Via S. Pansini, 5
80131 NAPOLI

SALVO GIUSEPPE
Via Regina Elena, 299
00100 ROMA

SANDIFORT KEESE
Via priv. Fantoli, 21/15
20138 MILANO

SANFILIPPO FRANCESCO
Corso dei Mille, 191
90100 PALERMO

SANNA MANLIO
Via Scano, 27
09100 CAGLIARI

SANTAGOSTINO MARIA ANTONIETTA
Via Minghetti
37044 COLOGNA VENETA

SANZI GAETANO
Via B. Chimizzi
88100 CATANZARO

SAVIO EMILIO
Via Brentana, 7
25100 BRESCIA

SAVOLDI ROBERTO
Via Gramsci
25100 BRESCIA

SCAGLIA JOLANDA
Via Cerri, 16
29100 PIACENZA

SCHARPE' S.
Univ. Antwerpen
Universiteitsplein, 1
2610 WILRIJK – Belgium

SCHIAVON SANTE
Via E. Vendramin, 7
35100 PADOVA

SCHMIDT ELLEN
Div. of Gastroenterology and Hepatology
Dept. of Int. Med. Medical School
HANNOVER

SCOGNAMIGLIO SALVATORE
P.zza Immacolata, 10
80129 NAPOLI

SCOTT I.V.
Dept. of Obstetrics and Gynaecology
University St. Mary's Hospital
Whitworth Park
MANCHESTER M13 OJH

SEMENZATO GIANPIETRO
Via Nazareth, 2, int. 3
35100 PADOVA

SERIO ANNA
Via Schiavello
87012 CASTROVILLARI

SGUIZZATO GIOVANNA
Via Piave, 2
35100 PADOVA

SHAVEL MIKEL
5/Mansard Court
07470 WAYNE U.S.A.

SICLARI VITTORIA
Via Lazzarini, 33
32100 BELLUNO

SIEDE WERNER
Briandring 4
FRANKFURT – Germ. Fed. Rep.

SILIPRANDI NORIS
Istituto Chimica Biologica dell'Università
35100 PADOVA

SIMONE MARIO
Via Strada Rivoltana
20090 RODANO

SIVINI CATERINA
Laboratorio Analisi
Ospedale Civile
31015 CONEGLIANO

van der SLIK WILLEM
University Hospital
Laboratory for Clinical Chemistry
Rijnsburgerweg, 10
LEIDEN – Holland

SMIDS PETRUS MARIA
v. Boetzelaer van Oosterhoutlaan 13
LEUSDEN – The Netherlands

SORACCO EDOARDO
Via Matteotti
44011 ARGENTA

SPANDRIO LUIGI
P.le Spedali Civili
25100 BRESCIA

SPANO' ALBERTO
P.zza Trieste, 18
80056 ERCOLANO

SPEZZANI LUCA
Via Mancini, 34
25017 LONATO

SPIEZIA MARIANO
Viale Fieschi
19100 LA SPEZIA

SPYKERS J.B.F.
Klin. Chem. Lab.
St. Lambertus Ziekenhuis
HELMOND – Holland

STATLAND B.E.
Dept. of Hospital Laboratories University
CHAPELL HILL – NORTH CAROLINA

STINSHOFF KLAUS
Boehringer-Biochemica
Bahnhofstrasse, 5
8132 TUTZING

STRAUB J.P.
Acad. Ziekenhuis Vrye
Universiteit
AMSTERDAM

SZASZ GABOR
Institut fur Klinische Chemie
and den Universitatskliniken
63 GIESSEN, Klinikstrasse 32b

SZILASI ALEX
Via Ambra d'Oro, 39
25100 BRESCIA

TACCONI ROSALIA
Via Roma, 15
62032 CAMERINO

TAGLIAZUCCHI ALBERTO
Via S. Amico, 10
17025 LOANO

TAMBUSCIO SERAFINO
Corso Europa
24067 SARNICO

TANGANELLI ENZO
Via F. Borromeo, 10
20062 CASSANO ADDA

TARANTINO MARIO
Via F. Ferruccio, 12
20145 MILANO

TARENGHI G.
Ames-Miles
Via F.L. Miles, 10
20040 CAVENAGO BRIANZA

TENTORI LEONARDO
Istituto Superiore di Sanità
Viale Regina Elena, 299
00161 ROMA

TERZANI GIULIANO
Via Nicolò Tartaglia, 16
00100 ROMA

TETTAMANTI GUIDO
Istituto Chimica Biologica dell'Università
Viale Saldini, 50
20100 MILANO

TIOZZO FILIPPO
c/o Istituto Behring
Via degli Scrovegni, 17
35100 PADOVA

TOMASI ALDO
Via Scanaroli
41100 MODENA

TOMASINI SERGIO
Via D. Michiel, 14
30100 VENEZIA-Lido

TOME' G. PAOLO
Via IV Novembre 152
55049 VIAREGGIO

TOPI GIANCARLO
Viale Pilsudski, 118
00197 ROMA

TRENTINI SAVERIO
I.L.
Via dell'Industria, 3
20037 PADERNO DUGNANO

TREVISAN MARIO
Via Garibaldi
35013 CITTADELLA

TRIMARCHI IVAN
Via Cavalieri della Stella, 7
98100 MESSINA

TRUZZI GIORGIO
Via Cadorna
46029 SUZZARA

VALENZA TOMMASO
Via R. Sanzio, 48
50065 PONTASSIEVE

VALOR PEREA RICARDO
Hospital Clinico
Facultad de Medicina
MADRID

VALSECCHI BRUNO
Via Stradone Farnese, 118
29100 PIACENZA

VANVREEDENDAAL MATTHYS
Charlotte plaate 3
ZOETEMEER – Holland

VANZETTI GIULIO
Via Ragusa, 8
20125 MILANO

VERCKO KSENIJA
Pavleka Micknine 64
41000 ZAGREB

VIDAL MARIELLA
Via Montegrappa, 15
31033 CASTELFRANCO

VILLA ANTONIO
Laboratorio Analisi
Ospedale Civile
36078 VALDAGNO

VILLANO STEFANIA
Via Palombini, 20
00165 ROMA

VIT ELISA
Via Castello 3722
30100 VENEZIA

WALID YASMINEH
University of
MINNEAPOLIS-MINNESOTA

WERNER MARIO
901 23rd Street
The George Washington University
Medical Center
20037 WASHINGTON

WILKINSON J.H.
Dept. of Chemical Pathology
Charing Cross Hospital Medical School
Fulham Palace Road
LONDON W6 8RF

WINKEL PER
University Hospital
COPENHAGEN

ZAMBONI MADDALENA
Via Mangano, 7
37100 VERONA

ZAMPETTI PIERLUIGI
Ospedale S. Verdiana
50051 CASTELFIORENTINO

ZANARDI GIORGIO
Via Cavour, 38
42017 NOVELLARA

ZANASI PIERLUISA
Via A. Volta, 16/b
27100 PAVIA

ZANGAGLIA ONORIO
Via E. Vendramini, 7
35100 PADOVA

ZAPPALA' ELIO
Via Decima
35044 MONTAGNANA

ZINI GUIDO
Via priv. Fantoli, 21/15
20138 MILANO

ZONTA FABIO
Via La Marmora, 13
34100 TRIESTE

AUTHOR'S INDEX

Finito di stampare
nel mese di gennaio 1978
presso le Grafiche Carollo e C. s.n.c.
di Motta di Costabissara (Vi)
per conto di Piccin Editore, Padova